Pyrethrum Flowers

Pyrethrum cinerariaefolium.

Pyrethrum Flowers

C B Gnadinger

2001

BIOTECH BOOKS
DELHI - 110035

First Indian Edition 2001

ISBN 81-7622-058-2

Published by : **BIOTECH BOOKS**
1123/74, Tri Nagar,
Delhi - 110 035
Phone: 7109765
e-mail: biotechbooks@yahoo.co.in

Printed at : **Chawla Offset Printers**
New Delhi - 110 052

PREFACE

TO SECOND EDITION.

More than three hundred references have been added to the bibliography of Pyrethrum since the first edition of this book was published in 1933.

In the present volume, data on production and new sources of pyrethrum have been brought up to date. Recent work on the isolation of the pyrethrins and on pyrethrum dermatitis is described. Working descriptions of fourteen chemical methods for assaying pyrethrum products are included; new biological assay methods are fully presented.

The decomposition of pyrethrum flowers and extracts in storage and the use of antioxidants to prevent decomposition are discussed. Additional information on live stock sprays is presented and brief descriptions of many recent investigations on the use of horticultural dusts and sprays are given. Much new material on the cultivation of pyrethrum in the United States is included in the final chapter.

C. B. GNADINGER.

PREFACE

TO FIRST EDITION.

Accurate knowledge of the chemical composition of pyrethrum flowers dates from the work of Staudinger and Ruzicka, published in 1924. In the preceding century, the secrecy surrounding the cultivation of the plant and the lack of information about its active principles resulted in the accumulation of a mass of misinformation and mistaken opinion, which persists, partially, even today.

Great progress toward revealing the facts about pyrethrum has been made in the last nine years. As a further step in that direction, an attempt has been made to describe in this book the results of the more recent investigations, including those conducted by the author in the laboratory of McLaughlin Gormley King Company.

It is hoped that the material presented will be of interest and value to all users of pyrethrum.

C. B. GNADINGER.

PREFACE

TO THE FIRST EDITION

Although [illegible] chemical composition was [illegible] until the work of Staudinger and Ruzicka was published in 1924. The prevailing ignorance surrounding the constitution of the flowers and the lack of information about its active principles resulted in the circulation of a mass of misinformation and misleading opinion which persists, partially, even today.

Great progress toward establishing the facts concerning [illegible] has been made during the last decade and in this volume an attempt has been made to describe in detail the more important results of the more recent investigations, including those conducted by the author in the laboratory of McLaughlin Gormley King Company.

It is hoped that the material presented will be of interest and value to all users of pyrethrum.

C. B. GNADINGER

ILLUSTRATIONS

TABLES

FIGURES

CONTENTS

PYRETHRUM FLOWERS

CHAPTER I

DESCRIPTION AND HISTORY OF PYRETHRUM

Pyrethrum is regarded as a section of the genus *Chrysanthemum,* family *Compositae.* This genus contains more than one hundred species, of which only a few, in the section *Pyrethrum,* are toxic to insects. The United States Department of Agriculture recognizes only three species of *Chrysanthemum* as being suitable for the manufacture of insect powder. These species are:

Chrysanthemum (Pyrethrum) cinerariaefolium (Treviranus) Boccone.

Chrysanthemum (Pyrethrum) roseum Web. and Mohr.

Chrysanthemum Marshalli Ascherson (synonym, *Pyrethrum carneum* Marschall Bieberstein).

Pyrethrum cinerariaefolium is the only species that is commercially important, the quantities of *Pyrethrum roseum* and *Pyrethrum carneum* imported into this country being negligible.

To the casual observer, *P. cinerariaefolium* resembles the ordinary field daisy; the two plants are, however, readily distinguishable. *P. cinerariaefolium* is a glaucous perennial, 18 to 24 inches high. The stems are unbranched and slightly hairy. The leaves are petioled and finely cut. The dried flower heads are hemispherical and consist of a short, rounded receptacle; a straw-colored involucre composed of three rows of scales; a disk composed of numerous yellow flowers; a circle of white or cream-colored ray flowers. The outer involucral scales are lanceolate and have a pronounced keel; their outer surface is light brown and hairy; the inner surface is smooth and lighter in color. The inner involucral scales are spatulate and longer than the outer ones, as well as lighter in color; they have a membranous margin. The disk florets are yellow, tubular, perfect and have a five-lobed corolla borne on the achene, which has five ribs and a toothed crown. The ray florets are ligulate, pistillate and the corolla is cream-colored or white; it is delicately veined and has three teeth at the tip. Commercial flowers vary from 6 to 24 mm. in width and from 0.070 to 0.300 gram in weight.

Powdered *P. cinerariaefolium* has a pleasant, characteristic odor which is more pronounced in the freshly prepared material; it is slightly sternutatory. It has at first an acrid, bitter taste, which is succeeded, after a few minutes, by a numbing sensation that affects the tongue and lips. This numbness is caused by the active principles of the plant and is similar to that caused by

FOLIAGE OF *Pyrethrum cinerariaefolium* WITH FLOWERS IN DIFFERENT STAGES OF MATURITY.

aconite root, although less intense. The color of the ground product depends on its age and the maturity of the flowers from which it is prepared. Newly harvested flowers, properly dried and ground, have a bright yellow color; after standing for some time, or when ground from old or poorly cured flowers, the color is dull and browner.

Pyrethrum roseum is widely grown for its ornamental flowers, whose ray florets are pink, carmine, rose, crimson or white. It blossoms earlier and less profusely than *P. cinerariaefolium* and is somewhat more resistant to disease and injury. Its dried flowers are easily distinguished from *P. cinerariaefolium* by the purple color of the ray florets, the ten-ribbed achenes and the brown margins of the involucral scales. *P. carneum* resembles *P. roseum*. It is seldom seen in this country. These two species yield a darker powder than *P. cinerariaefolium*. Other species of *Pyrethrum* reported to be toxic to insects are *P. achilleae, P. caucasicum, P. corymbosum, C. frutescens, C. myconis, P. par-*

thenium and *P. segetum.* Species said to be inert include *C. coronarium*, *C. indicum* and *C. leucanthemum.*

Pyrethrum flowers have occasionally been confused with pyrethrum root or pellitory, which was official in U. S. Pharmacopoeia IX. The latter plant, *Anacyclus pyrethrum,* (Linn), is a native of Africa, and its root was formerly used medicinally as a powerful irritant; it has no insecticidal use and is rarely used today.

Throughout the remainder of this book the term pyrethrum, unmodified, is applied to *P. cinerariaefolium.*

The use of pyrethrum flowers, for insecticidal purposes, apparently originated in Persia. Great secrecy is said to have surrounded the early use and preparation of the material, which was made from *P. roseum* and *P. carneum,* and this no doubt accounts in part for the difficulty of fixing the date of discovery of its activity. Various writers agree that the powder was introduced into Europe early in the nineteenth century, by an Armenian merchant, who discovered the secret of its preparation while traveling in the Caucasus. A second version is that the secret was revealed to the Russians by military prisoners. About 1840 a new species, *P. cinerariaefolium,* was produced in Dalmatia and rapidly superseded the Persian species in Europe. An interesting story traces the discovery of the effect of *P. cinerariaefolium* on insects to a German woman of Dubrovnik, Dalmatia, who picked a bouquet of the flowers for their beauty. When they withered she threw them into a corner where, several weeks later, they were found surrounded by dead insects. She associated the death of the insects with the insecticidal property of the flowers and embarked in the business of manufacturing pyrethrum powder. Although accounts of the discovery of the toxicity of pyrethrum to insects differ, it is fairly certain that pyrethrum was in use in Europe, as an insecticide, more than a century ago, and in Persia considerably earlier. The plant apparently was well known long before its insecticidal properties were recognized.

Pyrethrum powder was introduced into the United States about 1860; for some time no information was available as to its botanical source. Later, importations of the powder were almost entirely replaced by the whole flowers, which were powdered in this country, thus preventing the addition of powdered stems—in the country of origin. All of the pyrethrum imported was used in the form of a very fine powder, which lent itself admirably to all kinds of adulteration. Consumption of pyrethrum increased from 600,000 pounds in 1885 to 3,000,000 pounds in 1919. About 1913, the development of an inexpensive bellows-container popularized the use of the powder somewhat. About 1919, kerosene extracts of pyrethrum began replacing the powder for household purposes; nine years later the use of the powder had nearly ceased, although importation of pyrethrum reached a peak of 16,120,000 pounds in 1935. In 1924 the publication of the work of Staudinger and Ruzicka, on the identification of the active principles, awakened new interest in the product. In 1926 progress was made in adapting pyrethrum extracts for horticultural use, a field in which much work remains to be done. In 1929 it became possible to assay pyrethrum and produce a pyrethrum extract of standardized active principle content. During the next seven years, further progress in the chemistry of pyrethrum was made.

Although pyrethrum was used in large quantities each year, information concerning it was fragmentary and contradictory. As recently as 1924 each of the following statements was in circulation or could be substantiated by reference to the current literature:

"The best powder is made from all-closed Dalmatian flowers."

"Dalmatian flowers are superior to Japanese."

"The flowers should be harvested before the buds expand, the smaller, the better. Half-open flowers are not so desirable and open flowers are almost worthless."

"The best criterion for the value of pyrethrum is the determination of the ether extract."

"Pyrethrum acts by suffocation, clogging the breathing apparatus of insects."

"The toxic properties are due to a volatile oil."

"Flowers kept in sealed jars for five years do not deteriorate."

"Ground flowers are not injured by exposure in an open dish for eight months."

The first four of these opinions were generally accepted by the trade and formed the basis on which business was transacted. Each of these eight statements has since been proved incorrect.

CHAPTER II

COMMERCIAL SOURCES OF PYRETHRUM

Until 1914 nearly all of the pyrethrum used in this country was imported from Dalmatia. The World War cut off this source of supply entirely and enabled Japan to seize the market, which she still firmly retains. In 1926, for example, 4,400 tons of pyrethrum were imported into the United States, of which 3,600 tons, or about 82 per cent, came from Japan. In 1931 the proportion imported from Japan increased to about 91 per cent. The Kingdom of Yugoslavia, however, is attempting to meet Japanese competition by applying scientific methods to the cultivating, storing, grading and marketing of pyrethrum and by organizing and educating the producers. The complete reorganization of the Dalmatian pyrethrum industry is now under way.

DALMATIAN PYRETHRUM

The term "Dalmatian pyrethrum" has been applied to all *P. cinerariaefolium* grown on the eastern coast of the Adriatic Sea. This region, embracing Croatia, Dalmatia, and Hercegovina, was formerly a part of the Austro-Hungarian Monarchy. The Kingdom of Yugoslavia was formed in 1918 from certain territories that separated from Austria-Hungary and united with the kindred Balkan States, Serbia and Montenegro. For a time the boundaries and names of the old districts, Serbia, Montenegro, Bosnia, Hercegovina, Dalmatia, Croatia, Slovenia, Slavonia and Voivodina were preserved within the new Kingdom. In 1929, however, the Kingdom was re-divided into nine "Banovine" or Provinces and the old sub-divisions and names were discarded. The new Provinces are shown on the accompanying map together with the outline of the old Kingdom of Dalmatia.

The coast of Yugoslavia extends from the Italian frontier on the northwest to the Albanian border on the southeast. Running parallel with the coast is a large number of islands, which rise to 3,600 feet above sea level; these are the summits of a submerged mountain range. The mainland is deeply indented and rugged; along the coast the mountains rise abruptly from the sea, reaching a maximum height of about 6,000 feet. The mountain valleys are fertile and have frequent heavy rains, but many parts of the country suffer from drought and the seaward slopes are, in sections, barren and arid. Along the coast the climate is warm, sunny and healthful; the temperature varies from about –4° to 32° (25° to 90° F.), the mean average temperature being about 15° (59° F.). Back of the coast the high regions are extremely cold in winter and hot in summer. The temperature of eastern Zetska varies from –15° to 43° (5° to 109° F.)

YUGOSLAVIA. THE OLD BOUNDARY OF DALMATIA IS INDICATED BY THE DOTTED LINE THROUGH PRIMORSKA AND ZETSKA.

with a mean annual temperature of about 13° (55° F.). The rainfall along the coast is about 40 inches; in Zetska it is higher, being very heavy in the section formerly comprising Montenegro. Rain is rare in the summer months in Primorska.

Pyrethrum is grown in Savska, Primorska, and Zetska and on many of the islands. Primorska produces about 75 per cent of the entire crop.

The production of pyrethrum in Dalmatia has been carried on in a rather primitive manner. Pyrethrum thrives in a warm, dry climate; under these conditions it will grow on mountainous or waste land. A humid or heavy soil is said to be unfavorable. The seed are sown in hotbeds in late October or early November and the plants are transferred to the fields in March, generally on the protected slopes of the hills. From 20,000 to

A DALMATIAN PYRETHRUM FIELD.

45,000 plants are set out to the acre; they bloom in May and June, but the yield the first year is very scanty. The second harvest is good and the plants continue to bear for a number of years. On well drained hillsides plants produce for as long as 15 years. Where conditions are more primitive the fields are prepared in the late summer and the seed are sown broadcast in the fields about the first of November. In harvesting, the flower is cut off just below the head. If the weather is dry and warm, the flowers are dried in the shade, but if rain threatens they are dried in the sun as rapidly as possible. In some regions the flowers are first exposed to the sun for a short time and the drying is completed in the shade. The freshly-picked flowers lose from 65 to 75 per cent of their weight during drying, depending somewhat on the maturity of the flowers and the moisture content of the dried flowers, which varies from 7 to 11 per cent. The picking season extends from about the tenth of May to the fifteenth of June. A field will be picked over about three times at intervals of about a week, for example, on the eighth, fifteenth, and twenty-second of May. Occasionally a rainy season sets in at the harvest time; then the flowers dry very slowly, even under cover, and sometimes become moldy and discolored. Dalmatian flowers were formerly graded as:

HARVESTING PYRETHRUM NEAR SPLIT, YUGOSLAVIA.
(COURTESY OFFICE FOR FOREIGN TRADE, KINGDOM OF YUGOSLAVIA).

1, closed; 2, half-open; 3, mixed open and closed; 4, open. These grades were not sharply defined, closed flowers usually containing some half-open blossoms and the half-open grade usually consisting, in part, of closed and open flowers. Open flowers were generally true to name. The closed flowers were considered the most valuable, the open flowers being considered inferior. These grades have been discarded; the Yugoslav government is educating the producers to let the flowers mature and only three standard grades will be exported (see page 150).

At harvest time middlemen visit the growers and buy up the crop. Before the war the entire crop was exported through Trieste, which, like Dalmatia, was part of Austria. Today much of the crop is still exported through Trieste, which now belongs to Italy. Yugoslavia is making strong efforts to organize and develop an export trade from Split, its principal port, thus permitting government control of grading and shipping, which is impossible under present conditions.

The principal pyrethrum markets in Yugoslavia are Split, Dubrovnik, Šibenik and Šušak.

The production of pyrethrum in Yugoslavia from 1920 to 1934 was as follows (Table I):

TABLE I. PRODUCTION OF PYRETHRUM IN YUGOSLAVIA

Year	Area cultivated (acres)	Production (pounds)
1920	995	335,000
1921	1317	460,900
1922	1581	839,300
1923	2930	1,608,900
1924	4804	1,902,300
1925	6323	2,787,800
1926	6298	2,712,700
1927	5592	1,627,000
1928	4752	1,333,600
1929	4940	1,153,600
1930	6160	2,443,400
1931	5950	1,358,000
1932	5548	1,173,400
1933	5350	1,788,600
1934	5090	1,689,600

DRYING DALMATIAN PYRETHRUM.
(COURTESY OFFICE FOR FOREIGN TRADE, KINGDOM OF YUGOSLAVIA).

The quantities grown in the principal producing districts of Yugoslavia in 1931 were (Table II):

TABLE II. PRODUCTION IN YUGOSLAVIA BY DISTRICTS

Province	District	Area cultivated (acres)	Production (pounds)	Per cent of total production	Yield per acre (pounds)
Primorska	Split	2781	429,780	31.6	155
Primorska	Hvar	1030	150,090	11.1	146
Primorska	Šibenik	674	185,580	13.7	276
Primorska	Brač	304	189,760	13.9	624
Primorska	Makarska	208	55,540	4.1	267
Primorska	Korčula	27	13,220	1.0	489
Zetska	Dubrovnik	912	325,970	24.0	357
Zetska	Boka Kotorska	12	6,610	0.5	551
Zetska	Trebinje	2	1,410	0.1	705

Formerly the greater part of the Dalmatian crop was shipped to the United States; in later years most of it has gone to European countries. Importations into the United States from Yugoslavia in 1935 were negligible.

JAPANESE PYRETHRUM

The cultivation of pyrethrum in Japan is of comparatively recent origin. The plant is *P. cinerariaefolium*, identical with the Dalmatian pyrethrum, from which it cannot be distinguished either in whole or powdered form. Pyrethrum was introduced into Japan from England in 1881. Four years later cultivation was undertaken in Wakayama Prefecture from Dalmatian seed. About the same time seeds from California were sown at the farm of the Komaba Agricultural College, near Tokyo. In 1896 a few seedlings were planted in Hokkaido. Here the soil and climate proved favorable and cultivation was gradually extended.

Hokkaido (Yezzo) is the principal pyrethrum producing district; it formerly accounted for about 64 per cent of the total Japanese crop. The island is mountainous and pyrethrum is grown on the barren slopes, whose sterile soil is useless for other crops. The climate is severe during the four winter months and snow falls to great depths. The average annual temperature is about 7° (45° F.). The rainfall is scanty. The main island (Honshu) formerly produced about 25 per cent of the crop, principally in the Prefectures of Hiroshima, Okayama and Wakayama. The nearby Prefectures of Ehime and Kagawa, on the island of Shikoku, accounted for most of the remaining 11 per cent. The climate of these five Prefectures is milder than that of Hokkaido,

STRIPPING FLOWER HEADS FROM DALMATIAN PYRETHRUM.

the average annual temperature being about 15° (59° F.). The protecting heavy snows are lacking, however, and considerable frost damage to the roots of the pyrethrum plants occurs in winter. The rainfall is more abundant, the soil is sandy and well drained and the land is much more valuable than land in Hokkaido. The production in the different districts varies from year to year; moreover it is difficult to obtain accurate figures on the production of pyrethrum in Japan. In the last few years, production outside of Hokkaido has greatly increased, so that in 1935 it was estimated that only 40 per cent of the total crop was produced in Hokkaido.

In Hokkaido the sowing season is about the middle of May. The seeds are soaked in water and are then wrapped in sacking and buried in damp sand for four or five days. They are then mixed with dry sand and sown in well drained, sunny seed beds having carefully plowed, soft, sandy soil which has been freed from stones and clods. Fertilizer consisting of manure and superphosphate is worked into the beds before sowing; excessive use of fertilizer causes too rapid growth and is avoided. One pint of seed is used to 150 square yards of seed bed; this will yield seedlings enough for an area ten times as great. After sowing,

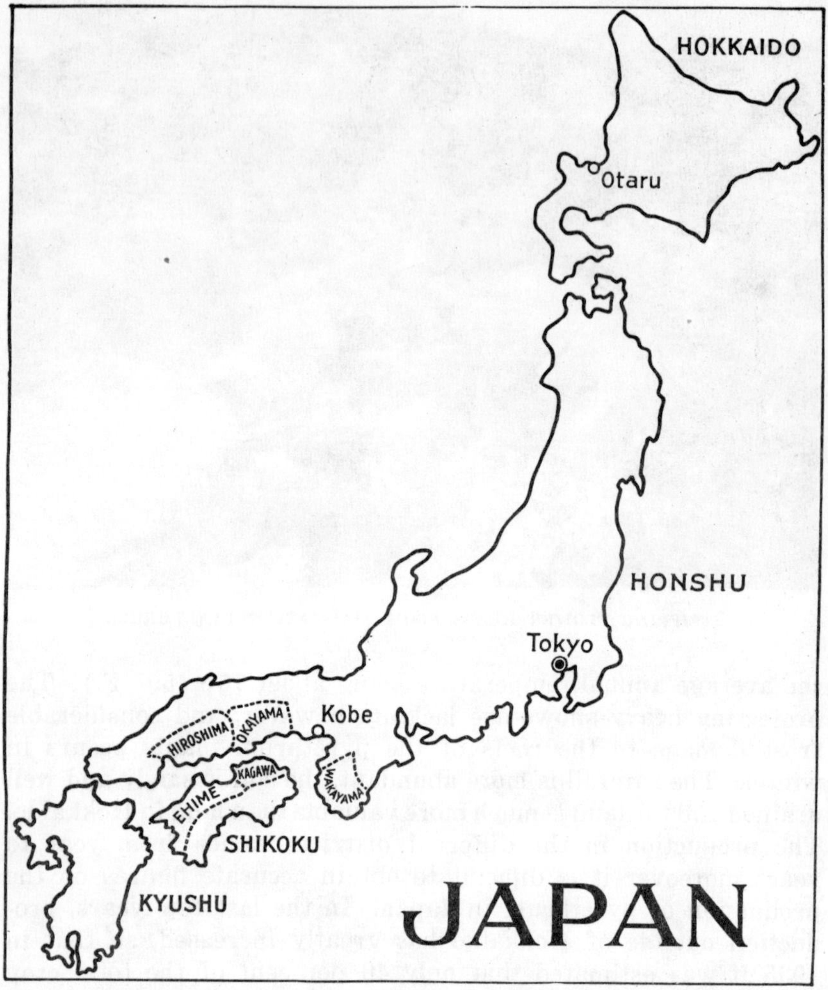

PRINCIPAL PYRETHRUM PRODUCING DISTRICTS OF JAPAN.
HOKKAIDO, HIROSHIMA, WAKAYAMA, OKAYAMA, EHIME AND KAGAWA.

the seeds are covered with earth or ashes, the beds are shaded with screens and in periods of drought they are carefully watered. The seedlings appear in about twelve days; when they are two to three inches high, fertilizer is added. About the first of October the seedlings reach a height of about four inches and are ready for transplanting. This must be done early enough to permit the roots to establish themselves firmly before cold weather, otherwise they will winter-kill. The field is carefully plowed, weeded, manured, and leveled. The seedlings are planted in rows

A PYRETHRUM FIELD, WAKAYAMA, JAPAN.

at intervals of 7 to 12 inches between plants and 1 to 2 feet between rows. The rows are raised or ridged to prevent water collecting around the roots. If the plants are set too deeply, few flowers are produced. No crop is expected in the following spring. Even in the second year the crop is very light, rarely exceeding 33 pounds per acre. The third, fourth and fifth years produce the largest crops after which the yield falls off. At this time the roots may be subdivided but this is less satisfactory than replacing with new plants. After the crop is harvested, manure is applied between rows in the autumn. Fish-scrap is also used as fertilizer, but the excessive use of nitrogenous fertilizer is said to result in all leaves and no flowers.

The harvesting period extends for 14 to 18 days in a given locality and the flowers are picked when they are about 70 per cent open. The picking season varies somewhat from year to year. On the mainland the harvest occurs late in May or early in June, but in Hokkaido it is generally in the first half of July. Occasionally heavy rain will delay the harvest until the last of July or first part of August, as in 1932. On the smaller farms the flower heads are picked, without stems, by hand; girls can gather 40 to 80 pounds a day. Larger producers cut the flowers with sickles and remove the flower heads with a comb or hackle.

STRIPPING FLOWER HEADS FROM FRESHLY HARVESTED FLOWERS. THE METAL STRIPPER HAS TEETH LIKE A COMB. (JAPAN).

Seeds for the next year are collected about two weeks after the petals have faded, from young, early flowering plants.

Drying generally requires about 5 to 7 days and is accomplished by spreading the flower heads on straw mats, in the sun, during the day, and placing them indoors at night. Sometimes the flowers with stems attached are suspended from drying frames. Flowers are also dried by exposing them to the sun for one day and then completing the process by storing on shelves in well ventilated sheds, where they are turned twice a day. In some cases artificial heat is used, temperature 65° (149° F.), but this is too expensive for general use.

After the flowers are thoroughly dried so that they can easily be crumbled between the fingers, they are packed in straw bags or gunny sacks. The former are used on the main island and Shikoku and the latter in Hokkaido. The flowers are collected from the farms at local inspection centers where they are classified by official inspectors as 1st, 2nd, 3rd, 4th or 5th grade. A discussion of this method of grading is given in Chapter VII. The classification is under the supervision of a board composed of two producers, two dealers, two producer-dealers

DRYING PYRETHRUM, WAKAYAMA, JAPAN.

and four government experts. Flowers from the mainland districts are not officially separated into the five grades. The average yield of dried flowers in Hokkaido for a period of eighteen years was 350 pounds per acre. Most of the crop is moved from the farms in August. Kobe is the great export center for pyrethrum, despite strenuous efforts to establish a similar center at Otaru, on Hokkaido. In Kobe the flowers are stored in bags. Just prior to exporting they are placed in matting sacks, called "ampera", each holding 112 pounds of flowers. Four of these are placed in a gunny sack and compressed in a hydraulic press, under a pressure of three tons, into a bale about 2 feet square by 3 feet high, which is bound with iron straps and weighs 448 pounds net. Experiments are being made to discontinue the use of the ampera.

"The following figures* represent a combination of producing cost estimates obtained from several trade sources and should be considered as approximate only. They are based upon the cultivation of one tan (0.245 acre) of pyrethrum flowers resulting in a harvest during the second and third years after planting.

* U. S. Consular Report, Jan., 1932.

TABLE III. COST OF PRODUCING JAPANESE PYRETHRUM

Item of expense	Western Japan		Hokkaido	
Seeds	Yen 0.60		Yen 1.00	
Ground rent	" 43.00		" 9.30	
Wages:				
Cultivation	" 40.00		" 24.50	
Harvesting	" 20.00		" 14.40	
Drying	" 8.00		" 4.80	
Fertilizer	" 20.00		" 8.00	
Mats, tools, et cetera	" 8.00		" 8.00	
Productions costs	Yen 139.60	($48.66)	Yen 70.00	($24.50)
Average total yield for two crops	Lbs. 430		Lbs. 270	
Average cost per pound	Yen 0.324	(11 cents)	Yen 0.26	(9 cents)

"To the above farmers' costs must be added about one sen to cover cartage, inspection and freight to Kobe from western Japanese points. Approximately 2.5 sen per pound is required to cover charges in bringing Hokkaido flowers to Kobe and a loss of two per cent in weight must also be reckoned on shipments from Hokkaido. Without profit to any of the parties concerned, the Kobe cost of western Japanese pyrethrum flowers is about 33.4 sen (11.5 cents) per pound as against 29.8 sen (10 cents) per pound for Hokkaido flowers."

BAGS OF PYRETHRUM IN WAREHOUSE OR "GO-DOWN", KOBE.

The production of Pyrethrum in Japan from 1911 to 1935 was as follows (Table IV):

TABLE IV. PRODUCTION OF PYRETHRUM IN JAPAN

Year	Acres cultivated	Production in pounds
1911	380	280,000
1912	1,260	1,036,000
1913	2,190	1,830,000
1914	2,710	2,124,000
1915	2,990	2,174,000
1916	6,790	4,335,000
1917	10,500	6,783,000
1918	9,500	5,964,000
1919	6,270	4,008,000
1920	5,930	3,879,000
1921	6,230	3,740,000
1922	7,710	4,073,000
1923	10,740	4,842,000
1924	18,640	8,094,000
1925	29,080	13,575,000
1926	33,730	15,993,000
1927	28,170	10,613,000
1928	26,800	11,624,000
1929	30,046	12,523,000
1930	39,532	16,029,000
1931	40,002	12,454,000
1932	44,900	13,440,000
1933	38,300	14,336,000
1934	50,900	17,763,000
1935	71,845	28,555,000

The production in the various Prefectures of Japan for 1935 was (Table V):

TABLE V. PRODUCTION IN JAPAN BY PREFECTURES (1935)

No.	Prefecture	Acres cultivated	Production in pounds	Per cent of total production	Yield per acre, pounds
1	Hokkaido	52,200	11,546,000	40.4	221
2	Hiroshima	5,000	4,681,000	16.4	936
3	Ehime	4,470	3,628,000	12.7	811
4	Wakayama	3,460	3,113,000	10.9	900
5	Kagawa	2,930	2,607,000	9.1	890
6	Okayama	3,080	2,374,000	8.3	770
7	Yamaguchi	700	604,000	2.1	862

Although Hokkaido produced 40 per cent of the crop, in 1935, the yield per acre was less than one-third that in the other Prefectures. In middle Japan the yields are about as follows:

Age of plant (years)	No. of flowers per plant	Pounds of flowers per acre Fresh	Dried
2	30 to 40	620	124
3	100 to 150	2480	496
4	200 to 300	3720	744

According to unofficial figures, the areas under cultivation and production for 1929, 1934 and 1935 were approximately as follows (Table VI):

TABLE VI. JAPANESE PRODUCTION BY PREFECTURES FOR THREE YEARS

Prefecture	1929 Acres cultivated	1929 Production pounds	1934 Acres cultivated	1934 Production pounds	1935 Acres cultivated	1935 Production pounds
Hokkaido	23,275	7,564,000	38,360	6,848,000	52,200	11,546,000
Hiroshima	2,083	1,762,000	3,400	3,024,000	5,000	4,681,000
Wakayama	1,225	854,000	2,300	2,195,000	3,460	3,113,000
Okayama	1,948	1,164,000	2,130	1,478,000	3,080	2,374,000
Ehime	907	594,000	2,920	2,307,000	4,470	3,628,000
Kagawa	608	585,000	1,790	1,501,000	2,930	2,607,000
All others	......	......	......	......	700	604,000
Totals	30,046	12,523,000	50,900	17,353,000	71,840	28,553,000

The annual conference of the Combined Insect Flower Merchants' Association meets each April, in Japan, and issues estimates of the acreage under cultivation and the production for the coming year. Price estimates are also made and the minimum, maximum and average estimates are published. The acreage estimates are usually more accurate than the estimates of production, the latter usually being high; the price estimates are likely to be misleading.

The estimated and actual production, in long tons for several years were:

	1931	1932	1933	1934	1935	1936
Estimated production	7,118	7,287	7,307	8,010	11,330	15,000
Actual production	5,600	6,000	6,400	7,930	12,750	......

In 1934 new plantings were greatly increased, especially on the mainland, where yields per acre are several times as great as on Hokkaido. This increased cultivation produced, in 1935, the largest crop of flowers on record. As a result, the price of pyrethrum reached an all time low of 7 cents per pound, C.I.F. New York, in 1936.

In Hokkaido flowers are marketed by the Hokkaido Growers' Cooperative Association. In general the Japanese grower sells his crop to brokers who in turn sell to the large dealers or exporters. Occasionally producers sell direct to dealers or exporters, who establish agencies at the principal producing centers. It is not unusual for flowers to go through eight to ten hands in passing from grower to exporter. Some American importers have established commercial agents in Japan to purchase pyrethrum direct from producers and brokers. Others prefer to deal with Japanese exporters. In either case the goods are paid for before leaving Japan. A further discussion of more modern methods of purchasing pyrethrum is given on page 187. Japan consumes about 33 per cent of her pyrethrum crop and exports the remainder, of which the United States consumes about 90 per cent.

Consumption of Pyrethrum in the United States

With the development of new methods of using pyrethrum, about 1925, importations into the United States increased sharply. The quantities brought into this country from 1920 to 1935 were* (Table VII):

Table VII. Importations of Pyrethrum into United States

Year	Amount imported pounds	Total value dollars	Av. value, cents per pound
1920	5,055,000	2,482,000	.491
1921	2,236,000	785,000	.351
1922	3,594,000	1,198,000	.333
1923	2,974,000	1,398,000	.471
1924	3,945,000	1,604,000	.407
1925	6,435,000	1,237,000	.192
1926	9,853,000	1,219,000	.124
1927	8,993,000	1,220,000	.136
1928	13,689,000	3,692,000	.269
1929	9,013,000	2,061,000	.229
1930	8,536,000	1,333,000	.156
1931	4,521,000	522,000	.115
1932	12,111,000	1,309,000	.108
1933	10,434,000	1,523,000	.146
1934	10,591,000	2,059,000	.195
1935	15,578,000	....	....

Since 1920 the import price of pyrethrum has fluctuated from 7 to 55 cents per pound; the average for the last 15 years was about 24.0 cents per pound.

The comparative value of the various grades of Dalmatian, Kenya and Japanese Pyrethrum is fully discussed in Chapter VII.

Other Sources

Pyrethrum has been grown in commercial quantities in Algeria, Australia, Brazil, Bulgaria, California, China, France, Italy, Kenya, Persia, Russia, Spain and Switzerland. The com-

* Information supplied by U. S. Department of Commerce.

COMPRESSING AND BALING JAPANESE PYRETHRUM, KOBE.

bined production of all of these sources is quite small. Production in France is said to have been 600 tons in 1934. In China about 1500 acres of pyrethrum are under cultivation. It is estimated that about 100 tons of flowers will be available for export from China in 1936.

Pyrethrum was introduced into Kenya Colony, East Africa, in 1928 *(48, 55)*.* Since then the area under cultivation has gradually increased; in 1934 there was an exportable surplus of about 80 tons. The seed are planted in shaded beds and the seedlings are transplanted to the field when about four months old. The plants are spaced 15 inches by 15 inches. Artificial fertilizer is not used. The plant flourishes at elevations of 6000 to 8000 feet. The flowers are dried on mats, in the sun, which generally requires about six days. The yield in Kenya is said to be 450 to 600 pounds of dried flowers per acre. In certain parts of Kenya the plants bloom for about

CONSTRUCTION OF COMPRESSED BALE OF JAPANESE PYRETHRUM. LEFT, COMPLETE BALE; CENTER, OUTER WRAPPING REMOVED SHOWING FOUR SECTIONS; RIGHT, QUARTER BALE WITH MATTING REMOVED.

nine months of the year and picking takes place every three or four weeks. Production in Kenya, available for export, has been estimated at about 200 tons, for 1935, 600 tons for 1936 and 1500 tons for 1937. Because of their higher pyrethrin content, at the present time Kenya flowers command a premium of 12

*Numbers (*italic*) in parentheses refer to bibliography, page 316.

to 15 per cent above the Japanese price. Kenya is rapidly becoming one of the most important sources of pyrethrum.

Importations into the United States from all sources, according to the U. S. Department of Commerce, were:

Source	1934	1935
Japan	10,094,000	15,203,000
Yugoslavia	299,000	154,000
Italy	113,000	66,000
United Kingdom	42,000	147,000
Russia	38,000	5,000
Brazil	5,000	.....
China	.....	2,000

CHAPTER III

ACTIVE PRINCIPLES OF PYRETHRUM

The search for the active principles of pyrethrum began about the middle of the nineteenth century. In spite of the fact that a large number of investigators attacked the problem, little more was known about the toxic constituents in 1909 than had been discovered sixty years earlier.

EARLY INVESTIGATIONS

The earliest recorded investigation was that of Ragazzini *(225)*, 1854, who failed to reach any conclusion. Heller and Kletzinsky *(394)*, 1863, believed the toxicity was due to ethereal oil and santonin. Hanamann *(394)* thought the ethereal oil was toxic, but failed to find santonin. Rother *(759)*, 1876, found that the active principle was a glucoside and then *(760)* decided that it was not. Hager *(383)*, 1878, found two active materials, one similar to trimethylamine and the other a resin. Dal Sie *(206)*, 1879, claimed to have isolated a free volatile acid as the most active constituent. Textor *(886)*, 1881, discovered that the soft resin soluble in benzine was toxic to flies. Gillette *(316)*, 1889, proved the active principle was soluble in ether. Hirschsohn *(432)*, 1890, showed that the active principle could be extracted with alcohol, ether, chloroform and benzine. Eymard *(257)*, 1890, proved that the essential oil was inert and isolated an active resin. Schlagdenhauffen and Reeb *(781)*, 1890, claimed to have isolated a toxic acid. Thoms *(897)*, 1890, thought the essential oil was the active material. Durrant *(245)*, 1897, also considered the volatile oil toxic together with an acid resin. Gerard *(307)*, 1898, agreed that the essential oil was toxic. Sato *(771, 772)*, 1905-7, isolated an active "sirupy resin." Fujitani *(289)*, 1909, prepared an active purified extract of pyrethrum, which he considered an ester. Siedler *(810)*, 1915, failed to isolate any active constituent. Roark *(608)*, 1917, showed that the essential oil was inert and that petroleum ether completely extracted the active principle and concluded that it consisted of esters. Yamamoto *(978)*, 1918, obtained a purified extract of pyrethrum containing a highlv unsaturated ester, which was very toxic to flies. On saponification he obtained an acid which he partially identified but could not separate the alcoholic constituent. La Forge *(608)*, 1924, obtained a purified extract which yielded a thick sirup when the solvent was evaporated. This sirup was purified by distillation in vacuum. The distillate, saponified, yielded the same acid found by Yamamoto; the alcoholic constituent could not be isolated.

Work of Staudinger and Ruzicka

When every other attempt to solve the problem had failed, Staudinger and Ruzicka *(841 to 850)* published a series of papers describing one of the finest investigations in the history of plant-chemistry. They had succeeded in isolating the two active principles of pyrethrum, and had solved the problem of their chemical nature. The work was done from 1910 to 1916; it was published in 1924.

Staudinger and Ruzicka completely extracted 100 kg. of powdered Dalmatian flowers with petroleum ether, which was distilled, yielding 5 to 6 kg. of thick, dark liquid. This residue was treated with 2 kg. of slightly warm methanol and the mixture was chilled to separate the resin and fat. The methanol solution was easily separated from the viscous mass and the residue was again treated with 2 kg. of methanol and the process was repeated twenty times using gradually decreasing quantities of methanol. The residue was a black, pitch-like body. The brown methanol solutions were each diluted with 5 per cent of water and again chilled, whereby resin and fat were precipitated. The solutions were filtered and the residue from each was added to the next succeeding solution, with which it was further extracted. The filtered solutions were not distilled, but were evaporated in vacuum in order to prevent decomposition of the active esters through substitution of methanol for their alcoholic component. After removing the methanol, the separated oil was taken up in low boiling petroleum ether; the active material dissolved in the petroleum ether while a dark resinous mass remained. The petroleum ether extracts contained large quantities of fatty acids, whose separation constituted the most disagreeable part of the isolation of the active principles. The petroleum ether solutions were shaken 10 hours with concentrated potassium carbonate solution; thereby the fatty acids were converted into soaps which dissolved in the petroleum ether. These soaps could not be removed by washing with water because emulsions formed which could not be broken up. The soaps were easily separated by shaking the petroleum ether solutions with 50 per cent calcium chlorid solution, thus converting the potash soaps into calcium soaps, which remained in the petroleum ether. Then on evaporating the petroleum ether and extracting the residue several times with methanol, the calcium salts remained undissolved. The methanol was removed in vacuum, the greenish yellow oil was dissolved in petroleum ether and the solution was washed with 25 per cent potassium carbonate solution to remove traces of acids. The petroleum ether was distilled, leaving 470 g. of yellow, crude oil, which contained the active materials almost completely, although contaminated with other compounds of unknown composition.

This oil was optically active and highly unsaturated, taking up large amounts of bromine. It was oxidized by potassium permanganate, chromic acid and ammoniacal silver solution. The active principles could not be separated by distillation with steam or superheated steam. The oil was rendered inert by reduction with hydrogen. Saponification with alcoholic potash did not give a satisfactory separation of the products. Satisfactory crystalline derivatives could not be obtained by treating the oil with hydrazine, phenylhydrazine, hydroxylamine, nitrophenylhydrazine, thiosemicarbazide or semioxamazide.

The crude oil was further purified by distillation in high vacuum, yielding a light ethereal oil, with a strong odor of insect powder, which was inert, and about 60 per cent of a yellow, viscous, strongly toxic oil boiling between 120° and 150°. This was not homogeneous but contained at least 50 per cent of impurities. The residue from the distillation was an inert resin.

Both the crude oil and the distilled oil were rendered inert by alkali, indicating the presence of esters. It was observed that the distilled oil could be dissolved in petroleum ether and washed with potassium permanganate solution or chromic acid solution without destroying the toxicity to insects. Both oils contained a carbonyl group which reacted with semicarbazide hydrochloride in methanol solution, forming insoluble semicarbazones. Methyl semicarbazide also formed a crystalline derivative. About half of the oil did not form a semicarbazone and this half was inert. The semicarbazone was separated in crystalline form; it was not homogeneous (m.p. 60°-100°) but was a mixture of two substances.

These semicarbazones eventually proved to be derivatives of the active principles. When the semicarbazone mixture was heated seven hours at 100° with 15 per cent oxalic acid solution, semicarbazide was split off and the ketones were regenerated, yielding an active oil. One of the semicarbazones (VI*) was isolated from the mixture in nearly pure condition (m.p. 120°-123°) and yielded a highly active oil.

When the semicarbazone mixture (m.p. 60°-100°) was saponified with sodium hydroxide in methanol at 0°, a new semicarbazone melting at 200° was obtained (VII). This was the semicarbazone of the alcoholic component of the active esters. A small quantity of a semicarbazone melting at 300° was also obtained. When the saponification was conducted without cooling to 0°, a large quantity of the semicarbazone melting at 300° was obtained with only traces of the semicarbazone of 200° melting point. Also, when the saponification was made with ethyl alcohol instead of methanol, even at low temperature, only a little of the 200° semicarbazone was obtained together with a large amount of a semicarbazone resembling the 300° semicarbazone but melting at 260°. Saponification with aqueous sodium hydroxide solution yielded no semicarbazone of 200° melting point, but only the higher melting semicarbazones. The semicarbazones melting from 230° to 300° were condensation or alteration products (VIII) of the semicarbazone melting at 200°.

When the semicarbazone melting at 200° was shaken, in the cold, for three to four weeks with benzene and aqueous potassium bisulfate solution the ketone-alcohol was split off and recovered, pure, from the washed benzene solution. This ketone-alcohol, the alcoholic component of the active esters of pyrethrum, was called pyrethrolone (I). It was a colorless, viscous, levorotatory oil, without characteristic odor, insoluble in water, slightly soluble in petroleum ether and miscible with alcohol, ether and benzene. It did not crystallize at $-80°$; it distilled at 111° to 112° (0.05 mm.) and decomposed completely at high temperatures. It had the formula $C_{11}H_{16}O_2$.

Pyrethrolone did not form a crystalline urethane. It reduced potassium permanganate, Fehling's solution and ammoniacal silver nitrate and added one molecule of bromine, in carbon bisulfide. With *p*-nitrophenylhydrazine it formed an osazone (IX) which was colored blue by alcoholic sodium hydroxide. With acetic anhydride, pyrethrolone formed an acetate boiling at 104°-105° (0.5 mm.). Pyrethrolone acetate formed an impure semicarbazone melting at 143°-145°, Pyrethrolone, treated

*Roman numerals in parentheses refer to graphic formulas, pages 31 and 32.

with sodium dissolved in methanol, yielded dehydropyrethrolone (X) as a result of auto-oxidation. The methyl ether of pyrethrolone (XI) was formed by heating pyrethrolone semicarbazone with methanol and sulfuric acid. It was a colorless liquid boiling at 82°-83° (0.25 mm.), soluble in petroleum ether; its semicarbazone melted at 182°-183°. Pyrethrolone methyl ether was also obtained by treating pyrethrolone with dimethyl sulfate, but was not obtained by the action of methanol on pyrethrolone in the presence of sulfuric acid. Pyrethrolone methyl ether was instantly oxidized by permanganate solution. The ethyl ether of pyrethrolone was formed by the action of ethyl alcohol and sulfuric acid on pyrethrolone semicarbazone. It boiled at 102°-103° in absolute vacuum and yielded a semicarbazone melting at 178°-180°.

Pyrethrolone in alcohol was reduced with hydrogen; it took up two molecules of hydrogen forming the saturated ketone-alcohol, tetrahydropyrethrolone $C_{11}H_{20}O_2$ (XII). Further reduction with hydrogen yielded a ketone, tetrahydropyrethrone, $C_{11}H_{20}O$ (XIII).

Tetrahydropyrethrone, the parent substance of pyrethrolone, was shown to be 3-methyl-2-amylcyclopentanone. It was best obtained by reduction of pyrethrolone acetate. It formed a semicarbazone, m.p. 194°. Oxidation with excess permanganate in the cold (XIV) gave caproic acid, levulinic acid and succinic acid, the latter a secondary oxidation product of levulinic acid.

The carbonyl group was found to be in the ring and not in the side chain. This was proved by forming the oxime of tetrahydropyrethrone (XV) which was converted into the isoxime (XVI) by warming with sulfuric acid. The isoxime was found to be a lactam and not an acid amide. The lactam was identified by treatment with hydrochloric acid, yielding γ-methyl-δ-aminodecylic acid (XVII) which was treated with sodium nitrite and dilute sulfuric acid and distilled with steam, forming the lactone of γ-methyl-δ-hydroxydecylic acid (XVIII) which on oxidation with permanganate gave caproic and levulinic acids.

Final proof of the formula of tetrahydropyrethrone was obtained by synthesizing the compound. This was accomplished by condensing the ethyl ester of levulinic acid (XIX) with the ethyl ester of α-bromheptylic acid (XX) by means of zinc at 150°. The products of the reaction were distilled in vacuum yielding among other compounds, ethyl-β-methyl-α-amylbutenedicarboxylate, from which the diethyl ester was prepared (XXI). This was condensed, by sodium in xylol, to a methyl-amyl-cyclopentenone derivative (XXII) which was converted to 3-methyl-2-amyl-Δ-2, 3-cyclopentenone (XXIII). The latter on reduction with hydrogen, over nickel, at 250° yielded 3-methyl-2-amylcyclopentanone, identical with tetrahydropyrethrone (XXIV).

Since pyrethrolone yielded tetrahydropyrethrone on reduction with hydrogen, it now remained to prove only the position of the double bonds to establish the formula of pyrethrolone.

Oxidation of pyrethrolone by permanganate (XXV) yielded only acetic acid, eliminating the possibility of the presence of ethyl or propyl groups and certain other groups in the side chain, and locating the double bonds in the side chain. With ozone pyrethrolone yielded a mixture of mono- and di-ozonide. The mono-ozonide (XXVI) gave a mixture of acetaldehyde and acetic acid on decomposing with water, establishing the position of one double bond. No crystalline decomposition products could be obtained. Ozonides were also prepared from pyrethrolone acetate. The non-volatile products obtained on decomposing the ozonide with water yielded malonic acid on oxidation with hydrogen peroxide, indicating the presence of a CH-CH_2-CH- group.

The exact structure of the side chain was not conclusively proved but the only formula that conforms to all of the findings is that shown on page 31. Pyrethrolone is a methyl-pentadienyl-cyclopentanolone. The allene grouping found in the side chain was unexpected and has not been found in any other natural product.

Pyrethrolone could be obtained only by saponifying the mixed semicarbazones isolated from the crude or distilled oil (page 25). When the crude or distilled oil was saponified with potassium hydroxide in methanol, only small amounts of pyrethrolone could be obtained together with the methyl ether of pyrethrolone. Saponification with ethyl alcoholic potash yielded neither pyrethrolone nor its methyl ether but only altered pyrethrolone.

The saponification of the semicarbazone mixture yielded, besides pyrethrolone, three acids. One of these was a colorless, liquid, dextrorotatory acid, boiling at 135° (12 mm.) having the composition $C_{10}H_{16}O_2$. It was called chrysanthemum monocarboxylic acid (II). It was soluble in petroleum ether and other organic solvents, insoluble in water and volatile with steam. In carbon bisulfide it took up one molecule of bromine but yielded no crystalline derivative. The methyl ester boiled at 86°-87° (10 mm.) and had an odor of pyrethrum flowers. The chloride boiled at 85° (12 mm.). The anilide melted at 101°; the amide at 131°. The anhydride boiled at 125° (15 mm.). Reduction of chrysanthemum monocarboxylic acid with hydrogen yielded 2, 2-dimethyl-3-isobutyltrimethylene-1-carboxylic acid (XXVII). Oxidation of the chrysanthemum acid with ozone yielded *l*-trans-caronic acid (XXVIII) and acetone.

Finally the chrysanthemum acid was synthesized from tetramethylbutadiene and diazoacetic acid ethyl ester. A mixture of cis-chrysanthemum acid and trans-chrysanthemum acid was obtained. Partial separation was effected by cooling the petroleum ether solution of the mixture to −80°. The cis-acid crystallized, m.p. 115°-116°; on oxidation with ozone it yielded cis-caronic acid and acetone. The trans-acid was separated, after evaporating the petroleum ether, by vacuum distillation, b.p. 145°-146° (13 mm.). It yielded acetone and trans-caronic acid on oxidation with ozone (XXIX). Chrysanthemum monocarboxylic acid is dimethyl-isobutenyl-trimethylene-carboxylic acid.

The second acid obtained from the saponification of the semicarbazone mixture was a solid, melting at 164°, having the formula $C_{10}H_{14}O_4$. It was dextrorotatory and non-volatile with steam. This acid was called chrysanthemum dicarboxylic acid (XXX). It was easily oxidized by permanganate and added bromine slowly, but was not reduced by hydrogen. The acid-chloride boiled at 151° (12 mm.) and melted at 48°-50°. The anilide melted at 204°-205°. Oxidation with ozone yielded *l*-trans-caronic acid and pyruvic acid (XXX). This acid does not occur in the flowers but was obtained by the saponification of its monomethyl ester, which does occur in the active oil and which was the third acid found in the products of saponification of the semicarbazones. Chrysanthemum dicarboxylic acid monomethyl ester, $C_{11}H_{16}O_4$, (III) was a colorless, heavy, dextrorotatory liquid boiling at 129°-130° (0.3 mm.), easily soluble in organic solvents. The chloride boiled at 114° (0.5 mm.). Oxidation with ozone yielded *l*-trans-caronic acid and the methyl ester of pyruvic acid, fixing the position of the $-COOCH_3$ group.

The inert portion of the original crude oil contained small amounts of the methyl esters of the chrysanthemum acids, possibly due to the action of the methanol used in the isolation process on the pyrethrins. The partial synthesis of the pyrethrins was carried out by treating the chlorides of the crysanthemum acids in benzene with a solution of pyrethrolone and quinoline in benzene, in the cold. Pyrethrolone was esterfied with chrysanthemum monocarboxylic acid and yielded a highly active ester. This ester when diluted with flour had the same activity as pyrethrum powder similarly diluted. It was called pyrethrin I (IV). Pyrethrolone esterfied with chrysanthemum dicarboxylic acid did not yield an active ester, but with chrysanthemum dicarboxylic acid monomethyl ester, the resulting compound was almost as active as pyrethrin I. This ester, from pyrethrolone and chrysanthemum dicarboxylic acid methyl ester, was called pyrethrin II (V).

The partially synthesized pyrethrins I and II did not yield semicarbazones of sharp melting points, probably because of changes in the highly unsaturated side chain of the pyrethrolone. The crystalline semicarbazone of pyrethrin II could not be isolated, either from the synthetic product or from the mixture of natural semicarbazones.

Pyrethrolone was esterfied with the synthetic cis- and trans-chrysanthemum monocarboxylic acids. The ester from the synthetic *dl*-trans-acid was much more active than that from the synthetic cis-acid, but was not as active as the natural pyrethrin I, which contains the *l*-trans chrysanthemum acid.

Pyrethrin I ($C_{21}H_{30}O_3=330$) was a viscous liquid boiling at about 150° in absolute vacuum. Pyrethrin II ($C_{22}H_{30}O_5=374$), also a heavy liquid, decomposed when distilled in absolute vacuum.

The word "pyrethrin," applied by Staudinger and Ruzicka to the active principles, should not be confused with the term "pyrethrine," used by certain French writers and manufacturers to designate oleoresin of pyrethrum. Some confusion has also arisen because the term "pyrethrin" was applied to the active principles of pellitory root (page 3) as early as 1876. This use of the term was apparently not known to Staudinger and Ruzicka when they named the active principles of pyrethrum flowers. Shepard *(799)* has discussed the origin of the name "pyrethrin" and has correctly pointed out that it is not feasible to change the nomenclature used by Staudinger and Ruzicka, because pellitory is a drug of slight and waning importance, whereas pyrethrum flowers are widely used and are becoming more important.

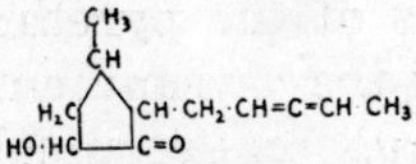

I. Pyrethrolone.

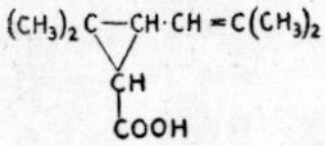

II. Chrysanthemum monocarboxylic acid.

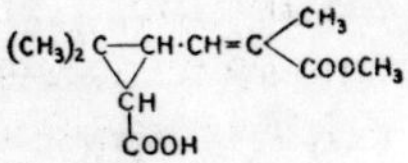

III. Chrysanthemum dicarboxylic acid monomethylester.

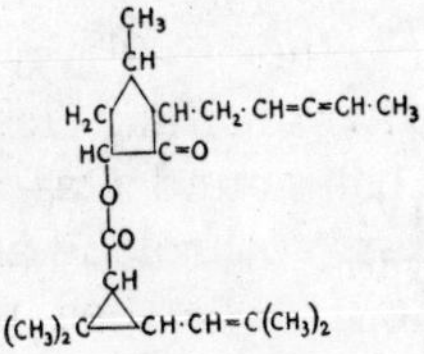

IV. Pyrethrin I.

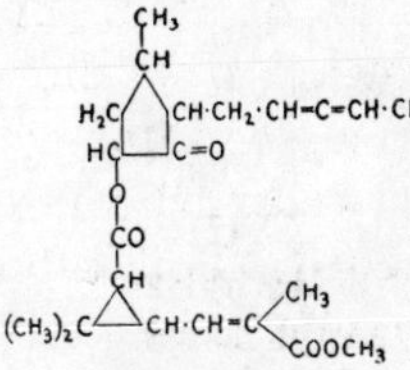

V. Pyrethrin II.

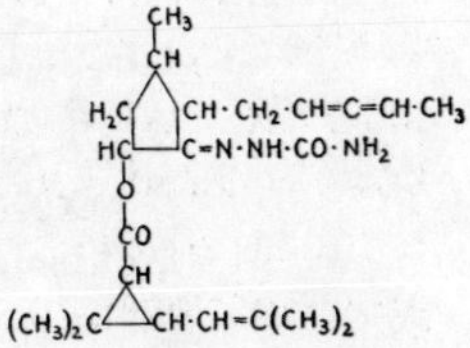

VI. Semicarbazone of pyrethrin I.

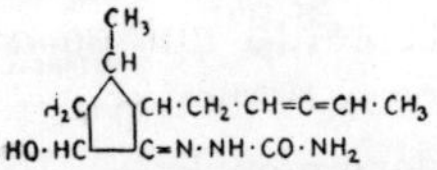

VII. Pyrethrolone semicarbazone.

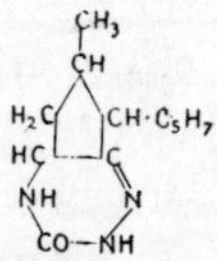

VIII. Altered pyrethrolone semicarbazone.

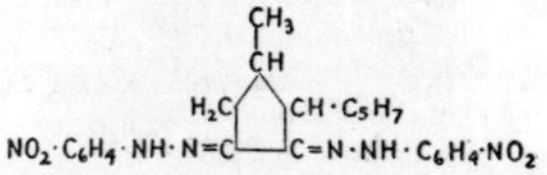

IX. p-Nitrophenyl osazone of pyrethrolone.

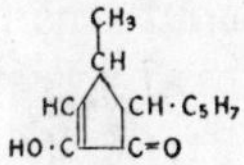

X. Dehydro-pyrethrolone.

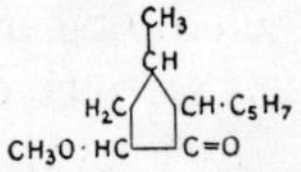

XI. Pyrethrolone methyl ether.

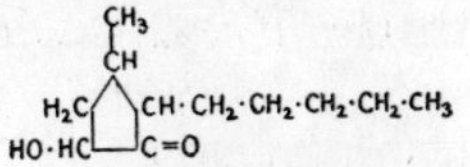

XII. Tetrahydro-pyrethrolone.

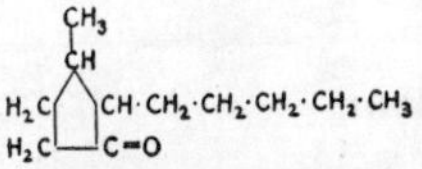

XIII. Tetrahydro-pyrethrone.

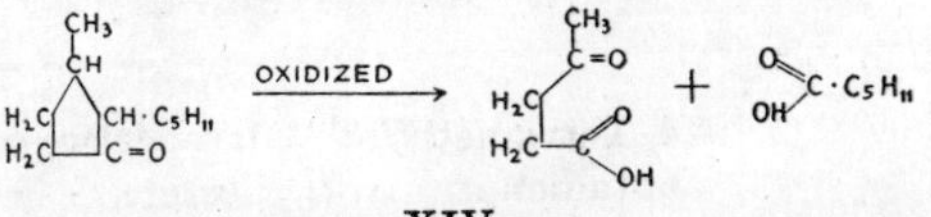

Tetrahydro-pyrethrone.

XIV. Levulinic acid. Caproic acid.

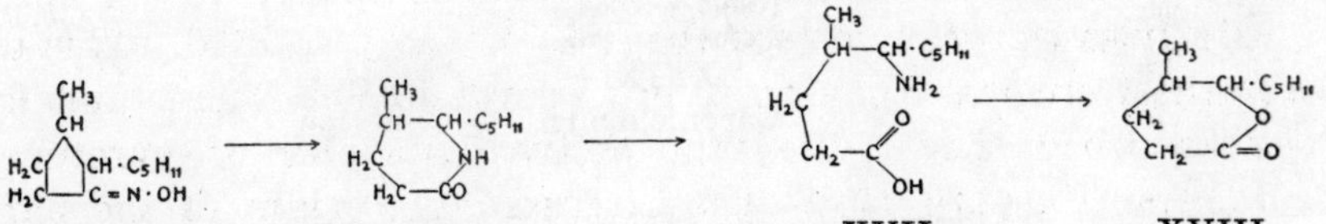

XV. Oxime of tetrahydro-pyrethrone

XVI. Isoxime.

XVII. γ-Methyl-δ-amino-decylic acid

XVIII. Lactone of γ-methyl-δ-hydroxydecylic acid.

GRAPHIC FORMULAS OF PYRETHRINS AND RELATED COMPOUNDS.

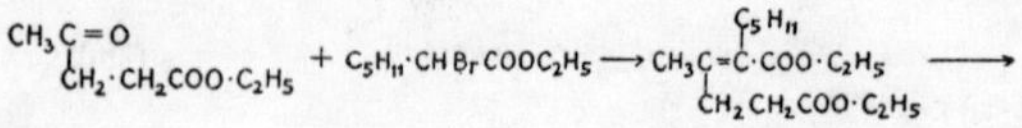

XIX. Levulinic ester. XX. Brom-heptyl ester. XXI. Diethyl-β-methyl-α-amyl-butene-dicarboxylate

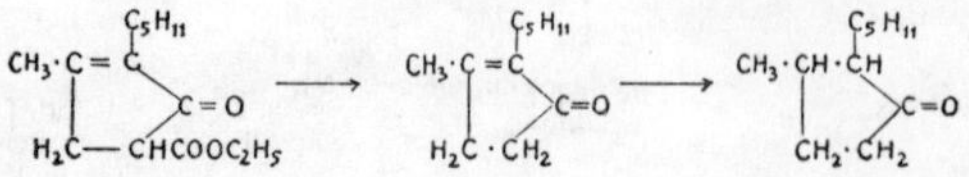

XXII. Not isolated. XXIII. 3-Methyl-2-amyl-Δ-2,3-cyclopentenone. XXIV. Tetrahydro-pyrethrone.

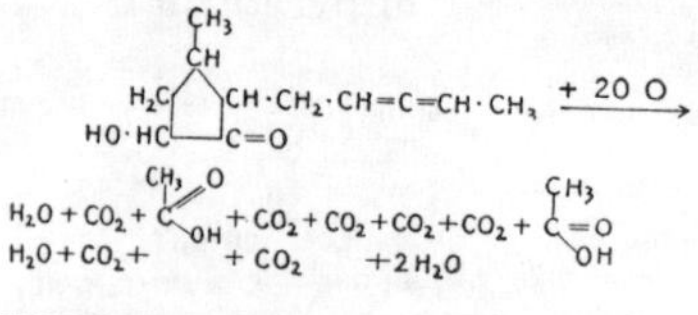

XXV. Oxidation of pyrethrolone by permanganate.

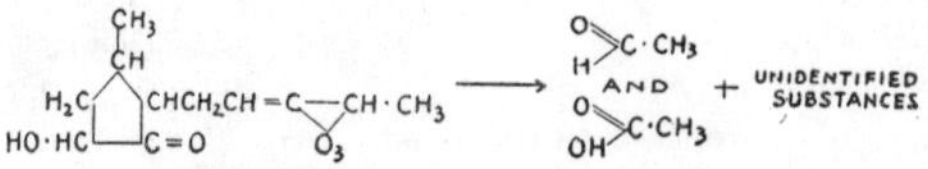

XXVI. Pyrethrolone ozonide.

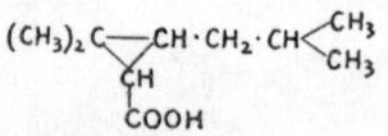

XXVII. 2,2,Dimethyl-3-isobutyl-trimethylene-1-carboxylic acid.

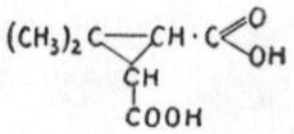

XXVIII. Trans-caronic acid.

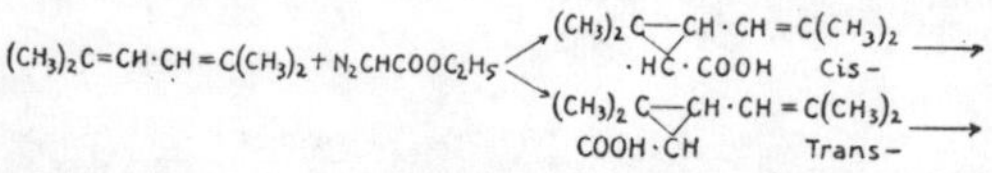

1,1,4,4, Tetramethyl-butadiene. Diazoacetic ester. Chrysanthemum monocarboxylic acids.

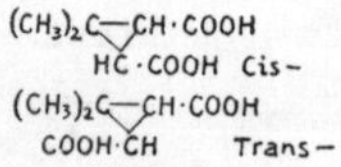

XXIX.

Caronic acids.

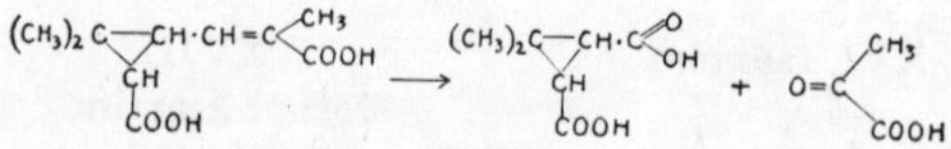

XXX.

Chrysanthemum dicarboxylic acid. *l*-Trans-caronic acid. Pyruvic acid.

GRAPHIC FORMULAS OF PYRETHRINS AND RELATED COMPOUNDS.

That part of the crude oil which did not form semicarbazones was inert to insects and had nothing to do with the activity of the pyrethrum. No toxic compounds could be found except the pyrethrins, nor could any semicarbazone be isolated from the petroleum ether extract except the semicarbazones of pyrethrins I and II and their decomposition products.

Pyrethrolone was esterified with 32 acids, including saturated aliphatic acids, unsaturated aliphatic acids, aromatic acids, acids of the terpene series and other trimethylene carboxylic acids.

Chrysanthemum monocarboxylic acid was combined with 82 compounds including alcohols, thio-alcohols, amines, phenols, aliphatic-aromatic alcohols, alicyclic compounds, terpenes, ketone-alcohols, cyclopentanolone and cyclopentenolone derivatives.

A few of the 104 compounds synthesized were slightly active; none approached the pyrethrins in toxicity. The hope that in the pyrethrins the chrysanthemum acids could be replaced by another acid, or the pyrethrolone by another alcohol was not fulfilled. Small changes either in the alcohol or the acid part of the pyrethrin molecule affected the toxicity greatly and more often than not destroyed it.

The pyrethrins are highly unsaturated both in the acid and alcohol parts of the molecule. Their toxicity depends not only on composition, but on the spatial arrangement of the atoms. For these reasons, the commercial production of synthetic pyrethrins is highly improbable.

The tests for toxicity were generally made on roaches, but lice, flies, bees and butterflies were also used, with the same results.

According to Audiffren *(85)*, chrysanthemum monocarboxylic acid gives a color reaction with Denigès reagent. The test is made as follows: to 1 cc. of dilute solution of chrysanthemum acid add 1 cc. of Denigès reagent; then add 0.4 to 0.5 cc. of concentrated sulfuric acid so as to form a layer in the bottom of the test tube. Mix after a few seconds; a pink or red color forms and soon turns violet and then green. After 24 hours a yellow precipitate is formed. The test is sensitive to 1 part in 100,000. Denigès reagent is prepared by mixing 5 g. of yellow mercuric oxide with 44 cc. of water; while stirring add 20 cc. of sulfuric acid and then add 40 cc. additional water and stir until dissolved.

This reaction is also described by Seil *(794)* who states that it will detect 0.5 mg. in 10 cc. The dicarboxylic acid gives no color reaction but on long standing yields a precipitate like that produced by the mono-carboxylic acid. According to Seil, the chrysanthemum acids, in aqueous solution, give the following reactions:

Reagent	Monocarboxylic acid Cold	Hot	Dicarboxylic acid Cold	Hot
Calcium chloride	—	—	—	—
Barium chloride	—	—	—	—
Silver nitrate	—	—	—	—
Ammoniacal silver nitrate	—	—	—	—
Zinc chloride	—	—	—	—
Lead acetate	+	+	+	+
Copper sulfate	+	+	+	+
Ferric chloride	+	+	+	+
Mercuric chloride	—	Reduced	—	Reduced
Fehling solution	—	.	—	—
Potassium permanganate	Reduced	Reduced	Reduced	Reduced
Bromine	Absorbed	....	Absorbed	....

In addition to the chrysanthemum acids, Ripert *(745)* claims to have isolated the following compounds from pyrethrum flowers: a resin acid, protocatechuic, isovaleric, caproic, lauric, palmitic, oleic and linoleic acids. Most of these acids are said to occur in both the free and combined states.

Gnadinger, Evans and Corl have shown that pyrethrins occur in the growing flowers as well as in the dried blossoms. Fairly mature flowers were picked and immediately dropped into acetone, where they were allowed to macerate for seven days. The carefully filtered acetone extract was diluted with water and applied to flies, roaches, and ants. All of these insects showed characteristic pyrethrum poisoning. Control solutions, containing acetone and water, in the proportions present in the diluted pyrethrum extract, were not toxic.

The relative toxicity of pyrethrins I and II is discussed in Chapter VI.

Active Principles of Japanese Pyrethrum

Staudinger and Ruzicka had isolated the active principles from Dalmatian pyrethrum. Yamamoto *(980)* was unable to isolate the pyrethrins, as semicarbazones, from Japanese pyrethrum. He suggested that the active principles of the Japanese flowers might be different from the pyrethrins of Staudinger and Ruzicka. In 1929, however, Gnadinger and Corl *(343)*, using a modification of the procedure of Staudinger and Ruzicka, proved that the active principles of Japanese pyrethrum are identical with those in Dalmatian flowers. They also succeeded in isolating the semicarbazone of pyrethrin II direct from the flowers and converted it into nearly pure pyrethrin II.

Isolation of Pyrethrins By Physical Means

Wilcoxon and Hartzell *(968)* have devised a purely physical method for isolating the pyrethrins. The following is a complete description of their procedure:

"About 500 g. of pyrethrum flowers were extracted with low boiling petroleum ether in a continuous extractor, and the solvent removed at low temperature under the water pump. The crude extract was dissolved in warm methanol and on cooling the solution for a few minutes, resins separated and were filtered off. The filtrate was treated with norite charcoal, which removed some of the pigments. The methanol was then removed under the water pump, yielding an oil which contained from 50 to 60 per cent of total pyrethrins. This oil was then submitted to a systematic fractionation between petroleum ether and 80 per cent methanol according to the scheme illustrated in Figure I. Each circle in the diagram represents a separatory funnel containing petroleum ether and 80 per cent methanol. The original sample, after being shaken in funnel No. 1, separates into two layers which are drawn off. The petroleum ether layer goes to funnel No. 2 where it is shaken with fresh 80 per cent methanol, while the methanol layer goes to funnel No. 3, where it is shaken with fresh petroleum ether, and this is continued as shown by the arrows in the diagram. It was

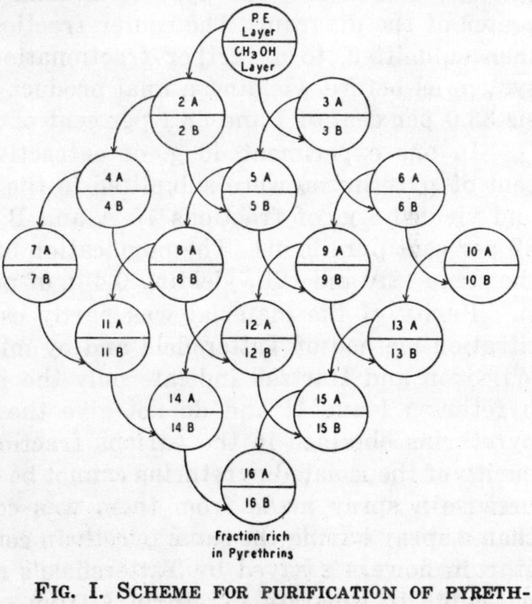

FIG. I. SCHEME FOR PURIFICATION OF PYRETHRINS BY IMMISCIBLE SOLVENTS. (WILCOXON AND HARTZELL).

found that the impurities tended to concentrate at the ends while the center fraction became richer in the pyrethrins. Since the solubility of the pyrethrins in petroleum ether is considerably greater than in 80 per cent methanol, about 50 cc. of the latter were used with 25 cc. of petroleum ether. The following experiment is typical of many performed:

"A sample of flowers containing 0.98 per cent of pyrethrins I and II was extracted and submitted to the preliminary purification yielding a product which contained 23.4 per cent of I and 38.3 per cent of II, total 61.7 per cent. This product was then treated according to the scheme outlined above, and various fractions analyzed. Fraction 16A and B contained 25.2 per cent I, and 61.9 per cent II, total 87.1 per cent.

"The analysis of some of the other fractions is shown below:

Fraction	Pyrethrin I	Pyrethrin II	Total
11A	36.9	35.5	72.4
14A	39.0	41.7	80.7
13B	14.3	47.2	61.5
15B	19.7	66.1	85.8

It may be seen that the pyrethrins tend to concentrate in the center of the diagram. The center fractions, 16 A and B, were then submitted, to a further fractionation following the same system as before, yielding a final product which gave on analysis 33.0 per cent of I and 68.1 per cent of II."

In one experiment 40 g. of extractive, containing 60 per cent of pyrethrins, were submitted to the fractionating process and yielded 3 g. of fractions 16 A and B combined, containing 88 per cent pyrethrins. The purification of this was repeated to the stage 12A and 12B, yielding 0.4 g. of pure pyrethrins.

Purity of the material was partly established by the acid titration method of Tattersfield and by micro ultimate analysis. Wilcoxon and Hartzell indicate only the relative percentage of pyrethrins I and II and do not give the actual weight of the pyrethrins obtained in the various fractions. The proof of the purity of the isolated pyrethrins cannot be considered conclusive, because a spray made from them was considerably less toxic than a spray having the same pyrethrin content, made from pyrethrum flowers assayed by Tattersfield's acid titration method. Attempts to separate pyrethrin I from pyrethrin II were not successful.

Ripert and Gaudin *(748)* have also employed methanol and petroleum ether for separating pyrethrin I from pyrethrin II.

Starting with a mixture of pure pyrethrins, in about equal proportions, 50 g. of the mixture were dissolved in one liter of pentane. This solution was washed five times with 500 cc. of methanol at 82°. The alcohol was diluted with water and the pyrethrins extracted with pentane (petroleum ether). This petroleum ether solution was dried over sodium sulfate and chilled to –50°, which precipitated the pyrethrin II. After dissolving the pyrethrin II in pentane and again precipitating it by chilling, it was dissolved in chloroform. The pyrethrin II was 98.9 per cent pure. The petroleum ether solutions, rich in pyrethrin I, were chilled to –120°. The pyrethrin I remained in solution in the petroleum ether from which it was obtained, 96 per cent pure.

La Forge and Haller *(538)* have devised a method for isolating pyrethrin II. The following is their description of the procedure:

"The starting material, a commercial concentrated petroleum ether extract, consisted of a thick brown liquid with a total pyrethrin content of about 30 per cent, in which the pyrethrins were present in about equal amounts. After removal of the remaining solvent by warming under reduced pressure (an essential step), 450 grams of the material were divided into nine 50-gram portions.

"To each portion 75 cc. of glacial acetic acid were added, and the mixture was stirred until practically all had dissolved. The same quantity of acetic acid to which 15 cc. of water had been added was mixed with the solution, causing the separation of a semisolid mass of material. The containers were then cooled in an ice bath for about half an hour, after which the contents were filtered separately on cold Büchner funnels. The solid material consisted of fats and waxes with comparatively little occluded pyrethrins. The green filtrates were combined and after addition of decolorizing carbon were again filtered. The acetic acid solution was then divided into two or three portions and strongly diluted with water, and the separated oil was extracted with petroleum ether. A small quantity of insoluble dark syrup was discarded. The petroleum ether solution was washed with water, a little sodium chloride being added to break slight emulsions. After the solution had been dried with sodium sulfate, most of the solvent was recovered by distillation on the steam bath; the remainder was removed under reduced pressure. The same method was used in isolating all pyrethrum concentrates referred to below from their acetic acid solutions.

"The quantity of concentrate obtained from 450 grams of the original material was 140 grams. It contained, besides other impurities, fatty acids, which were removed by dissolving the material in 400 cc. of freshly distilled aniline and agitating the solution for about 10 hours at $-10°$ with a solution of 140 grams of potassium carbonate in 600 cc. of water. The aniline solution readily separated from the slimy soap solution, which was drawn off and on acidification yielded the extracted fatty acids. The aniline solution was washed three times with saturated bicarbonate solution and then, with constant stirring, was slowly run into 400 cc. of concentrated hydrochloric acid and 400 grams of cracked ice. These procedures were usually carried out in two operations with half the quantities stated. The separated oil was taken up in a suitable quantity of petroleum ether. A small quantity of insoluble material was discarded. The residue obtained from the washed and dried petroleum ether solution consisted of 90 grams of yellow oil containing 38.9 per cent pyrethrin II and 29.2 per cent pyrethrin I. Only slight variations from these values were observed in similar runs.

"The 90 grams of partly purified material obtained by the process just described were dissolved in 200 cc. of glacial acetic acid and 400 cc. of petroleum ether in a separatory funnel, and 20 cc. of water were added in small portions with vigorous shaking.

"The addition of water caused the solution to separate into two layers, a petroleum ether fraction (P) and an acetic acid fraction (A). The latter was washed with six 40-cc. portions of petroleum ether, yielding an acid fraction (A-1).

"The combined petroleum ether washings were added to the petroleum ether fraction (P), and the resulting solution was washed with six 40-cc. portions of acetic acid containing 10 to 12 per cent of water. These acid washings were combined (A-2).

"The acid-washed petroleum ether solution (P-1) yielded about 35 grams of a product containing 43 per cent pyrethrin I and 14.4 per cent pyrethrin II. This was set aside for the isolation of pyrethrin I.

"The acid solution A-2 was extracted once with 240 cc. of petroleum ether and then with two 60-cc. portions of the same solvent. The combined petroleum ether washings were designated P-2, and the washed acid solution was designated A-3.

"The two acetic acid solutions A-1 and A-3 were combined,

and the dissolved material was isolated by the addition of water and extraction with petroleum ether. The quantity of concentrate so obtained was 30 grams. It contained 72.2 per cent pyrethrin II and 14.2 per cent pyrethrin I. The material will be referred to as of stage I.

"This product was dissolved in 100 cc. of acetic acid and 200 cc. of petroleum ether, and 14 cc. of water were added in small portions, with vigorous shaking after each addition. The separated acid solution was washed with four 20-cc. portions of petroleum ether. The washings and the separated petroleum ether layer were added to solution P-2.

"The dissolved material in the acid solution (A-4) was isolated by dilution with water and extraction with petroleum ether and 20 grams of material containing 76.5 per cent pyrethrin II and 7.4 per cent pyrethrin I were obtained. (stage II).

"At stage III, which is reached by repeating the solution in acetic acid-petroleum ether, addition of water and subsequent washing of the acid layer with petroleum ether, using corresponding quantities of solvents, the yield was about 14 grams of material containing 82.7 per cent pyrethrin II and 6 per cent pyrethrin I. It is necessary to repeat the process whenever at stage III the pyrethrin II content falls below 80 per cent, but it is not possible by further repetition to raise it much above 83 per cent when the proportions originally present were of the order given.

"All petroleum ether extracts except P-1 were combined with P-2. The quantity of material obtained from these combined extracts was 25 grams. When treated in a second series of operations according to the procedure described, it yielded 8 grams of concentrate containing at stage III about the same percentages of the two pyrethrins as the corresponding final product from the first series.

"As the separation and concentration of pyrethrin II proceeds, the concentrate becomes progressively more viscous and lighter in color and more difficultly soluble in petroleum ether. A concentrate of more than 70 per cent of pyrethrin II separates in part even from a moderately dilute petroleum ether solution in cooling, but this is of no advantage in purification, as the separated product carries the impurities with it.

"From 8 to 10 grams of solvent-free concentrate containing about 80 per cent pyrethrin II and 3 to 8 per cent pyrethrin I were distilled in a molecular still, of suitable size, of the type described by Hickman in the Journal of Franklin Institute, 213, 119, 1932. The still was heated in a small oil bath kept at the constant temperature of boiling chlorobenzene (131°), the pressure as recorded by the mercury gage being about 10^{-6} mm. The distillation usually required 8 to 10 hours.

"In the earlier trial runs a number of fractions was collected and analyzed to determine at which points the distillation should be interrupted. In later distillations three fractions were usually collected. The first fractions contained most of the pyrethrin I and the more volatile impurities. The second fractions generally consisted of 80 to 90 per cent pyrethrin II and were combined for redistillation. The third fractions were practically 100 per cent pyrethrin II. There was always more than 50 per cent loss due to polymerization in the still, and the still residue consisted of a glassy mass almost insoluble in all solvents. The yield of pure pyrethrin II was in no case more than 20 per cent of the quantity of material distilled, but an additional quantity was obtained on redistillation of the second fractions.

"The analytical results and physical constants for one sample are given below:

"Anal. calcd. for $C_{22}H_{30}O_5$: C, 70.54; H, 8.08; CH_3O, 8.28. Found: C, 70.63, 70.84; H, 7.69, 7.69; CH_3O, 8.10. Calcd. to pyrethrin II (mol. wt. 374), 97.72. $n\frac{30}{D}=1.5247$; $n\frac{20}{D}=1.5285$. Rotation: 0.135 g. in 2.28 cc. ether at 20° C. rotated –0.34°; $l=0.96$ dm.; $[\alpha]\frac{20}{D}=-6.0$; 0.235 g. in 2.28 cc. benzene at 20° C. rotated –0.42°; $l=0.96$ dm.; $[\alpha]\frac{20}{D}=-4.2$.

"None of the samples showed a pyrethrin II content of appreciably more than 100 per cent as determined by the methoxyl method of Haller and Acree *(385)*, indicating that none of them contained methyl pyrethrolone, alkyl esters of the chrysanthemum acids, or other methoxyl containing impurities. These may have been present in the concentrates before distillation, but, being of much lower molecular weight, they would be removed in the more volatile fractions."

The pyrethrin II so obtained by La Forge and Haller was a colorless, very viscous liquid. It was slightly levorotatory instead of dextrorotatory, as reported by Staudinger and Ruzicka.

Action of Pyrethrins on Cold-blooded Animals

The exact nature of the action of the pyrethrins on insects has not been determined. When an insect, such as a cockroach, is dusted with pyrethrum powder, it behaves normally for a few seconds but soon begins to stagger; in a minute or two it can scarcely walk and begins to struggle vigorously, finally turning on its back where it remains, helpless, until death. The time required to kill depends on the dose applied and the vitality of the insect and may be as long as 24 to 48 hours. House flies treated with pyrethrum extract can fly for a considerable time after they have lost control of their legs. With most insects death from pyrethrum poisoning is accompanied by violent and sometimes prolonged struggles.

Fujitani *(289)* found that fish are susceptible to the pyrethrins but protozoa are very tolerant. This has been confirmed by Kemper *(516)*, who also found that frogs, lizards, eels, dogfish, goldfish, tench and other species of fish, crabs and clams were killed by adding powdered pyrethrum to the water in which they were kept. Zeigler *(987)*, using oleoresin of pyrethrum prepared with ether, found that pyrethrum was toxic to frogs when introduced into the stomach. Turtles were killed by injections into the skin of the neck. Juillet *(496)* decided that pyrethrum is a neuro-muscular poison and is more effective by mouth than when applied to the integument. Saling *(764)* concluded that the pyrethrins are nerve poisons, as contrasted with respiratory or blood poisons. Krüger *(532)* investigated the action of pyrethrum on the transparent larvae of *Corethra plumicornis* and observed changes in the hypodermis, muscles and nerve fibers. Gaudin *(302)* found that echinoderms are practically unaffected by pyrethrins. He also investigated the action of pyrethrins on other organisms *(303)*.

Hartzell and Wilcoxon *(406)* have showed that death of the rose chafer, *Macrodactylus subspinosus*, Fab., can be caused by applying pyrethrum extract to the body without permitting it to come in contact with the head, thorax, spiracles, or other vulnerable parts. They also found that pyrethrum extract was toxic to representative species of ten orders of insects, spiders, centipedes, millipedes, sow bugs, snails, earthworms and frogs. The extract caused death in all cases where applied to the integument. The toxic action was much quicker when the pyrethrins were injected into the body cavity. Recently Wilcoxon and Hartzell *(968)* have studied the action of the pyrethrins on the yellow meal worm, *Tenebrio molitor* L. They found that death may ensue from the external application of pyrethrins under conditions where no tracheal penetration takes place. By experiments with dyes soluble in the pyrethrins they showed that the pyrethrins can penetrate the integument, at least in certain regions. These workers suggest that the low but appreciable solubility of the pyrethrins in water (or body fluids) may explain how the pyrethrins reach the nerve ganglia when applied externally. They were able to detect histological changes in nerve ganglia of insects killed by pyrethrum, by means of a staining technique using toluidine blue and they concluded that destruction of nerve tissue is a primary factor in causing death.

Buchmann *(142)*, however, has concluded that the pyrethrins do not form true solutions in water but exist instead in colloidal solution, the size of the colloidal particles lying between 0.01 and 0.001 microns.

Hartzell *(405)*, continuing his study of the nerve lesions caused by pyrethrin poisoning, concluded that "death is caused by the destruction of the cells of the central nervous system accompanied by paralysis." He also found lesions in the nerves of grasshoppers killed by pyrethrins.

The effects of controlled applications of pyrethrins to various parts of meal-worm larvae and roaches have been studied by O'Kane *(650)*.

Although pyrethrum is deadly to most insects, the growing pyrethrum plant is attacked by several pests. Riley *(734)* reports both leaves and flowers of pyrethrum plants are attacked by the rose chafer, which Hartzell and Wilcoxon used in the experiments mentioned above. The writer has seen the growing flower heads of first year plants infested with species of thrips, which are killed by pyrethrum sprays, and has found a small, unidentified bookmite, which was not affected by powdered pyrethrum, in shipments of dried flowers of high pyrethrin content from both Japan and Yugoslavia. The writer has also seen cadelle, the larvae of *Tenebroides mauritanicus*, living in finely powdered pyrethrum for five or six days without apparent ill effects; the pyrethrum powder contained 0.92 per cent pyrethrins.

Effect of Pyrethrins on Warm-blooded Animals

Yamamoto *(978)* found that pyrethrins had no effect on dogs when introduced into the stomach. Zeigler *(987)* reported that guinea pigs showed no toxic symptoms from subcutaneous injections or from applications of powdered pyrethrum to the mouth and nostrils. Dogs were not affected by feeding the oleoresin in capsules nor by subcutaneous injections, but intravenous injections produced convulsions sometimes resulting in death. Chevalier *(167)* found that intraperitoneal injections of 10 to 15 milligrams of pyrethrins per kg. of body weight caused nervous disturbances in guinea pigs. Intravenous injections of 6 to 8 milligrams of pyrethrins per kg., in dogs, caused death.

Shimkin and Anderson *(801)* have also investigated the action of pyrethrins on mammals.

A number of old cases have been reported where illness or death of human beings was said to have been caused by pyrethrum. These instances are not well authenticated because the purity of the powder was not established. Staudinger and Ruzicka experienced no ill effects from their years of work on pure pyrethrins. The writer has known large doses of oleoresin of pyrethrum to be administered in capsules to chickens without any ill effect, and has known workmen to grind the flowers day after day, for years, without experiencing any symptoms of toxic action. Workmen have smeared their arms with the oleoresin and spilled highly concentrated acetone and kerosene extracts on hands and arms without any effect.

The writer has handled pure pyrethrins in considerable quantities without ill effect and has taken doses of about 50 milligrams by mouth, with no effect except the numbing of the tongue and lips. Audiffren *(85)* found chrysanthemum monocarboxylic acid in the urine of patients to whom pyrethrum had been administered as a vermifuge.

It is well known that some workers are affected with dermatitis when handling pyrethrum or its extracts. A number of these cases have been reported by McCord, Kilker and Minster *(600)*. Allergic reactions have also been observed among those picking the fresh flowers. Feinberg *(264, 265)* states that pyrethrum apparently contains three toxic principles: 1, the active insecticidal esters, the pyrethrins; 2, the substance that causes the dermatitis, "probably an oleoresin," 3, "the specific allergen, which may give rise to symptoms of allergy, particularly in the respiratory tract."

Garratt and Bigger *(295)* reported the occurrence of asthma, due to the use of pyrethrum in the patient's bed. Cases of pyrethrum allergy have also been reported by Ramirez *(704)*.

Feinberg *(265)*, in a series of 225 patients sensitive to ragweed pollen, obtained cutaneous reactions with extract of pyrethrum in 104, or 46 per cent of the cases. He states that "sensitiveness to pyrethrum is a group sensitiveness, occurs only in ragweed sensitive individuals, and is probably present in the majority of the latter." He concludes that the allergenic substances of ragweed pollen and pyrethrum are closely related and warns that ragweed sensitive people may be susceptible to allergic attacks, at any time of the year, as the result of exposure to pyrethrum products.

Sweitzer and Tedder *(861)* reported only four cases of allergic sensitivity to pyrethrum in a series of 618 patients treated.

Schwartz *(790)* found 20 cases of dermatitis in a factory manufacturing pyrethrum-oil sprays. Patch tests, on the most susceptible individual, with all the ingredients used in the plant, showed that he was markedly hypersensitive to pyrethrum flowers. The perfumes and oils used gave negative or slight reactions.

The belief that pyrethrum is toxic to cold-blooded animals and non-toxic to warm-blooded animals seems to be well founded.

CHAPTER IV

EVALUATION OF PYRETHRUM BY CHEMICAL METHODS

Although the insecticidal properties of pyrethrum had been known for more than a century, no satisfactory method for determining the toxic principles was devised until 1929. The earlier chemical methods of analysis were used solely for the detection of adulterants or foreign materials and afforded no idea of the amount of toxic principles present in the flowers. The physiological tests on insects were the only methods, at that time, for comparing the relative toxicity of different samples and these were inaccurate and unsatisfactory even under the best conditions.

The identification of the pyrethrins by Staudinger and Ruzicka in 1924, opened the way for the development of chemical methods for determining the percentage of pyrethrins in the flowers.

Methods of Staudinger and Harder

The first chemical assay methods were published by Staudinger and Harder *(840)* in 1927. The pyrethrins, it will be recalled from the preceding chapter, are esters and form insoluble semicarbazones; each of these characteristics was used as the basis for a method of assaying pyrethrum. The two methods were known as the acid method and the semicarbazone method.

For the acid method Staudinger and Harder extracted 500 g. of pyrethrum powder with petroleum ether for about 18 hours. The petroleum ether was evaporated in a current of carbon dioxide, the last part being removed in vacuum. The residue was gently warmed with 50 cc. of methanol and cooled in freezing-mixture, whereupon resin and fat separated. The methanol solution was filtered and the insoluble material was again treated with 50 cc. of methanol, as above, and then six times with 30 cc. portions of methanol. The methanol solutions were diluted with 10 per cent of water and placed in freezing-mixture for 2 hours; a further separation of resin and fat occurred. The methanol solution was filtered and concentrated by removing the solvent in vacuum. The solution was saponified by heating for 8 hours with a five-fold excess of methyl-alcoholic soda. The solvent was removed in vacuum with gentle warming, water was added and the alkaline solution was extracted several times with ether to remove neutral material. The alkaline solution was acidified with dilute sulfuric acid and the chrysanthemum monocarboxylic acid was volatilized by distilling with steam for 1 to 2 hours. The monocarboxylic acid in the distillate was determined by titration with 0.1N sodium hydroxide, (using phenolphthalein as indicator) and from it the pyrethrin I content was calculated. The chrysanthemum dicarboxylic acid remained in the residue from the steam distillation. This residue was treated with animal charcoal and extracted for 18 hours with ether in a Kutscher-Steudel extractor. The ether was evaporated, the dicarboxylic acid was titrated with 0.1N soda solution and the pyrethrin II content was calculated from the titration.

The semicarbazone method followed the procedure just described to the point where the methanol solutions were distilled in vacuum; here, instead of saponifying the residue with alcoholic soda, the solvent was completely removed and the crude oil obtained was dissolved in 50 cc. methanol and treated with a solution of 1.2 g. semicarbazide hydrochloride and 2 g. sodium acetate in 8 cc. of water. After 24 hours the solvent was removed in vacuum. The residue was washed 4 to 5 times with 100 cc. portions of water and then dissolved in ether and again washed with water several times. The ether was evaporated and the nitrogen in the semicarbazones was determined by the Kjeldahl method. From the nitrogen content of the semicarbazones the pyrethrin content was readily calculated.

The following analyses published by Staudinger and Harder (Table VIII) are given to show the degree of accuracy which they obtained with the two methods:

TABLE VIII. COMPARISON OF RESULTS OBTAINED BY STAUDINGER AND HARDER

Kind of powder	Acid method Pyrethrin I %	Acid method Pyrethrin II %	Total %	Semicarbazone method Total pyrethrins %
Open Dalmatian G	.157	.155	.31	...
Open Dalmatian G	.142	.112	.25	...
Open Dalmatian G	.141	.110	.25	...
Swiss—A	.252	.156	.41	.27
Swiss—A	.256	.134	.39	...
Swiss—A	.204	.134	.34	...
Swiss—A	.270	.102	.37	...
Swiss—E	.162	.181	.34	.44
Powder H—B	.107	.146	.25	...
Powder H	.327	.183	.51	...
Powder R	.239	.017	.25	.23
Powder S	.338	.086	.42	.34

These were obviously not satisfactory quantitative methods since concordant results were not obtained on the same sample by the acid method, which also did not yield the same results as the semicarbazone method. By the acid method pyrethrin I and pyrethrin II were estimated separately while the semicarbazone method determined the total pyrethrin content. Inspection of Staudinger and Harder's analyses shows that the ratio of pyrethrin I to pyrethrin II in different samples of flowers, is not constant.

Early Methods of Tattersfield and His Associates

Tattersfield, Hobson and Gimingham *(880)* began by isolating the mixed pyrethrin semicarbazones by the method which Staudinger and Ruzicka had used. Also, like Staudinger and Ruzicka, they were unable to isolate pyrethrin II direct from the flowers and therefore they saponified the mixed semicarbazones, obtaining:

(A) pyrethrolone
(B) chrysanthemum monocarboxylic acid
(C) chrysanthemum dicarboxylic acid methyl ester

By esterifying (A) and (B) they synthesized pyrethrin I and from (A) and (C) they obtained pyrethrin II, following the same procedure Staudinger and Ruzicka *(850)* had used. Staudinger and Ruzicka had noted that the partially synthesized pyrethrins prepared in this way did not yield semicarbazones of definite melting point and suggested that a change might have taken place in the side chain or that isomers might be present. Tattersfield also has recognized that his pyrethrin II was possibly not pure.

Continuing his work with the partially synthesized pyrethrins, Tattersfield developed modifications of Staudinger and Harder's ester or acid method and also their semicarbazone method. Tattersfield's original acid method *(880)* gave slightly low results on samples very rich in pyrethrins. It was slightly modified to insure complete saponification of the pyrethrins *(590)*. The procedure finally adopted by Tattersfield is as follows:

For samples of poor quality—below 0.7 per cent total pyrethrins—10 g. are extracted with petroleum ether (b. p. below 40°) in a Soxhlet extractor. For samples between 0.7 and 1.5 per cent, 5 g., and for samples above 1.5 per cent pyrethrins, 2.5 g. are used.

"After taking down to a small bulk in a current of carbon dioxide and evaporating the solvent in a vacuum desiccator, the residue is extracted with four lots of 2.5 cc. each of gently warmed methanol, (previously purified by refluxing with caustic soda for several hours and then distilling), each of which is cooled and filtered into a 100 cc. Kjeldahl flask through a pad of cotton-wool. A final washing with 2.5 cc. cold methanol is made, a few drops of phenolphthalein in methanol are added and then, drop by drop till just alkaline, a solution of caustic potash in methanol of 1*N* strength. A further 5 cc. are added and the mixture is refluxed for a full 8 hours. The methanol is taken off in partial vacuum with gentle warming (the temperature not being allowed to rise above 25°), and the residue is dissolved in water. The solution is acidified with 1*N* sulfuric acid and the volatile acid is distilled with steam. The distillation apparatus used is similar to the original Pregl *(701)* micro-Kjeldahl still (Fig. II), except that the rubber connections with the condenser are eliminated so that the glass tube connects directly with a small worm condenser and leads to a receiver marked at the 50 cc. level. The volume in the distillation flask should not be allowed to exceed 30 cc. Two lots of 50 cc. are distilled and the acids in the first distillate are extracted with two lots of 50 cc. of petroleum ether, each extract being washed with 20 cc. of distilled water. The two extracts are combined, evaporated on a water bath after addition of 20 cc. of distilled water, and the residue is titrated while warm with 0.02*N* soda, the sides of the flasks being washed down towards the end with a little neutral methanol. Phenolphthalein is used as indicator. From the titration the amount of pyrethrin I can be determined. The second distillate of 50 cc.

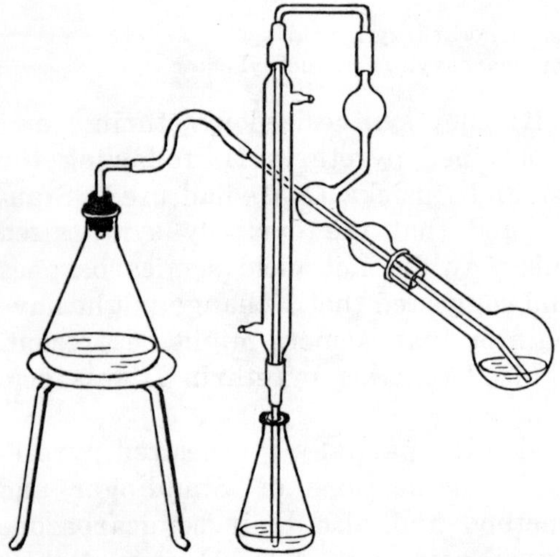

FIG. II. PREGL'S APPARATUS FOR DETERMINING NITROGEN.

may be extracted with petroleum ether; it should not show more than a trace of titratable acid. The hot aqueous residue in the distillation flask is treated with 0.2 g. of calcium sulphate and, after standing overnight, filtered through a cotton-wool plug, washed three or four times with water and extracted exhaustively with sodium treated ether in the apparatus shown in Fig. III. In a rapid extractor 20 hours extraction appears to be the minimum time necessary for complete extraction of the dicarboxylic acid in the case of samples of high pyrethrin content. After adding 20 cc. of distilled water the ether is evaporated, the aqueous layer is heated to boiling, cooled and filtered through a cotton-wool plug and the filtrate, after heating to boiling, is titrated with 0.02*N* soda.

1 cc. 0.02*N* alkali=3.36 mg. monocarboxylic acid= 6.60 mg. pyrethrin I
=1.98 mg. dicarboxylic acid= 3.74 mg. pyrethrin II."

According to Tattersfield, considerable care is necessary to obtain accurate results, especially in determining pyrethrin II. The process is rather prolonged.

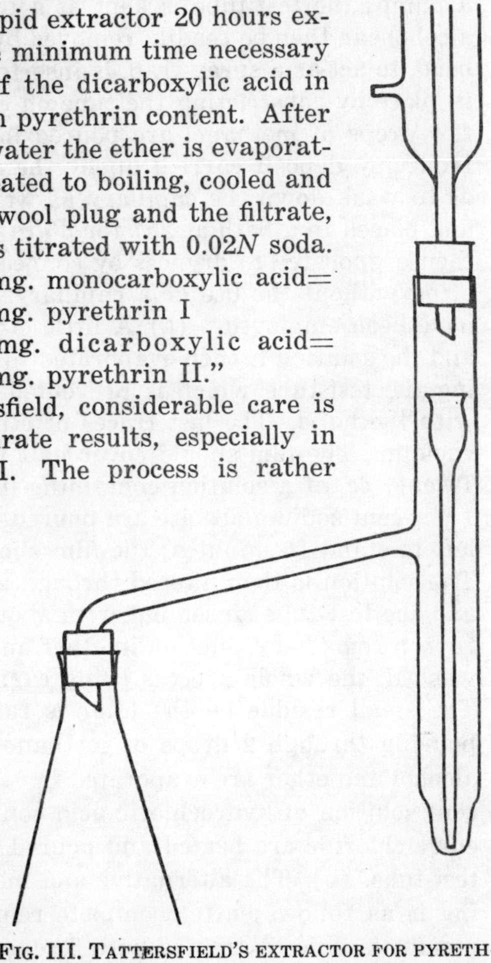

FIG. III. TATTERSFIELD'S EXTRACTOR FOR PYRETHRIN II (X ¼). ETHER IS BOILED IN THE FLASK, CONDENSES, FLOWS THROUGH THE INNER TUBE, PERCOLATES THROUGH THE AQUEOUS LAYER AND RETURNS TO THE FLASK.

The adaptation of Staudinger and Harder's semicarbazone method by Tattersfield, as a micro method, was considerably more difficult. The semicarbazone method, as finally adopted follows *(880)*:

"10 g. of the powdered flower-heads are extracted completely with low boiling petroleum ether and the residue, after evaporation of the solvent, is extracted with absolute methanol, in exactly the same way as for the acid method. The extracts are filtered successively into a small measuring cylinder and brought to a volume of 10 cc. The solution is mixed and 4 cc. are pipetted into a hard glass test-tube; 30 mg. of semicarbazide hydrochloride and 50 mg. of sodium acetate are added and the solution is concentrated to about 0.5 cc. This is done by inserting a rubber stopper having a capillary tube reaching to the bottom of the liquid and a side-tube which is connected to a pump; the test-tube is kept at a temperature of 25° and the alcohol can then be readily removed by evacuation. A small glass bead, to act as a spray-trap, is inserted in the side-tube and held in place by constricting the tube on either side. At the finish a few drops of methanol are poured into the side-tube, to free it from any deposit carried up by the spray, and so manipulated as to wash down the capillary as well. The test-tube is corked and placed in a bath at 25° for 36 to 48 hours. The methanol is then evaporated to dryness by connecting to a pump and evacuating without the use of a capillary. Two methods of washing have been employed. (a) A little ether is added to the residue and the solution is then evaporated to a film by tilting and rotating the test-tube which is prevented from cooling by warming with the hand. The last traces of ether are removed in a good vacuum. The film should cover half the length of the test-tube. Twenty cc. of a solution containing 0.1 per cent acetic acid and 1 per cent sodium acetate are poured carefully into the tube and left to stand 10 minutes; the film should be completely covered. The solution is then filtered through a small plug of cotton-wool and the test-tube rinsed out with about 2 cc. of water. The film is then remade by solution in ether and evaporation and is again washed, the whole process being carried out three times in all. The small residue on the filter is returned to the test-tube by pouring through 2 drops of methanol and 1 cc. of ether. The alcohol and ether are evaporated in vacuum. Five cc. of a 15 per cent solution of hydrochloric acid containing 5 per cent of mercuric chloride are heated and poured through the filter into the test-tube. (b) The alternative and more rapid method of washing is as follows: after complete removal of the methanol, the residue is twice extracted with 5 cc. of ether and the ether solution poured into a small separating funnel. The residue in the test-tube is washed four times with 10 cc. of water, the washings are successively filtered through a cotton-wool plug into the separating funnel and used to wash the ether solution with gentle rotation. The ether solution is washed twice more with small amounts of water, returned to the test-tube and evaporated. Five cc. of the 15 per cent hydrochloric acid solution containing 5 per cent of mercuric chloride are heated and used to wash into the test-tube the insoluble material left in the separating funnel and the filter, the cotton-wool plug in the latter being previously loosened.

"The test-tube is fixed to a small reflux condenser and, after the addition of a little pumice, the solution is boiled for seven hours. Ammonia is then determined by Pregl's revised micro-Kjeldahl method *(701)*. The solution used for rendering alkaline is prepared by diluting a 40 per cent caustic soda solution with an equal volume of saturated sodium thiosulfate solution, which is necessary for the decomposition of the mercury ammonium complex. The acid used for absorbing the ammonia is $0.02N$ and $0.02N$ alkali is used for the titration. Using the mean molecular weight of 350 for the pyrethrins, 1 cc. of $0.02N$ acid is equivalent to 70 mg. of mixed pyrethrins I and II."

The following analyses reported by Tattersfield *(880)* show the excellent agreement between his acid and semicarbazone methods (Table IX):

TABLE IX. COMPARISON OF RESULTS OBTAINED BY TATTERSFIELD'S METHODS

Sample No.	Total pyrethrin content	
	Acid method %	Semicarbazone method %
1	0.71	0.73
2	0.59	0.59
3	0.88	0.88
4	0.81	0.84
5	1.01	1.05
6	0.69	0.73
7	1.13	1.13
8	0.71	0.77

Tattersfield's semicarbazone method is so long and tedious that it is hardly usable as a practical method.

Tattersfield and Hobson *(878)* later described a short method for determining pyrethrin I. This method was offered in the belief that pyrethrin I is ten times as toxic as pyrethrin II and, therefore, the pyrethrin II content is of no importance. A detailed discussion of this question is given in Chapter VI.

The following is the procedure for Tattersfield's short acid method for pyrethrin I *(878)*:

"10 g. of the ground pyrethrum are extracted in a Soxhlet apparatus by means of petroleum ether (boiling range 40° to 50°), which is kept vigorously boiling over a carbon-filament lamp. The extraction is continued until the ether draining over is colorless. The petroleum ether solution which should have a volume of approximately 50 cc. is then poured into a long-necked flask of 100 cc. capacity, subsequently used for distillation, the extraction flask being rinsed once with a little petroleum ether, 4 to 5 cc. of 1*N* caustic soda in methanol added and the mixture vigorously refluxed on the water-bath for 1½-2 hours. The mixture is then acidified with 1*N* sulfuric acid and distilled in steam in the apparatus previously described (Fig II). Petroleum ether distills first and until it is completely removed, a flame is not placed under the distillation flask. To prevent risk of fire the worm-condenser is attached to the receiver by means of a cork with two holes, one of which contains a glass tube with a length of rubber tubing to carry any inflammable vapor away from the flame. Distillation is continued until 50 cc. of aqueous distillate stand below the petroleum ether in the receiver, after which another flask is attached and distillation continued until a further 50 cc. have distilled. The whole of the first distillate is then transferred to a fairly large separating-funnel and vigorously shaken for one minute, the aqueous layer is separated and the petroleum ether layer, after washing once with water, is run off into a flask containing 20 cc. of water, to which a few drops of alcohol and phenolphthalein have been previously added together with just enough alkali to render the liquid a faint pink. Titration is carried out with 0.02*N* soda until the aqueous layer is distinctly alkaline after vigorous shaking in the corked flask. The second 50 cc. portion of distillate is added to the first aqueous fraction (which has already been extracted once), vigorously shaken in a separating-funnel with 50 cc. of petroleum ether, the washed petroleum ether layer added to the titration flask and the titration finished as before; very little additional 0.02*N* soda is usually required. After deducting a blank which should be determined for the petroleum ether (about 0.2 cc. 0.02*N* soda), the monocarboxylic acid and pyrethrin I content can be calculated, the following factors being used:

1 cc. 0.02*N* alkali=3.36 mg. monocarboxylic acid
1 cc. 0.02*N* alkali=6.60 mg. pyrethrin I"

Copper Reduction Method of Gnadinger and Corl

Within a few days of the publication of Tattersfield's first methods Gnadinger and Corl *(343)* published a rapid quantitative method based on an entirely different principle. They had used a modification of Staudinger and Ruzicka's method for isolating the pyrethrins and had isolated 220 grams of the mixed semicarbazones of pyrethrin I and pyrethrin II. From the mixed semicarbazones, the semicarbazone of pyrethrin I was separated in nearly pure condition (m.p. 117°-119°) by repeated crystallization of the crude mixture from benzene and alcohol. The pure semicarbazone of pyrethrin II could not be isolated, but after fifty-one crystallizations from benzene, alcohol and mixtures of benzene and petroleum ether, a semicarbazone was obtained that melted at 56°-59°. Both semicarbazones crystallized from benzene in white needles. By heating the semicarbazones with oxalic acid solution they were converted into the corresponding pyrethrins. The crude pyrethrins were extracted from the oxalic acid solution with petroleum ether and washed free from acids with 1 per cent sodium hydroxide solution. Further purification was effected by washing with 3 per cent chromic acid solution and 3 per cent potassium permanganate solution, which did not attack the pyrethrins appreciably. The petroleum ether was then removed in vacuum and the residue was dissolved in alcohol, whereby a small amount of insoluble matter was precipitated. The alcohol sclution was filtered and distilled in vacuum, the residue was dissolved in petroleum ether and the solution was filtered and distilled in vacuum at a maximum temperature of 55° until constant weight was attained.

Ten grams of semicarbazone of pyrethrin I (m.p. 117°-119°) yielded 3.53 g. of pure pyrethrin I; 10 g. of semicarbazone of pyrethrin II (m.p. 56°-59°) yielded 2.76 g. of nearly pure pyrethrin II. Analyses:

Calculated for pyrethrin I: C, 76.31; H, 9.16
Found : C, 75.95; H, 9.32
76.36; 9.06
Calculated for pyrethrin II: C, 70.54; H, 8.08
Found : C, 72.71; H, 8.18

The insecticidal activity of the isolated pyrethrins was determined on roaches *(Blatta germanica)*. Pyrethrin I was slightly more toxic than pyrethrin II, as Staudinger and Ruzicka had noted.

Staudinger and Ruzicka mentioned that pyrethrolone reduces alkaline copper solution. It was found that the pyrethrins have the same property and the idea of using the reaction for determining them, at once suggested itself. The copper reducing power of the pyrethrins was found to be considerably less than that of dextrose and other reducing sugars, and for this reason the gravimetric sugar methods could not be applied satisfactorily. Of the other methods, the colorimetric method of Folin, for determining dextrose in blood, was the most promising. In this method a measured amount of specially treated blood is heated with an alkaline copper solution, in a tube designed to prevent oxidation of the precipitated cuprous oxide. A tube containing a known amount of dextrose is treated in precisely the same manner. When the heating is finished, the Folin phosphomolybdate reagent is added to the tubes and a deep blue color is developed by the action of the cuprous oxide on the phosphomolybdate solution. The intensities of the colors are compared and the amount of dextrose in the blood is calculated in the usual way.

In a long series of experiments this method was adapted to the determination of pure pyrethrins in alcoholic solution. A special size of Folin sugar tube was used because oxidation of the reduced copper took place rapidly when test tubes were used as suggested by Benedict. Several modifications of Folin's reagent were tried and a large number of alkaline copper solutions were investigated. The effects of varying the temperature and time of the reaction were noted. More than 700 experiments were made before the best procedure for conducting the test was determined.

Comparisons of the colors produced by known amounts of the pyrethrins with those obtained with known amounts of dextrose, indicated that dextrose could be used as a standard, as in the Folin method. The ratio of the pyrethrin used to the copper reduced was not a constant. A number of comparisons were made between different quantities of pyrethrins and a constant amount of dextrose; the results of these experiments were plotted. The curves were not linear functions, but more nearly conformed to the parabolic equation used by Allihn in his work on dextrose. In Table X the amounts of dextrose equivalent in copper reducing power to different quantities of pyrethrin I and pyrethrin II are given. These ratios were obtained by comparing the color produced by 2 mg. of dextrose with the colors yielded by the different weights of pyrethrins; the method by which these comparisons were made is described on page 58.

TABLE X. COMPARISON OF COPPER REDUCING POWER OF DEXTROSE AND PYRETHRINS

Pyrethrin I mg.	Weights having equivalent copper reducing power Dextrose mg.				Avg. mg.	Pyrethrin II mg.	Dextrose mg.		Avg. mg.
5.0	0.735	0.767	0.787	0.797	0.772	...	....	....	....
7.5	1.166	1.175	1.228	1.246	1.204	7.5	0.973	1,003	0.988
12.5	2.020	2.036	2.046	2.061	2.041	12.5	1.636	1.642	1.639
15.0	2.222	2.222	2.253	2.292	2.247	15.0	1.956	1.980	1.968
17.5	2.469	2.500	2.500	....	2.489	17.5	2.253	2.272	2.263
20.0	2.730	2.826	2.836	....	2.797	20.0	2.402	2.539	2.471

By substituting the values in Table X in the equation $y=a+bx+cx^2$, (where y=mg. of dextrose, x=mg. of pyrethrin, and a, b and c are constants) and applying the method of least squares, two equations were obtained.

Equation I (for pyrethrin I): $y= -0.2536+0.2237x -0.00365x^2$

Equation II (for pyrethrin II): $y= -0.0101+0.1437x -0.0009x^2$

From these equations the dextrose equivalent to the quantities of pyrethrins in Table X was calculated; the observed and calculated values are compared in Table XI, where the ratio of the reducing power of pyrethrin I to that of pyrethrin II also appears.

TABLE XI. OBSERVED AND CALCULATED RESULTS AND COMPARATIVE REDUCING POWER OF PYRETHRINS I AND II

Pyrethrin mg.	Dextrose equiv. to Py. I Obs., mg.	Calcd., mg.	Diff., mg.	Dextrose equiv. to Py. II Obs., mg.	Calcd., mg.	Diff., mg.	Ratio* Py. I : Py. II
5.0	0.772	0.773	0.001	....	....	....	...
7.5	1.204	1.218	0.014	0.988	1.017	0.029	1.22
12.5	2.041	1.972	–0.069	1.639	1.645	0.006	1.24
15.0	2.247	2.280	0.033	1.968	1.943	–0.025	1.14
17.5	2.489	2.543	0.054	2.263	2.229	–0.034	1.10
20.0	2.797	2.760	–0.037	2.471	2.504	0.033	1.13
Avg.			–0.0007			0.0018	1.166

* Calculated from dextrose equivalent to given weight of pyrethrin I and pyrethrin II; theoretical ratio, 1.133 (374/330).

It was expected that the amounts of dextrose equivalent to a given weight of the two pyrethrins would be inversely proportional to the molecular weights of the pyrethrins. This is shown to be approximately the case in Table XI. Direct colorimetric comparisons were made between equal weights of pyrethrin I and pyrethrin II. The colorimetric readings were almost directly proportional to the molecular weights of the pyrethrins.

By substituting different values for y in Equation I, the weights of pyrethrin I equivalent to amounts of dextrose from 0.750 to 2.875 mg. were calculated. These calculations were made at intervals of 0.125 mg. of dextrose and other values were interpolated at intervals of 0.025 mg. In this way the second column of Table XII was formed.

TABLE XII. COPPER REDUCING POWER OF DEXTROSE AND PYRETHRINS*

Dextrose	Pyrethrin I	Pyrethrin I and II	Dextrose	Pyrethrin I	Pyrethrin I and II	Dextrose	Pyrethrin I	Pyrethrin I and II
mg.	mg.	mg.	mg.	mg.	mg.	mg.	mg.	mg.
0.750	4.87	5.19	1.475	9.07	9.68	2.200	14.31	15.26
0.775	5.01	5.34	1.500	9.23	9.85	2.225	14.52	15.49
0.800	5.14	5.48	1.525	9.39	10.02	2.250	14.73	15.71
0.825	5.28	5.63	1.550	9.55	10.19	2.275	14.95	15.94
0.850	5.41	5.77	1.575	9.72	10.37	2.300	15.17	16.18
0.875	5.55	5.92	1.600	9.88	10.54	2.325	15.40	16.42
0.900	5.69	6.07	1.625	10.04	10.71	2.350	15.62	16.66
0.925	5.82	6.21	1.650	10.21	10.89	2.375	15.85	16.91
0.950	5.96	6.36	1.675	10.38	11.07	2.400	16.09	17.16
0.975	6.10	6.51	1.700	10.55	11.25	2.425	16.33	17.42
1.000	6.24	6.66	1.725	10.72	11.43	2.450	16.57	17.68
1.025	6.38	6.81	1.750	10.89	11.62	2.475	16.81	17.93
1.050	6.52	6.95	1.775	11.07	11.81	2.500	17.05	18.19
1.075	6.67	7.11	1.800	11.24	11.99	2.525	17.31	18.46
1.100	6.81	7.26	1.825	11.42	12.18	2.550	17.58	18.75
1.125	6.95	7.41	1.850	11.60	12.37	2.575	17.85	19.04
1.150	7.10	7.57	1.875	11.78	12.57	2.600	18.11	19.32
1.175	7.25	7.73	1.900	11.96	12.76	2.625	18.38	19.61
1.200	7.39	7.88	1.925	12.15	12.96	2.650	18.67	19.91
1.225	7.54	8.04	1.950	12.33	13.15	2.675	18.95	20.21
1.250	7.69	8.20	1.975	12.52	13.35	2.700	19.25	20.53
1.275	7.84	8.36	2.000	12.71	13.56	2.725	19.56	20.86
1.300	7.99	8.52	2.025	12.90	13.76	2.750	19.86	21.18
1.325	8.14	8.68	2.050	13.10	13.97	2.775	20.20	21.55
1.350	8.29	8.84	2.075	13.30	14.19	2.800	20.55	21.92
1.375	8.44	9.00	2.100	13.49	14.39	2.825	20.89	22.28
1.400	8.60	9.17	2.125	13.69	14.60	2.850	21.24	22.65
1.425	8.76	9.34	2.150	13.90	14.82	2.875	21.59	23.03
1.450	8.92	9.51	2.175	14.11	15.05			

* Third column calculated for a mixture of equal parts of pyrethrins I and II.

The proportion of the two pyrethrins in the flowers varies (page 117). It was assumed that the pyrethrins occur in equal amounts and from the values for pyrethrin I, in Table XII, the corresponding values for a 1:1 mixture of pyrethrins I and II were calculated, using the average molecular weight 352 and multiplying the figures for pyrethrin I by the factor 1.06666, (third column, Table XII). The error due to this assumption is negligible. The copper reducing power of the pyrethrins is inversely proportional to their molecular weights, that is, 330 g. of pyrethrin I is equivalent in reducing power to 352 g. of 1:1 mixture of pyrethrins I and II, or 374 g. of pyrethrin II. From this, 1 mg. of pyrethrin I is equivalent in reducing power to 1.067 mg. of 1:1 mixture of pyrethrins I and II; 1 mg. of pyrethrin II is equivalent in reducing power to 0.941 mg. of 1:1 mixture of pyrethrins I and II.

Let us apply these equivalents to three samples of pyrethrum, each containing 1 per cent total pyrethrins, but having different ratios of pyrethrin I to pyrethrin II; furthermore, let the variations in the ratio of I to II be the greatest so far reported (page 118).

No.	Pyrethrins I	Pyrethrins II	Ratio I:II	Amount of 1:1 mixture of pyrethrins I and II equivalent in reducing power to Pyrethrin I	Amount of 1:1 mixture of pyrethrins I and II equivalent in reducing power to Pyrethrin II	Total pyrethrin content which would be shown by copper reduction method
	%	%		%	%	%
1	0.50	0.50	1:1.00	0.53	0.47	1.00
2	0.61	0.39	1:0.64	0.65	0.37	1.02
3	0.29	0.71	1:2.45	0.31	0.67	0.98

The maximum error, due to the greatest variations in the ratio of pyrethrin I to pyrethrin II so far reported, would be 0.04 per cent.

The semicarbazones of the pyrethrins had little or no copper reducing action. The by-products of the semicarbazone conversion, with oxalic acid, were carefully examined for copper reducing compounds; none was found that resisted the purification with sodium hydroxide, chromic acid and permanganate. The pyrethrins were altered by prolonged heating at 90°; they became insoluble in petroleum ether and lost their activity but the copper reducing power was slightly increased. Saponification of pyrethrin I with alcoholic sodium hydroxide solution lowered the copper reducing power.

The semicarbazone of pyrethrolone was prepared from the mixed semicarbazones of pyrethrin I and II by saponification at 0°, with a methanol solution of sodium hydroxide. The pyrethrolone semicarbazone melted at 203°; it was converted into the ketone-alcohol by prolonged shaking with benzene and potassium bisulfate solution.

The copper reducing power of pyrethrolone was slightly less than that of pyrethrin I. Pyrethrolone was dissolved in petroleum ether and washed with 3 per cent potassium permanganate solution; 98.5 per cent of the pyrethrolone was oxidized and removed from the petroleum ether. The residue from the evaporation of the petroleum ether solution had very little copper reducing action.

Having found that the amount of pure pyrethrins in alcoholic solution could be determined accurately, it remained to apply the method to pyrethrum flowers and to eliminate, by appropriate means, any other copper reducing material that might be present.

Preliminary tests with petroleum ether extracts yielded percentages of pyrethrins of about the magnitude expected. Washing the petroleum ether extracts with 1 per cent sodium hydroxide or with concentrated sodium bisulfite solution did not remove any material that reduced alkaline copper solution.

Staudinger and Ruzicka found that the pyrethrins in petroleum ether solution were not oxidized by chromic acid solution. Washing the petroleum ether extracts of pyrethrum flowers with chromic acid solution did not decrease the amount of copper reducing material appreciably. No petroleum ether soluble compounds that reduced copper could be found, excepting the pyrethrins.

The residues from the purification of the semicarbazones of the pyrethrins were carefully examined for petroleum ether soluble copper reducing compounds but none was found.

No other compounds that formed semicarbazones could be isolated from the petroleum ether extract excepting the pyrethrins; this agrees with the findings of Staudinger and Ruzicka.

Application of the method to daisy flowers *(Chrysanthemum leucanthemum)* showed no pyrethrins present and tests on pyrethrum stems yielded only traces of pyrethrins. Finally, the pyrethrins were isolated from 2500 g. of pyrethrum flowers and weighed as semicarbazones. The same flowers were assayed by the copper reduction method. The gravimetric method yielded 0.66 per cent of pyrethrins; the copper reduction method 0.88 per cent. In view of the unavoidable losses in separating the semicarbazones, the agreement is good.

It is believed, therefore, that the following method determines only the toxic principles in pyrethrum flowers.

Reagents:

1. Petroleum ether, 90-99 per cent distilling between 20° and 40°; maximum boiling point, 60°. Petroleum ether which does not meet these specifications should not be used.

2. Aldehyde-free alcohol.—Allow 95 per cent alcohol, containing 5 g. of *m*-phenylenediamine hydrochloride per liter, to stand for 24 hours with frequent shaking. Boil under a reflux condenser for at least 8 hours, allow to stand overnight and distil, rejecting the first 10 per cent of distillate and the last 5 per cent of solution. Store in a dark place in well-filled bottles, tightly corked.
3. Basic lead acetate solution.—Dissolve 20 g. of Horne's basic lead acetate in sufficient recently boiled water to make 1 liter.
4. Alkaline copper solution.—Dissolve 2.500 g. of purest copper sulfate, $CuSO_4.5H_2O$, in about 100 cc. of water, warming gently; cool when dissolved. Dissolve 5.000 g. of highest purity sodium potassium tartrate and 7.500 g. of purest sodium hydroxide (at least 96 per cent NaOH) separately in about 100 cc. of cold water. Transfer the solutions to a 500-cc. volumetric flask, mix and dilute to the mark. This solution should not be used after it is three days old.
5. Folin's reagent.—Dissolve 150 g. of sodium molybdate, $Na_2MoO_4.2H_2O$, in 300 cc. of water. Filter through a 15-cm. quantitative filter paper into a 1-liter flask and wash with 75 cc. of water. Add 0.1 to 0.2 cc. of bromine and shake until the bromine is dissolved. Let stand 1 hour, then add with shaking 225 cc. of 85 per cent phosphoric acid. Add 150 cc. of sulfuric acid (1. vol. of concentrated acid mixed with 3 vols. of water and cooled). Remove the liberated bromine by means of a moderately rapid current of air; the aeration requires about half an hour. Finally add 75 cc. of 99 per cent acetic acid, mix and dilute to a volume of 1 liter.
6. Standard dextrose solution.—Dissolve 1.0000 g. of pure anhydrous dextrose, accurately weighed, in water and transfer to a 200-cc. volumetric flask. Add 40 cc. of aldehyde-free alcohol, mix and dilute to 200 cc. with water. Transfer 10 cc. of this solution to a 250-cc. volumetric flask by means of a pipet, add 210 cc. of aldehyde-free alcohol and dilute to 250 cc. with water. Ten cc. of this solution contain 2 mg. of dextrose. The strong dextrose solution is stable for months; the dilute solution should be made fresh each week.

Apparatus:

1. Constant-temperature water bath, set at 78°, corrected, and controlled within ± 0.2°. An oil bath should not be used instead of a water bath.
2. Colorimeter of Duboscq or Klett type, preferably with artificial illuminator.
3. Folin sugar tubes, blown to contain 15.5 cc. to base of constriction. When heated to 78° the surface of the liquid must fall between points A and B, Fig. IV. The internal diameter of the constricted portion should be the same for all tubes in a set.

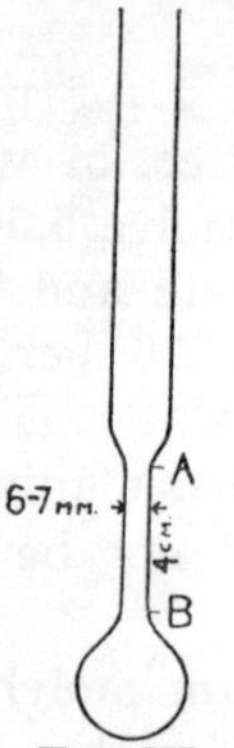

FIG. IV. MODIFIED FOLIN TUBE. BULB CONTAINS 15.5 CC. TO POINT B.

Determination:

Extract 15 g. of ground pyrethrum flowers (about 30 mesh) for 5 hours with petroleum ether in a Soxhlet extractor. For samples very low in pyrethrins 20 g. may be used and for high test samples 10 g. are sufficient. Cool the petroleum ether solution, which should measure less than 100 cc., to about 20° and let stand for at least half an hour, or allow to stand overnight at room temperature. Filter through a quantitative filter paper into a 400-cc. beaker, add a few grains of ignited sand and evaporate at a temperature not exceeding 75°. As soon as the last traces of petroleum ether are driven off, transfer the residue immediately with 5 or 6 portions of boiling 95 per cent aldehyde-free alcohol to a 100-cc. volumetric flask (previously marked at the 80-cc. point), using sufficient boiling alcohol to make the volume 80-85 cc. To the hot solution add from a pipet 15 cc. of basic lead acetate solution and make to the mark with hot alcohol. Shake vigorously, cool at once to 20° and again make to the mark with alcohol. Filter and to the filtrate add about 1 g. of anhydrous sodium carbonate. Let stand for 10 to 15 minutes, shaking frequently, and filter. Immediately pipet 10 cc. of the clear filtrate into a Folin tube and add, also from a pipet, 6 cc. of alkaline copper solution. Mix thoroughly, keeping the solution in the bulb of the tube. Measure 10 cc. of standard dextrose solution (2 mg. of dextrose) with a pipet into a second tube and add 6 cc.

of copper solution. Place the tubes upright in the constant-temperature bath, set at 78° corrected, and heat for exactly 45 minutes. Remove from the bath and place in water at 20° for 3 minutes. Add 10 cc. of Folin reagent from a pipet and let stand for 3 minutes; then stopper the tubes, mix thoroughly, transfer to 100-cc. volumetric flasks and make to the mark with water. Filter through a Gooch crucible fitted with a heavy asbestos pad, using gentle suction, and discard the first 10 to 20 cc. Do not use filter paper. The dextrose solution need not be filtered. Compare the solutions at once in the colorimeter and from the readings calculate the dextrose equivalent to the unknown solution in the usual way. Reference to the third column of Table XII will give the amount of pyrethrins, in milligrams, in the unknown solution, or milligrams of pyrethrins in one-tenth of the weight of flowers taken for extraction. The following example shows the method of calculation:

Weight of flowers taken......................	15 g.
Colorimetric reading of standard dextrose......	20 mm.
Colorimetric reading of unknown..............	24 mm.
Dextrose equivalent to unknown, $\frac{20\times 2}{24}$, or	1.667 mg.
Pyrethrins equivalent to dextrose, Table XIII....	11.01 mg.
Aliquot of flowers taken, $15\times\frac{10}{100}$, or 1.5 g.	
Pyrethrins in 1.5 g. = 11.01 mg., or 0.734 per cent.	

In making the color comparisons the standard dextrose is set at 20 mm. The unknown will then read between 14 and 50 mm., for quantities between 23 and 5 mg. of pyrethrins. If the reading is less than 12 mm., the entire amount of copper may have been reduced, and the determination should be repeated using a smaller aliquot. It is also desirable, if the reading is more than 40 mm., to run a duplicate determination using 40 g. of flowers.

Before making comparisons it is advisable to be sure that the colorimeter is not out of adjustment. This is easily done by filling both colorimeter cups with the blue solution from the standard dextrose and setting them at the 20 mm. mark. If the instrument is properly adjusted the colors will, of course, match. If the lighting is uneven, or if air bubbles collect on the tips of the plungers, the colors will not be the same. An air bubble on the tip of a plunger may cause an error of 2 mm.;

the solution will have a grayish color. After adjusting the instrument so that the colors match, when the dextrose standard is placed in both cups and set at 20 mm., leave the left cup in position, remove the right cup, rinse it several times with the unknown solution, fill and read; remove the right cup to be sure no air bubbles are trapped on the plunger tip and repeat the reading.

When assaying freshly harvested flowers, difficulty is sometimes experienced in obtaining a clear filtrate of the alcoholic solution which has been clarified with lead acetate. The slight cloudiness of the filtrate does not affect the accuracy of the results. In such cases 2 g. of anhydrous sodium carbonate should be used to remove the excess of lead.

Blank determinations should be run on every new lot of reagents. No difficulty was experienced in obtaining blanks equivalent to 0.05 mg. of dextrose, and for this reason no correction for the blank was made in calculating Table XII. Reagents that yield a blank equivalent to more than 0.10 mg. of dextrose should be rejected. High blanks may be due to impurities in any of the reagents, but the sodium potassium tartrate, sodium molybdate and aldehyde-free alcohol especially should be carefully tested. If the blank is between 0.05 and 0.10 mg. of dextrose, when compared with the standard dextrose solution, the error due to the blank is negligible.

The petroleum ether extract should not be heated longer than necessary to drive off the solvent. Generally only 5 or 10 minutes' heating above 60° will be required if the petroleum ether meets the specifications given.

The proportion of alcohol and water in the contents of the Folin tubes is carefully adjusted; if more alcohol is used the copper salts will be precipitated; if the percentage of water is increased the pyrethrins will be thrown out of solution. In either case incorrect results will be obtained. The Folin tubes should be dried before using and the measurements should be made with pipets. Smaller aliquots than 10 cc. can be used for a determination, but sufficient aldehyde-free alcohol (80.5 per cent) must be added to the tube to make the total volume of the pyrethrin solution 10 cc. Larger aliquots than 10 cc. cannot be used.

It is essential that the bath be maintained at 78° corrected, ±0.2°, and a stirrer must be provided to insure even temperature and circulation around the tubes, which should be immersed to a depth of 8 to 10 cm. Variations in time of heating or in temperature will yield results that are not comparable with Table XII. After removing the tubes from the bath they should be treated as nearly alike as possible; therefore it is inadvisable to run more than 3 or 4 tubes at a time.

The dextrose used as standard in this work was of the highest purity obtainable and was dried for 30 days over sulfuric acid. All volumetric glassware, weights and thermometers used were Bureau of Standards certified. Colorimeter readings were made with artificial light and a blue glass screen.

The foregoing method was applied to pyrethrum flowers and stems and to daisy flowers. The pyrethrin content varied, in the samples examined, from 0.40 to 1.21 per cent; duplicate determinations agreed within 0.03 per cent or less. Stems contained only one tenth as much pyrethrins as the flowers; daisy flowers contained no pyrethrins. The range of the pyrethrin content was greater than reported by Staudinger and Harder (0.4 to 0.6 per cent), or by Staudinger and Ruzicka (0.3 per cent). Tattersfield at first reported from 0.59 to 1.29 per cent pyrethrins, but later assayed flowers containing as high as 2.30 per cent. The writer has examined American grown flowers containing 2.07 per cent pyrethrins.

Ripert *(745)* has stated that the lead acetate purification of the Gnadinger-Corl method causes low results, since the fatty acids and their esters are precipitated in the process, removing pyrethrins from the solution. Ripert does not present data to support his statement. That the lead acetate treatment does not introduce an appreciable error can readily be shown by making duplicate determinations on the same sample of flowers, using the regular lead acetate solution in one determination and substituting distilled water for lead acetate solution in the other. An experiment of this type showed:

with lead acetate solution	0.98% pyrethrins
using water instead of lead acetate	0.98% pyrethrins

In a second experiment the lead acetate solution was replaced with aldehyde free alcohol; analysis showed but slight difference:

with lead acetate solution	1.22% pyrethrins
with alcohol instead of lead acetate	1.28% pyrethrins

Tatu *(883)* has suggested that the cuprous oxide obtained in the Gnadinger-Corl method be determined volumetrically instead of colorimetrically. He transfers the cuprous oxide to a filter and washes it with water, taking precautions to prevent oxidation by the air; he then dissolves it in ferric sulfate solution acidified with sulfuric acid and titrates the resulting ferrous sulfate with standard permanganate solution.

Raupp *(705)* has also described a volumetric modification of the Gnadinger-Corl method. The determination described on page 60 is followed to line 37; from this point Raupp proceeds as follows:

> Dissolve 7 g. of copper sulfate in sufficient water to make 400 cc.; dissolve 35 g. of sodium potassium tartrate and 10 g. of sodium hydroxide in sufficient water to make 100 cc. Measure 20 cc. of the copper sulfate solution and 5 cc. of the alkaline tartrate solution into a 200 cc. Erlenmeyer flask and boil for 2 minutes; cool, add 20 cc. of the alcoholic pyrethrin solution, prepared as described above, boil again for 2 minutes and cool. In a second 200 cc. Erlenmeyer flask, measure 20 cc. of the copper sulfate solution and 5 cc. of the alkaline tartrate solution, boil for 2 minutes and cool. To each of the flasks add slowly a solution of 1 g. potassium iodide, 20 cc. of water and 10 cc. dilute sulfuric acid (1:5). Mix and titrate with 0.1*N* sodium thiosulfate solution, using starch solution as indicator. From the difference in the two titrations the amount of copper reduced by the pyrethrins is calculated; from this, the dextrose equivalent to the copper is obtained from Allihn's table and from the dextrose the pyrethrins are obtained from the Gnadinger-Corl table; the percentage of pyrethrins in the flowers is then calculated.

Later Methods of Martin and Tattersfield

Martin and Tattersfield have adapted the principle of the Gnadinger-Corl method to the analysis of minute amounts of material. Their method is intended for assaying such small amounts as 0.5 g. of pyrethrum, or even a single flower head. According to its authors, considerable care and some practice in the technique are required for the satisfactory application of this method. Their description of the method is as follows *(590)*:

"Reagents:

1. Alkaline ferricyanide solution.—Dissolve 1.649 g. of potassium ferricyanide and 28.6 g. of anhydrous sodium carbonate in sufficient water to make 1 liter of solution.
2. Ferrocyanide precipitant.—Dissolve 5 g. of potassium iodide, 10 g. of hydrated zinc sulfate and 50 g. of sodium chloride in water and dilute to 200 cc.

3. Acetic acid.—Dilute 3 cc. of glacial acetic acid to 100 cc. with water.
4. Zinc sulfate solution.—Dissolve 1.25 g. hydrated zinc sulfate in water and dilute to 250 cc.
5. Starch solution.—Dissolve 1 g. soluble starch and 20 g. sodium chloride in water and dilute to 100 cc.
6. Aldehyde-free absolute alcohol.—Prepare as described on page 59, using absolute alcohol instead of 95 per cent alcohol.
7. Thiosulfate solution, 0.005*N*.—Dissolve 1.243 g. sodium thiosulfate in recently boiled water and dilute to 1 liter; protect with a soda-lime tube. The thiosulfate solution is most readily standardized by titration of 0.005*N* solution of potassium iodate, prepared by weighing exactly 0.1784 g. of the dried solid and making up to a volume of 1 liter. Ten cc. are delivered into a small flask by means of a carefully calibrated pipet. The addition of 5 cc. of 2 per cent potassium iodide solution, followed by 3 cc. of 3 per cent acetic acid, liberates the iodine equivalent of the potassium iodate, which is then titrated with the thiosulfate solution."

"Determination:

0.5 g. of ground flowers is extracted with petroleum ether (b.p. below 40°) in a Soxhlet apparatus, and the solvent is removed by gentle warming in a current of carbon dioxide, final traces being removed in a vacuum desiccator. The residue is extracted on a boiling water bath with 5 successive portions of 4 cc. each of aldehyde-free absolute alcohol. To the hot solution 1 cc. of 0.1*N* sodium hydroxide and 4 cc. of dilute zinc sulfate are added, the solutions are mixed, and warmed on the water bath for a few minutes, precipitation of proteins being thus effected. The solution is cooled to 20°, made up to 25 cc. with aldehyde-free alcohol, shaken and allowed to stand. The final protein-free extract is obtained by filtering through a small Whatman No. 1 filter paper. For the estimation, 2 cc. of extract (delivered from a fine nozzle and standardized pipet) are heated for 45 minutes in a boiling water bath with 10 cc. of alkaline ferricyanide solution (accurately delivered), in a Folin tube, as modified by Gnadinger and Corl, Fig. IV. The solution is then cooled, washed into a small conical flask, and excess of ferrocyanide precipitant immediately added. The iodine equivalent of the remaining ferricyanide is liberated by the addition of 10 cc. of 3 per cent acetic acid, and, by means of a micro-burette, titrated with thiosulfate, using starch as indicator toward the end of the reaction. A blank determination on 10 cc. of ferricyanide is carried out, using 2 cc. of 80 per cent alcohol. From the difference in the thiosulfate titrations the amount of pyrethrins in 2 cc. of the filtrate may be read directly from the graph, Fig. V. The pyrethrins in 25 cc. of the extract and, therefore, in 0.5 g. of material used, can thus be readily obtained."

The graph (Fig. V) was constructed by preparing solutions of pyrethrins of different strengths from a sample of pyrethrum which had been assayed by Tattersfield's acid method and the Gnadinger-Corl method. The above method was applied to these known solutions and from the results the graph was plotted.

The preceding method has been modified by Martin and Tattersfield so as to be applicable to a single flower head (about 0.1g.). The reagents used are the same, excepting the alkaline ferricyanide solution, which is made to contain 3.30 g. of potassium ferricyanide and 57.2 g. of anhydrous sodium carbonate in 1 liter of solution. The procedure is the same as described above, excepting that 0.1 g. of pyrethrum is taken for extraction, 10 cc. of the pyrethrum solution are taken instead of 2 cc. as before, and 5 cc. of the alkaline ferricyanide just described are used. The tube is heated at 78° for 45 minutes in a constant-temperature bath fitted with a stirrer. A blank determination is made using 5 cc. of ferricyanide solution and 10 cc. of aldehyde-free alcohol. The test and the blank are titrated with 0.005N thiosulfate solution and from the difference between the titrations the pyrethrins in the

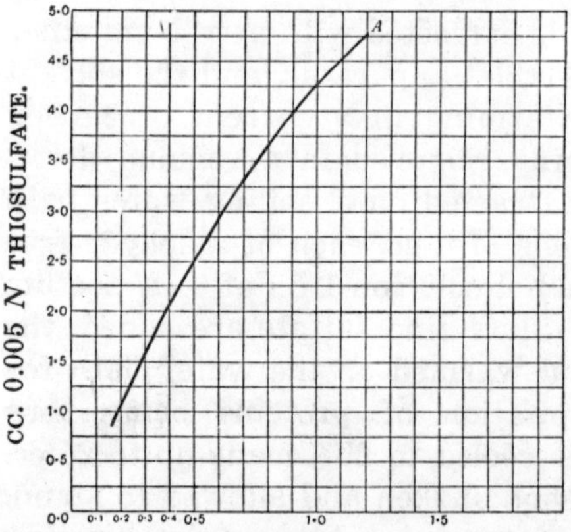

FIG. V. TATTERSFIELD'S GRAPH FOR DETERMINING PYRETHRINS. (METHOD ON PAGE 64).

10 cc. of solution taken are determined from a second graph, Fig. VI. This graph was prepared in a similar way to that in Fig. V. Knowing the amount of pyrethrins in 10 cc. of the solution, the percentage in the powder is easily calculated.

SEIL'S ACID METHOD

Seil *(794)* has modified the acid method, making it shorter and easier to operate. Moreover he has adapted the acid method to the analysis of oil extracts of pyrethrum (see page 223). Seil's method for pyrethrum flowers is as follows:

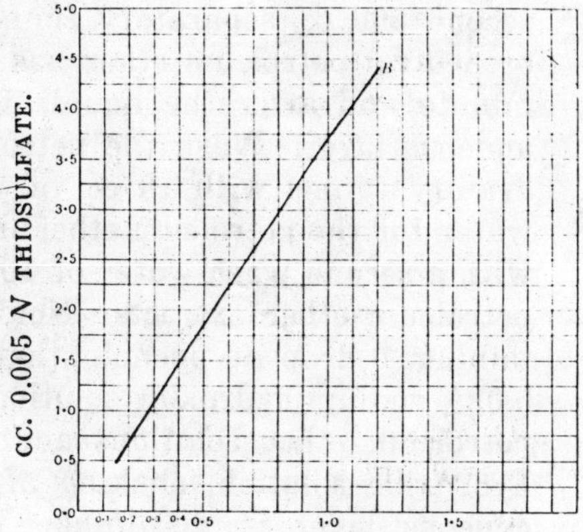

MG. PYRETHRINS I AND II IN 10 CC. OF FILTRATE.

FIG. VI. TATTERSFIELD'S GRAPH FOR DETERMINING PYRETHRINS IN SINGLE FLOWER HEADS. (METHOD ON PAGE 66).

> Extract 12.5 g. of ground flowers (20 to 30 mesh) in a Soxhlet extractor with low boiling petroleum ether. After the extraction is complete, recover the petroleum ether on a water bath; add 10 to 15 cc. of 0.5N ethyl alcoholic sodium hydroxide and reflux the mixture for 1 to 2 hours. Transfer the alkaline alcoholic solution with water to a 600 cc. beaker, adding sufficient water to bring the volume of liquid in the beaker to 200 cc. Add a few glass beads and remove the alcohol by boiling, taking care to avoid boiling over. When the volume is reduced to 150 cc. cool the solution and transfer to a 250 cc. volumetric flask to which 1 g. of filter-cel has previously been added. Mix thoroughly, add 10 cc. of 10 per cent barium chloride solution, make to the mark with water and mix again. After the precipitate has settled filter through a fluted paper, transfer 200 cc. of the clear filtrate to a 500 cc. Erlenmeyer flask, add 1 cc. of concentrated sulfuric acid and distil with steam, using an efficient trap and condenser. Continue the distillation until the liquid in the distilling flask is between 15 and 20 cc., collecting the distillate (usually about 250 cc.) in a 500 cc. Squibb sep-

aratory funnel. Allow the distilling flask, containing the dicarboxylic acid, to cool.

To the separatory funnel add 50 cc. of neutral petroleum ether, and shake thoroughly for one minute. After the liquids have separated, draw off the aqueous layer into a second 500 cc. separatory funnel to which a second 50 cc. of neutral petroleum ether has been added. Shake for one minute and after the liquids have separated, discard the aqueous layer. Wash the petroleum ether in the first separatory funnel with 10 cc. of water using the same wash water for the petroleum ether in the second funnel. Repeat with a second wash water of 10 cc. as before. Combine the petroleum ether extracts. Neutralize 15 cc. of water containing 1 drop of phenolphthalein indicator solution with 0.02*N* sodium hydroxide solution and add it to the combined petroleum ether solutions and titrate with 0.02*N* sodium hydroxide solution, shaking after each addition until the aqueous layer is just pink. Each cc. of 0.02*N* sodium hydroxide solution consumed is equal to 0.0066 g. of pyrethrin I. The 200 cc. aliquot taken corresponds to 10 g. of sample. Therefore the number of cc. consumed times 0.066 gives the percentage of pyrethrin I.

Filter the solution containing the dicarboxylic acid through a Gooch crucible, washing the flask with a little water. Make the clear filtrate alkaline with bicarbonate of soda and transfer to a separatory funnel; wash it twice with chloroform. Wash the first chloroform extract with water, using the same wash water for the second chloroform extract. Combine the aqueous solutions, acidify strongly with hydrochloric acid, saturate with sodium chloride and extract with 50 cc. of ether, shaking for about 1 minute. Repeat the extractions with three more portions of ether, using 50 cc. for the second and 25 cc. each for the third and fourth extracts. Wash the ether of the first extraction with 10 cc. of water using the same wash water for the other ether extracts successively. Repeat with a second wash with 10 cc. of water as before. Combine the ether solutions, draw off any water, separating, and filter the ether into a flask. Recover the ether on a water bath and dry the residue at 100° for 10 minutes. Add 2 cc. of neutral alcohol, warm gently, then add 20 cc. of distilled water and heat to dissolve the acid. If a residue remains undissolved, cool and filter through a Gooch crucible. Add a drop of phenolphthalein indicator solution and titrate with 0.02*N* sodium hydroxide solution, of which 1 cc. is equivalent to 0.00374 g. pyrethrin II.

Ripert's Method

Ripert *(745)* claims to have proved that Tattersfield's and Seil's acid methods yield high results, because of the presence of certain fatty acids and their esters, as well as free chrysanthemum acids and the mono-methyl ether of chrysanthemum dicarboxylic acid. Ripert says that his method is applicable to all products containing pyrethrum. He does not give specific instructions for assaying pyrethrum flowers; it is not clear, for example, whether he uses petroleum ether or ethyl ether for extracting the ground flowers. Since altered or oxidized pyrethrins are much more soluble in ethyl ether than in petroleum ether, only petroleum ether should be used. Ripert appears to have overlooked this fact. With this modification, Ripert's method would be carried out as follows:

Extract 40 to 50 g. of ground flowers, with petroleum ether, in a suitable (Soxhlet) extractor. Evaporate the petroleum ether and dissolve the residue in 100 cc. of ethyl ether. Neutralize the ethyl ether solution exactly with normal aqueous potassium hydroxide. Draw off the aqueous solution, wash it with 50 cc. of ethyl ether. Combine the ethyl ether solutions and wash once with distilled water; discard the aqueous solutions. Transfer the ethyl ether solution to a 500-cc. flask, evaporate the ether, add 50 cc. of normal methyl alcoholic potassium hydroxide solution and boil under a reflux condenser for 1½ hrs. Evaporate the alcohol, under reduced pressure, on a steam bath. Dissolve the residue in hot distilled water, transfer to a 500-cc. separatory funnel (50 cc. of water should suffice) and cool. Rinse the saponification flask with two 50-cc. portions of ethyl ether and add to the aqueous solution in the separatory funnel. Finally rinse the flask with 10 cc. of ethyl alcohol, which is reserved. Shake the separatory funnel vigorously and add 50 cc. of 25 per cent sodium chloride solution. Generally the two layers separate readily; if they do not, add the 10 cc. of ethyl alcohol reserved above. Draw off the aqueous layer into another separatory funnel. Wash the ethyl ether with 20 cc. of water, adding 25 cc. of the sodium chloride solution after shaking; repeat this washing of the ethyl ether. Combine the three aqueous solutions in the second separatory funnel and add 10 cc. of saturated barium chloride solution. Add 100 cc. of ethyl ether, shake and allow the funnel to stand; the precipitate will collect at the interface of the two liquids. Filter the aqueous layer through a rapid filter paper, using filter-cel or kieselguhr if necessary; wash the ether layer with 25 cc. of water and use this water to wash the filter. Acidify the aqueous filtrate strongly with 4*N* hydrochloric acid and extract three times with 200 cc. of ethyl ether. Wash the combined ether extracts three times with 10 cc. portions of neutral saturated sodium chloride solution. Transfer the ethyl ether solution to a flask (which is to be used later for steam distillation) and evaporate the ether. Dissolve the residue in the flask in 5 cc. of neutral 95 per cent ethyl alcohol and titrate with 0.2*N* alcoholic potassium hydroxide solution, using phenolphthalein as indicator. This titration corresponds to the sum of the chrysanthemum mono- and di-carboxylic acids and other acids. Record the titration, add to the flask a volume of normal sulfuric acid equal to half the volume of the 0.2*N* alkali required for the titration and distil with steam. Conduct the distillation so that 100 cc. of distillate are collected in about 10 minutes; superheated steam may be used. After collecting 100 cc. of distillate, replace the receiver with another flask and collect a second 100 cc. of distillate. Transfer the first 100 cc. of distillate to a 250-cc. separatory funnel; rinse the receiver with 100 cc. of neutral petroleum ether, which is added to the separatory funnel; shake vigorously. Draw off the aqueous layer into the receiving flask and wash the petroleum ether solution with 25 cc. of the neutral sodium chloride solution; draw off the latter into a second flask and reserve. Transfer the petroleum ether to an Erlenmeyer flask containing 25 cc. of water, neutralized to phenolphthalein. Titrate with 0.02*N* sodium hydroxide solution, shaking vigorously after each addition of alkali until a faint pink color persists. Pour the petroleum ether into a separatory funnel, add the previously extracted first 100 cc. of distillate, shake the mixture and draw off the aqueous layer into an Erlenmeyer flask. To the petroleum ether add the second 100 cc. of distillate, rinsing the receiving flask with 25 cc. of neutral petroleum ether. Shake vigorously and draw off the aqueous layer into the flask containing the extracted first 100 cc. of distillate. Wash the petroleum ether with the 25 cc. of sodium chloride solution, reserved above, and add the latter to the flask containing the extracted distillates. Titrate the petroleum ether with 0.02*N* sodium hydroxide solution, exactly as in the first extraction. Titrate the combined aqueous distillates with 0.02*N* sodium hydroxide solution, using phenolphthalein as indicator. The following example shows the method of calculating the results:

Weight of flowers taken,	40 g.	
Total acids, first titration,	9.5 cc. 0.2*N* alkali, or	95.0 cc. 0.02*N* alkali
Required to neutralize pyrethrin I:		
1st 100 cc. of distillate	27.4 cc. 0.02*N* alkali	
2nd 100 cc. of distillate	3.0 cc. 0.02*N* alkali	
Total	30.4 cc. 0.02*N* alkali	
Required to neutralize the aqueous distillate	13.1 cc. 0.02*N* alkali	
Total for pyrethrin I plus distillate		43.5 cc. 0.02*N* alkali
Difference, required for pyrethrin II		51.5 cc. 0.02*N* alkali

Since 1 cc. 0.02*N* alkali is equivalent to 6.6 mg. pyrethrin I and 3.7 mg. pyrethrin II the sample contained:

$$\frac{30.4 \times 0.0066}{40} = 0.50 \text{ per cent pyrethrin I and,}$$

$$\frac{51.5 \times 0.0037}{40} = 0.48 \text{ per cent pyrethrin II.}$$

The analyst is advised to consult the original paper.

Haller-Acree Method for Pyrethrin II

The method of Haller and Acree *(385)* is based on the fact that methyl esters yield methyl iodide quantitatively, when treated with hydriodic acid. The methyl iodide is absorbed in an acetic acid solution of potassium acetate, to which bromine has been added, whereby it is converted into methyl bromide and iodic acid; the latter is determined volumetrically. Pyrethrin II is a methyl ester, but pyrethrin I is not, hence the method can be used for determining only pyrethrin II.

Reagents (analytical grade):

Petroleum ether, b.p. 30°-60°.

Hydriodic acid, sp. gr. 1.70, constant boiling, from which free iodine has been removed by heating to 100° and treating with a slight excess of 50 per cent hypophosphorus acid.

Potassium acetate solution, 20 g. dissolved in 200 cc. of glacial acetic acid.

Sodium acetate solution, 25 g. dissolved in 100 cc. of water.

Sulfuric acid solution, 10 cc. dissolved in 100 cc. of water.

Formic acid (90%).

Sodium thiosulfate solution, 0.05*N*.

Chloroform, phenol, bromine and potassium iodide.

Apparatus:

As shown in Fig. VII.

Boiling rods, made from glass tubes 60 mm. long, with 1 mm. bore; the tubes are sealed at one end and also 10 mm. from the other end; the open end is fire polished.

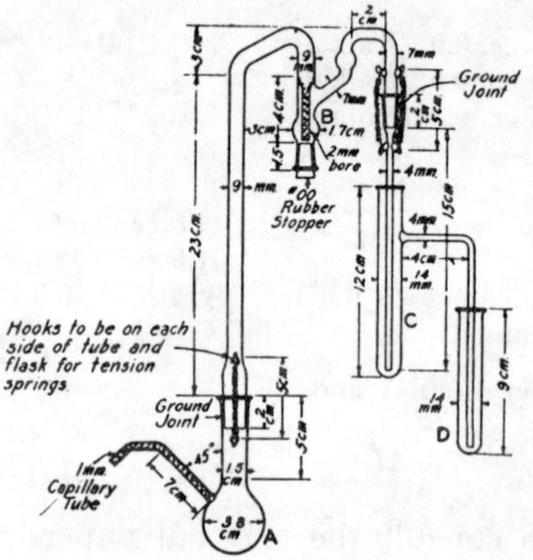

FIG. VII. SEMI-MICRO ZEISEL METHOXYL APPARATUS. (CLARK, J. AM. CHEM. SOC., 51, 1481, 1929).

Determination:

Extract 5 g. of ground pyrethrum in a Soxhlet extractor for seven hours with petroleum ether. Evaporate most of the petroleum ether on a water bath at 70° and transfer the residue quantitatively to the flask A, used for the methoxyl determination, by means of 20 cc. of chloroform, used in small portions. Place a boiling rod, open end down, in the flask and evaporate the chloroform on a water bath, removing the last traces under reduced pressure. Add 2.5 cc. of melted phenol and 5 cc. of hydriodic acid to flask A and connect it with the rest of the apparatus. Place a little water in scrubber B; phosphorus need not be used in B when the hydriodic acid has been treated as described above. Place in absorption tubes C and D, 10 cc. of the potassium acetate solution in glacial acetic acid, to which 20 drops of bromine have been added. Boil the mixture in flask A at such rate that the vapors do not rise more than half the length of the condenser and, at the same time, pass a slow stream of carbon dioxide through the capillary side arm of flask A. Continue the boiling for 1.5 hours, transfer the contents of the absorption tubes to a 250-cc. flask, containing 5 cc. of sodium acetate solution, and dilute to about 125 cc. Add 20 drops of 90 per cent formic acid to remove the excess bromine. When the liquid becomes water-white and the last traces of bromine are removed, add 2 cc. of dilute sulfuric acid, followed by 1 g. of potassium iodide. Titrate the liberated iodine with 0.05*N* sodium thiosulfate solution. Run a blank determination on the reagents and correct for the 0.05*N* sodium thiosulfate required to titrate the iodine liberated in the blank determination. The relation of the compounds formed may be outlined as follows:

$$1 \text{ mol. pyrethrin II} \rightarrow CH_3I \rightarrow HIO_3 \rightarrow 6I$$

from which it follows that 1 cc. of 0.05*N* thiosulfate solution is equivalent to 3.117 mg. of pyrethrin II. For a five gram sample, therefore, the number of cubic centimeters of 0.05*N* thiosulfate solution (corrected for the blank) multiplied by 0.0623 gives the percentage of pyrethrin II.

Gnadinger-Corl Acid Method

The following method was developed by the writer in the course of horticultural work on the development of high test strains of pyrethrum. The determination can be made on the same weighed sample that is used for determining pyrethrins by the copper reduction method. This method gives results somewhat higher than the copper reduction method.

Fifteen grams of ground flowers are extracted with petroleum ether as in the copper reduction method. The petroleum ether extract is cooled, filtered, evaporated, taken up in boiling aldehyde-free alcohol, treated with basic lead acetate solution and with anhydrous sodium carbonate, as described on page 60. Ten cc. of the solution, thus purified, are used for determining pyrethrins, by the copper reduction method, in the usual way; fifty cc. are taken for the following modified acid method:

Pipet 50 cc. of the alcoholic solution into a 500 cc. Erlenmeyer flask, add 5 cc. of potassium hydroxide solution (20 g. of 85% KOH in sufficient water to make 100 cc. of solution) and boil for 1½ hours under a reflux condenser. Remove the flask from the condenser, cool slightly, add about 0.25 g. of crushed pumice (10 to 20 mesh) and distil off the alcohol on a steam bath or hot plate until not more than 8 or 10 g. of solution remain. Cool, add 150 cc. of water and 25 cc. of sulfuric acid (5 cc. of concentrated acid diluted to 100 cc. with water) and distil with steam, collecting 200 cc. of distillate in a 500 cc. separatory funnel. The distillation should be conducted so that the volume of the solution remaining in the distillation flask is about 50 cc. Extract the distillate with two 50 cc. portions of petroleum ether, keeping the petroleum ether extracts separate. Wash the two petroleum ether extracts successively with a 15 cc. portion of distilled water; repeat the washing with a second 15 cc. of water. Transfer the two petroleum ether extracts to a glass stoppered flask, containing 15 cc. of recently boiled and cooled water, which has been neutralized with 0.02*N* sodium hydroxide, using phenolphthalein as indicator. Titrate with 0.02*N* alkali, shaking vigorously after each addition of alkali, until a faint pink color persists. Each cc. of 0.02*N* alkali is equivalent to 0.0066 g. of pyrethrin I. Since the 50 cc. aliquot taken corresponds to 7.5 g. of sample, the percentage of pyrethrin I is easily calculated.

Transfer the warm contents of the distillation flask to a 200 cc. volumetric flask, rinsing distillation flask and trap with water, cool to 20° and make to the mark with water. Add 1 g. of Celite analytical filter aid, mix thoroughly and filter. Pipet 150 cc. of the clear filtrate into a separatory funnel, add 35 g. of C. P. sodium chloride and 5 cc. of concentrated hydrochloric acid and extract with three 50 cc. portions of ethyl ether, keeping the three ether extracts in separate funnels. Wash the ether extracts successively with a 10 cc. portion of water; repeat the washing with a second 10 cc. of water. Combine the ether solutions, draw off any water that has collected and filter the ether through a dry paper into a 500 cc. Erlenmeyer flask, rinsing funnel and filter with a little ether. Add a few grains of sand to the ether solution, evaporate to dryness on a steam bath and dry at 100° for 10 minutes. Add 25 cc. of water to the residue in the flask, boil for 1 minute, cool and titrate with 0.02*N* sodium hydroxide solution, using phenolphthalein as indicator. One cc. of 0.02*N* alkali is equivalent to 0.00374 g. of pyrethrin II. The original aliquot of 50 cc. corresponded to 7.5 g. of flowers; the second aliquot of 150 cc., taken for the ether extraction, corresponds, therefore, to 5.625 g., from which the percentage of pyrethrin II can be calculated.

Ordinary 95% alcohol can be used instead of aldehyde-free alcohol, in the purification process, when the copper reduction determination is omitted.

Comparison of Results Obtained With Various Methods

By the end of January, 1931, eight methods, based on three different principles, were available for chemically assaying pyrethrum flowers. The first question, which naturally arose, was whether the results obtained by the different methods, applied to the same samples, would agree.

Shortly after the earlier methods were published the writer assayed two samples, one of high pyrethrin content and one low in pyrethrins. Portions of these carefully mixed, ground flowers were sent to Tattersfield, H. H. Richardson and the laboratory of a large oil company. The analyses made in the different laboratories showed excellent agreement, Table XIII.

TABLE XIII. ANALYSES OF PYRETHRUM SAMPLES BY DIFFERENT METHODS

Analyst	Sample	Tattersfield acid method Py. I. %	Py. II. %	Total py. %	Gnadinger-Corl method Total pyrethrins %
Tattersfield	82	0.39	0.60	0.99	0.92
		0.38	0.60	0.98	
N. J. A.	82	...	...	...	0.95
		...	...	...	0.95
Richardson	82	0.36	...	...	...
		0.38	...	...	...
Gnadinger	82	...	...	...	0.97
		...	...	...	0.97
		...	...	...	0.93
Tattersfield	42	0.20	0.24	0.44	0.40
		0.20	0.22	0.42	0.42
Gnadinger	42	...	...	...	0.39
		...	...	...	0.39

Martin and Tattersfield *(590)* later reported the analyses of 12 samples by the acid method and the copper reduction method (Table XIV).

TABLE XIV. COMPARISON OF TATTERSFIELD ACID METHOD WITH GNADINGER-CORL METHOD. ANALYSTS, MARTIN AND TATTERSFIELD

Sample	Tattersfield acid method Pyrethrins %	Gnadinger-Corl method Pyrethrins %
1	0.63	0.65
2	1.05	1.06
3	0.95	1.03
4	0.98	0.94
5	1.23	1.16
6	0.56	0.73
7	2.18	2.30
8	2.15	1.98
9	0.43	0.40
10	0.99	0.92
11	1.50	1.49
12	1.59	1.49

More recently Hartzell and Wilcoxon *(406)* published the analyses of 6 samples of flowers, using the two methods (Table

XV). Hartzell and Wilcoxon concluded that "The agreement between the method of Gnadinger and that of Tattersfield is quite good."

TABLE XV. COMPARISON OF TATTERSFIELD AND GNADINGER-CORL METHODS ANALYSTS, HARTZELL AND WILCOXON

Sample	Tattersfield acid method Pyrethrins %	Gnadinger-Corl method Pyrethrins %
1	1.01	1.03
2	0.79	0.81
3	0.75	0.73
4	0.52	0.59
5	0.98	0.82
6	0.67	0.64

The following results (Table XVI) were obtained on check samples by four different laboratories using the copper reduction method.

TABLE XVI. CHECK ANALYSES BY FOUR LABORATORIES USING THE COPPER REDUCTION METHOD

Lot	Lab. 1 Pyrethrins %	Lab. 2 Pyrethrins %	Lab. 3 Pyrethrins %	Lab. 4 Pyrethrins %
1	0.95	0.92	....	....
2	1.05	1.05	....	....
3	1.10	1.01	....	....
4	1.00	0.96	....	....
5	0.99	0.94	....	....
6	0.98	0.98	....	....
7	1.02	1.01	....	....
8	1.10	1.11	....	....
9	....	0.56	0.53	....
10	....	1.06	0.99	....
11	....	0.62	0.62	0.64
12	....	0.87	0.89	0.86
13	....	0.59	....	0.57
14	....	0.59	....	0.56
15	....	0.79	....	0.80
16	....	0.70	....	0.73
17	....	1.18	....	1.23

Thus it has been shown that results obtained with Tattersfield's semicarbazone method agree with those obtained by his acid method (page 51), which in turn check with the results by the Gnadinger and Corl method. Martin and Tattersfield *(590)* report good agreement between their ferricyanide methods, the acid method and the copper reduction method.

McDonnell *(605)* was unable to check the Tattersfield acid

method against analyses made for him by the writer, using the copper reduction method. In view of the satisfactory agreement obtained by other analysts, the difficulty would seem to be in McDonnell's analytical work rather than in the methods. Further evidence that McDonnell's analyses were incorrect is the extremely low proportion of pyrethrin I which he reported.

Comparison of Seil and Gnadinger-Corl Methods

Gnadinger and Corl *(352)* have compared the results obtained by applying the Seil method and the copper reduction method to pyrethrum flowers. The ages of the samples varied from one to fifty-five months. Analyses of a given sample by the two methods were made simultaneously. Blank determinations on all of the reagents employed in the two methods gave negligible corrections. The standard alkali used in the Seil method was standardized against U. S. Bureau of Standards benzoic acid and acid potassium phthalate. These analyses, given in Table XVII, show that, in every case, a higher pyrethrin content is obtained by the Seil method than by the copper reduction method. The samples were from 5.2 to 32.7 per cent higher in pyrethrins by the Seil method, averaging 12.8 per cent higher. The age of the flowers was apparently not an important factor, both old and new flowers showing higher pyrethrin contents by the Seil method.

Table XVII. Comparison of Seil and Gnadinger-Corl Methods (Gnadinger and Corl)*

No.	Kind of flowers	Crop year	Age when assayed mos.	Pyrethrin content: Seil method I %	Seil method II %	Total %	Gnadinger-Corl method %
1	Japanese	1930	55	0.31	0.46	0.77	0.58
2	Japanese	1930	46	0.28	0.40	0.68	0.62
3	Japanese	1933	12	0.38	0.51	0.89	0.80
4	Japanese	1933	11	0.43	0.47	0.90	0.81
5	Japanese	1933	11	0.44	0.49	0.93	0.87
6	Japanese	1934	7	0.42	0.56	0.98	0.92
7	Japanese	1934	7	0.41	0.65	1.06	0.93
8	Japanese	1934	7	0.44	0.64	1.08	0.95
9	Japanese	1934	3	0.43	0.64	1.07	0.92
10	Japanese	1934	1	0.41	0.60	1.01	0.96
11	Dalmatian	1934	7	0.28	0.39	0.67	0.57
12	American	1934	1	0.50	0.53	1.03	0.95

*Includes additional analyses, made since date of original publication.

Although the Seil method gives higher results than the copper reduction method when applied to pyrethrum flowers, with certain types of extracts concordant results are obtained (page 227). This is because the interfering substances are removed in the process of manufacturing the extracts.

Tattersfield *(875)* states that the Seil method gives substantially the same results as the Tattersfield acid method.

Tattersfield reports that only minute quantities of the free volatile and water-soluble acids, upon which Ripert lays so much stress, occur in pure pyrethrum flowers. Tattersfield also reports that Ripert's method yields higher results for pyrethrin II than either the Seil or Tattersfield methods.

Haller and Acree *(385)* have compared the pyrethrin II contents, found by their method, with those found by the acid methods of Seil and Tattersfield on the same samples. Their analyses, given below, show that the results obtained by the Haller-Acree methoxyl method are lower, in every case, than found by the Seil and Tattersfield acid methods. Sample 6 was 10 years old; the ages of the other samples are not stated.

Sample No.	Haller-Acree methoxyl method II %	Seil acid method II %	Tattersfield acid method II %
1	0.38	0.42	0.45
2	0.34	0.40	0.43
3	0.24	0.29	0.30
4	0.37	0.47	0.53
5	0.37	0.54	0.47
6	0.23	0.33	0.28

Unfortunately the analyses by the methoxyl method were made 2 or 3 months after the analyses by the Seil and Tattersfield methods. Hence it is quite possible that the lower results by the methoxyl method may be due, at least in part, to loss of pyrethrins by decomposition, during the period between analyses.

Gnadinger and Corl have compared the results obtained by their copper reduction method and their acid method (page 73). The comparisons, made on a limited number of samples, show slightly higher results by the acid method:

Description of flowers	Pyrethrin content: Gnadinger-Corl acid method I %	II %	Total %	Copper reduction method Total %
Tennessee, 1935 crop	0.97	0.76	1.73	1.76
Colorado, 1935 crop	0.75	0.74	1.49	1.36
Colorado, 1935 crop	0.68	0.52	1.20	1.20
Tennessee, 1935 crop	0.52	0.69	1.21	1.13
Tennessee, 1935 crop	0.49	0.76	1.25	1.11
Colorado, 1935 crop, immature	0.53	0.38	0.91	0.95

Interference of Altered Pyrethrins With Assay Methods

In the paper describing his acid method Tattersfield commented: "It is important to realize that after long exposure to damp conditions pyrethrum powder loses its toxicity and a genuine sample may be devoid of insecticidal value. Whether or not the chemical methods outlined will make it possible to detect such loss of toxicity, is a matter for further investigation." It has been suggested that the pyrethrins might decompose into the alcohol pyrethrolone and the two chrysanthemum acids. If this were the case neither the Tattersfield method nor the Gnadinger-Corl method would detect such deterioration. The pyrethrins do not decompose in this way, however; instead they form resinous compounds that are insoluble in petroleum ether; this is true of both pyrethrin I and pyrethrin II. The solubility of pyrethrins altered by exposure to air at room temperature has recently been more fully investigated. In petroleum ether, of the specifications given on page 58, the solubility of a mixture of altered pyrethrins (originally 66 per cent pyrethrin I and 33 per cent pyrethrin II) was found to be less than 3.6 mg. per 100 cc. of solution at 20°. Using a 15 g. sample, as in the Gnadinger-Corl method, the maximum error due to the presence of altered pyrethrins would be about 0.03 per cent.

Hartzell and Wilcoxon *(406)* found that pyrethrum exposed in air to sunlight, ultraviolet light and heat showed a loss of toxicity to aphids and in each case a loss in pyrethrin content, determined chemically, was also found, indicating that both the Tattersfield and Gnadinger-Corl methods detect deterioration of pyrethrum flowers (page 155). Gnadinger and Corl have shown that pyrethrum loses steadily in pyrethrin content during storage (page 157), indicating that the decomposed pyrethrins do not interfere appreciably with the accuracy of their method.

Presence of Interfering Compounds in the Flowers

Ripert *(743)* has condemned all of the methods for evaluating pyrethrum, both chemical and physiological. He claims to have isolated the methyl ether of pyrethrolone from pyrethrum flowers and states that this compound is present to the extent of 0.1 to 0.4 per cent and would interfere with the copper reduction method, being estimated as pyrethrins without having any toxicity to insects. In order to make clear the significance of this statement, it should be recalled that the pyrethrins are esters of the ketone alcohol pyrethrolone with the chrysanthemum acids, and the copper reducing action of the pyrethrins is due to the presence of the ketone group in the alcoholic component of the esters.

Ripert does not give the details of the method he used for isolating the methyl ether of pyrethrolone from the flowers and the proof of its identity which he presents is not conclusive. His statement that the compound would interfere in the copper reduction method is not supported by any evidence. When the copper reduction method was developed, the writer realized that pyrethrolone might be present in the flowers; accordingly pyrethrolone was prepared, at that time, and its properties were investigated. The possible presence of pyrethrolone methyl ether was not considered. Neither pyrethrolone nor its methyl ether was found in the flowers by Staudinger and Ruzicka; the thoroughness of their work makes it improbable that quantities of the methyl ether as great as Ripert reports would be overlooked. Gnadinger and Corl *(350)* have made the following report on an investigation of Ripert's claims.

"Staudinger and Ruzicka called attention to the fact that the pyrethrins in petroleum ether solution are not appreciably attacked by potassium permanganate solution. They also found that pyrethrolone in aqueous suspension and pyrethrolone methyl ether are immediately oxidized by permanganate solution. The writers had found that the pyrethrins in petroleum ether are scarcely attacked by permanganate, while pyrethrolone under the same conditions is completely oxidized. Hence this procedure might afford a means of separating the pyrethrins from pyrethrolone and possibly from pyrethrolone methyl ether.

"To determine the effect of permanganate solution on the pyrethrins, the semicarbazones of the pyrethrins were separated from 15 kg. of Japanese pyrethrum by a slight modification of Staudinger and Ruzicka's method. The mixed semicarbazones were converted into mixed pyrethrins I and II with oxalic acid solution. The crude pyrethrins were purified by the method described on page 124 and were immediately dissolved in sufficient petroleum ether to make the pyrethrin content about 3.5 g. per liter. This stock solution was kept in the dark. Analysis by Tattersfield's acid method showed that the mixed pyrethrins consisted of 51.3 per cent pyrethrin I and 48.7 per cent pyrethrin II.

"The extent to which the pyrethrins in petroleum ether are oxidized by permanganate was determined in the following manner: an aliquot of the stock solution of pyrethrins, in petroleum ether, was measured into a separatory funnel and diluted with sufficient petroleum ether to make the volume 100 cc.; 25 cc. of 2 per cent potassium permanganate solution were added and the funnel was vigorously shaken for one minute. Immediately 20 cc. of 5 per cent sodium bisulfite solution and 5 cc. of 15 per cent sulfuric acid were added and the funnel was gently shaken until the excess of permanganate was destroyed. In some cases 50 cc. of 2 per cent permanganate solution were used; in every case an excess of permanganate solution was present.

"The petroleum ether and aqueous layers were separated; the former was washed with 25 cc. of water and the aqueous layer was extracted with an equal volume of petroleum ether. This second petroleum ether extract was washed with the 25 cc. of water used to wash the original petroleum ether solution. The combined petroleum ether extracts were filtered into a 400-cc. beaker. A second aliquot of the stock solution of pyrethrins, equal to the first, was measured into another 400-cc. beaker and was diluted with petroleum ether to the same volume as the solution which had been treated with permanganate. The two petroleum ether solutions were then evaporated on a steam bath and the pyrethrin content of the residues was determined by the writer's method. The results of these experiments are given in Table XVIII.

TABLE XVIII. EFFECT OF WASHING PYRETHRINS IN PETROLEUM ETHER WITH 2% PERMANGANATE SOLUTION

Aliquot taken	Pyrethrins found: Not treated	Pyrethrins found: Treated with K Mn 0_4	Pyrethrins lost by K Mn 0_4 treatment	
cc.	mg.	mg.	mg.	%
50	185.4	174.1	11.3	6.1
40	141.9	137.2	4.7	3.3
35*	135.6	128.6	7.0	5.2
30	106.3	100.7	5.6	5.3
25*	85.9	84.0	1.9	2.2
20	70.2	66.8	3.4	4.8
10	34.6	33.6	1.0	2.9
Avg.	...	...	...	4.3

* 50 cc. K Mn 0_4 solution used.

"The aliquots taken were selected to correspond to the pyrethrin content which would be present in pyrethrum flowers when using 15 g. of flowers for assay. The loss of pyrethrins by the permanganate treatment varied from 2.2 per cent to 6.1 per cent with an average of 4.3 per cent.

"To determine the effect of permanganate solution on pyrethrolone, part of the mixed semicarbazones of pyrethrins I and II was saponified at 0° as recommended by Staudinger and Ruzicka. The resulting pyrethrolone semicarbazone was converted into pyrethrolone by prolonged shaking with benzene and sodium bisulfate solution. This pyrethrolone (0.898g.) was completely dissolved in petroleum ether in which it was but slightly soluble. The petroleum ether solution was washed with 3 per cent permanganate solution, which was instantly reduced. The washings with permanganate were continued until an excess of permanganate remained. The petroleum ether solution was then filtered and distilled in vacuum at 45° to constant weight; the residue weighed 0.012 g., that is, 98.7 per cent of the pyrethrolone was oxidized and removed from the petroleum ether solution, by the permanganate treatment.

"To determine the effect of permanganate solution on pyrethrolone methyl ether a second portion of the mixed semicarbazones of pyrethrins I and II was saponified at 0° and the pyrethrolone semicarbazone so obtained (m.p. 200°-203°) was used to synthesize pyrethrolone methyl ether, employing the procedure of Staudinger and Ruzicka. Thirteen grams of pyrethrolone semicarbazone yielded 4.7 g. of pyrethrolone methyl ether. A small portion of the pyrethrolone methyl ether was treated with semicarbazide hydrochloride; the semicarbazone of pyrethrolone methyl ether so obtained melted at 187°. This melting point is slightly higher than recorded by Staudinger and Ruzicka, (183°).

"The remainder of the pyrethrolone methyl ether was dissolved in petroleum ether, the concentration being 0.643 g. per 100 cc. Aliquots of this solution were treated with permanganate solution exactly as described above in the case of the pure pyrethrins, and the copper reducing power of the permanganate treated solution was compared with that of a solution of the same pyrethrolone methyl ether content which had not been treated with permanganate. Since no tables have been calculated for the copper reducing power of pyrethrolone methyl ether, it was necessary to express its reducing power in terms of the amount of dextrose having equivalent copper reducing power, as shown in Table XIX.

TABLE XIX. EFFECT OF TREATING PYRETHROLONE METHYL ETHER IN PETROLEUM ETHER WITH 2 PER CENT PERMANGANATE SOLUTION

Pyrethrolone methyl ether taken	Permanganate solution used	Dextrose equivalent in reducing power to pyrethrolone methyl ether: Not treated with $KMnO_4$	Treated with $KMnO_4$	Pyrethrolone methyl ether lost by $KMnO_4$ treatment
mg.	cc.	mg.	mg.	%
128.6	25.	13.84	6.25	54.
128.6	50.	13.84	0.00	100.
64.3	25.	6.41	0.00	100.

"In the first experiment the permanganate solution was completely reduced by about half the pyrethrolone methyl ether taken. When the amount of permanganate was doubled, or the quantity of pyrethrolone methyl ether was halved, the copper reducing power of the pyrethrolone methyl ether was completely destroyed. The amounts of pyrethrolone methyl ether taken were several times greater than would be encountered in assaying pyrethrum flowers containing the amount of pyrethrolone methyl ether Ripert claims to have found. It should also be noted that the copper reducing power of pyrethrolone methyl ether is somewhat less than that of the pyrethrins. This was also true of pyrethrolone.

"From the foregoing experiments it is quite clear that the pyrethrins can be separated with slight loss from pyrethrolone or pyrethrolone methyl ether by treating them, in petroleum ether, with 2 per cent potassium permanganate solution.

"According to Ripert *(743)*, methyl pyrethrolone is the form in which the reserve of pyrethrolone is stored by the plant, pending esterification with the chrysanthemum acids by metabolic processes. If this is the case one would expect to find larger quantities of methyl pyrethrolone in freshly-harvested flowers than in flowers several months old. In Table XX are given the analyses of samples of pyrethrum flowers assayed as soon as possible after harvesting. Two series of 15 g. samples of the ground flowers were extracted with petroleum ether in Soxhlet extractors. One series of petroleum ether extracts was assayed by the writers' method. Another series of petroleum ether extracts was first treated with 2 per cent permanganate solution exactly in the manner described above, before assaying.

"The average loss in pyrethrin content of the flowers by the permanganate treatment was 0.05 per cent equivalent to 5.2 per cent of the total pyrethrins present. This is only 0.9 per cent greater than the loss with pure pyrethrins. This indicates that pyrethrolone and pyrethrolone methyl ether are not present in the flowers shortly after harvesting.

TABLE XX. EFFECT OF PERMANGANATE TREATMENT ON PYRETHRIN CONTENT OF FRESHLY HARVESTED PYRETHRUM FLOWERS

Where grown	Harvested	Assayed	Pyrethrin content of flowers Not treated with K Mn 0_4	Treated with* K Mn 0_4	Loss of pyrethrins by K Mn 0_4 treatment	% of pyrethrin content
			%	%	%	
Colorado	7/14/32	7/21/32	1.04	1.00	0.04	3.8
Japan	6/15/32	8/ 9/32	1.00	0.94	0.06	6.0
Yugoslavia	5/27/32	7/19/32	0.84	0.79	0.05	5.9
Avg.	....	....	....	....	0.05	5.2

* 50 cc. 2 per cent K Mn 0_4 used.

"Ten additional samples of pyrethrum flowers from different sources were assayed 10 months after they were harvested. These samples were assayed by the copper reduction method both with and without the permanganate treatment. The results are compared in Table XXI.

TABLE XXI. EFFECT OF PERMANGANATE TREATMENT ON PYRETHRIN CONTENT OF FLOWERS 10 MONTHS OLD

Where grown	Pyrethrin content of flowers Not treated with K Mn 0_4 %	Treated with K Mn 0_4 %	Loss of pyrethrins due to K Mn 0_4 treatment %	% of pyrethrin content
Rose, Yugoslavia	0.70	0.59	0.11	15.7
Trogir, Yugoslavia	0.76	0.65	0.11	14.4
Trogir, Yugoslavia	0.71	0.59	0.12	16.9
Hokkaido, Japan	0.76	0.68	0.08	10.5
Hokkaido, Japan	0.79	0.69	0.10	12.6
Hokkaido, Japan	0.71	0.59	0.12	16.9
Hokkaido, Japan	0.85	0.72	0.13	15.3
Hokkaido, Japan	0.86	0.72	0.14	16.3
Hokkaido, Japan	1.01	0.85	0.16	15.8
Cyprus	0.75	0.63	0.12	15.0
Avg.	...	...	0.12	15.0

"The loss in pyrethrin content averaged 0.12 per cent of the weight of the flowers, equivalent to 15.0 per cent of the pyrethrins present. The losses were fairly constant for flowers from different countries and of different pyrethrin contents. Correcting for the loss due to the effect of permanganate on pure pyrethrins, the average loss in these ten month old flowers, due to other oxidizable material, is 10.7 per cent of the pyrethrins present. In the freshly harvested flowers the corrected loss was 0.9 per cent.

"Gnadinger and Corl *(347)* have shown that the pyrethrin content of pyrethrum flowers becomes gradually reduced as the flowers age, the loss amounting to about 30 per cent in one year. This loss of pyrethrins is due to oxidation or to a molecular rearrangement and is accelerated by light. The altered pyrethrins are almost insoluble in petroleum ether, the solubility being about 3.6 mg. per 100 cc. at 20°. Their copper reducing power is slightly higher than that of the pyrethrins. Whether they are oxidized in petroleum ether solution by permanganate was not determined because of their low solubility in that solvent. In ether, however, a considerable oxidation of altered pyrethrins occurs when the permanganate treatment is applied. An ether solution containing 88 mg. of pyrethrins, altered by exposure to air at room temperature, was washed with 25 cc. of saturated permanganate solution. The ether solution was evaporated and the residue was weighed; 62 mg. of altered pyrethrins were lost by oxidation. A second ether solution containing 176 mg. of altered pyrethrins was treated with permanganate in the same manner as the pure pyrethrins, as previously described. The loss by the permanganate treatment was 45 mg. The altered pyrethrins are apparently more readily attacked than pyrethrins I and II and it is quite probable that the altered pyrethrins in the samples reported in Table XXI were oxidized by the permanganate treatment. This would account, in part, for the fact that old flowers show a greater loss in pyrethrin content when treated with permanganate than new flowers similarly treated."

These experiments proved

1. The permanganate treatment separates pyrethrolone and its methyl ether from pyrethrins I and II with slight loss of the pyrethrins.

2. The pyrethrins in petroleum ether are not readily oxidized by dilute permanganate solution, the loss averaging about four per cent.

3. Pyrethrolone and pyrethrolone methyl ether, in petroleum ether, are instantly oxidized by dilute permanganate solution and removed from the petroleum ether.

4. Pyrethrins altered by exposure to air are, in ether solution, appreciably oxidized by permanganate solution.

5. Freshly harvested pyrethrum flowers assayed by the per-

manganate method show about the same loss of pyrethrins as solution of pure pyrethrins of the same strength.

6. Pyrethrum flowers ten months old show a somewhat greater loss by the permanganate method probably due in part to the oxidization of altered pyrethrins.

7. No evidence of any appreciable amount of pyrethrolone or pyrethrolone methyl ether was found in flowers from America, Yugoslavia or Japan.

Ripert, in a later paper *(745)* has described a quantitative method for determining methyl pyrethrolone in pyrethrum flowers. The semicarbazones of the pyrethrins are separated from the petroleum ether extract of the flowers in the usual way. One portion of the purified semicarbazones is used for a determination of chrysanthemum dicarboxylic acid; in a second portion the methoxyl content is determined by Zeisel's method. Ripert found that the methoxyl content was always higher than that calculated from the pyrethrin II content indicated by the chrysanthemum dicarboxylic acid present; this, he concluded, was due to the presence of methyl pyrethrolone.

Haller and Acree *(385)*, however, found that the pyrethrin II content of pyrethrum flowers, determined by their methoxyl method (page 71) was significantly lower than the pyrethrin II content determined by the Tattersfield and Seil acid methods. They point out that if methyl pyrethrolone were present, as claimed by Ripert, the results obtained by the methoxyl method would be higher than those obtained by the acid methods, since methyl pyrethrolone has a methoxyl content about twice that of pyrethrin II.

SUMMARY

The chemical methods for assaying pyrethrum flowers may be classified as follows:

Methods determining pyrethrins I and II separately:

- Staudinger and Harder, acid method, pyrethrins I and II.
- Tattersfield, Hobson and Gimingham, acid method, pyrethrins I and II.
- Tattersfield and Hobson, acid method, pyrethrin I.
- Seil, acid method, pyrethrins I and II.
- Ripert, acid method, pyrethrins I and II.
- Haller and Acree, methoxyl method, pyrethrin II.
- Gnadinger and Corl, acid method, pyrethrins I and II.

Methods determining pyrethrins I and II combined:

Staudinger and Harder, semicarbazone method.

Tattersfield, Hobson and Gimingham, semicarbazone method.

Gnadinger and Corl, copper reduction method.

Martin and Tattersfield, ferricyanide reduction method A.

Martin and Tattersfield, ferricyanide reduction method B.

Tatu, modified Gnadinger-Corl copper reduction method.

Raupp, modified Gnadinger-Corl copper reduction method.

The Seil method has gradually replaced the Tattersfield methods because it is easier to operate. Unfortunately the Seil method gives high results, as Gnadinger and Corl and also Haller and Acree have shown. The Gnadinger-Corl acid method also gives slightly higher results than their copper reduction method.

Many of the conflicting results obtained by different investigators are probably due to failure to consider the unstable nature of the pyrethrins. Comparisons of results obtained by different methods can properly be made only when the analyses are made simultaneously.

Pyrethrins which have been altered and rendered inert by exposure to air, light and heat, or by prolonged storage, are not determined by these methods because the altered pyrethrins are almost insoluble in petroleum ether. The errors introduced by the solubility of altered pyrethrins in petroleum ether are small. The acid methods appear to give high results for pyrethrin II.

Purchasers of pyrethrum, who buy on a guaranteed pyrethrin content, should ascertain by which method the pyrethrin content has been determined.

CHAPTER V

BIOLOGICAL METHODS FOR EVALUATING PYRETHRUM

The first biological tests with pyrethrum were made by the unknown discoverers of its toxicity to insects. The earliest recorded tests are those of Kalbruner *(505)*, 1874, who states that 4 grains of pyrethrum powder should kill flies in 2 or 3 minutes. Sayre *(775)*, 1880, used flies and tadpoles for testing powdered *Pyrethrum roseum*. Unger *(923)*, 1888, used roaches for determining the efficiency of *P. roseum* and *P. cinerariaefolium*. Slaus-Kantschieder *(816)*, 1913, shook 1 g. of powder in a 25 cc. flask and then introduced flies; first class powder killed the flies in 5 minutes. Kuraz *(534)*, 1915, considered the time required to paralyze flies as the best index of toxicity.

DUSTING EXPERIMENTS

All of the earlier investigators used powdered pyrethrum flowers applied to the insects as a dust. Probably the greatest source of error in such experiments, aside from variability in the vitality of the test insects, is due to the fact that most of the pyrethrins occurs in the achenes and even when fine powder is used, a large part of the pyrethrins is held within cell walls and is not available for contact with the insects. The fineness of the powder is another factor that was overlooked in the earlier methods.

Staudinger and Ruzicka diluted the pyrethrins with flour, thus obtaining powders of the same fineness with the pyrethrins uniformly distributed.

TATTERSFIELD'S METHOD USING *Aphis rumicis*

Tattersfield *(880)*, 1929, employed a colloidal suspension of pyrethrins (prepared by diluting an alcoholic extract of pyrethrum with water) for comparing the toxicity of different samples of flowers to aphids. This was a distinct improvement over the use of dusts because the liquids can be applied more evenly and furthermore the pyrethrin particles are uniformly distributed throughout the liquid and are readily brought in contact with the insect. The effect of this is plainly shown by the fact that Tattersfield obtained kills of 46 to 69 per cent, on specially resistant aphids, with aqueous suspensions containing 2 to 3 mg.

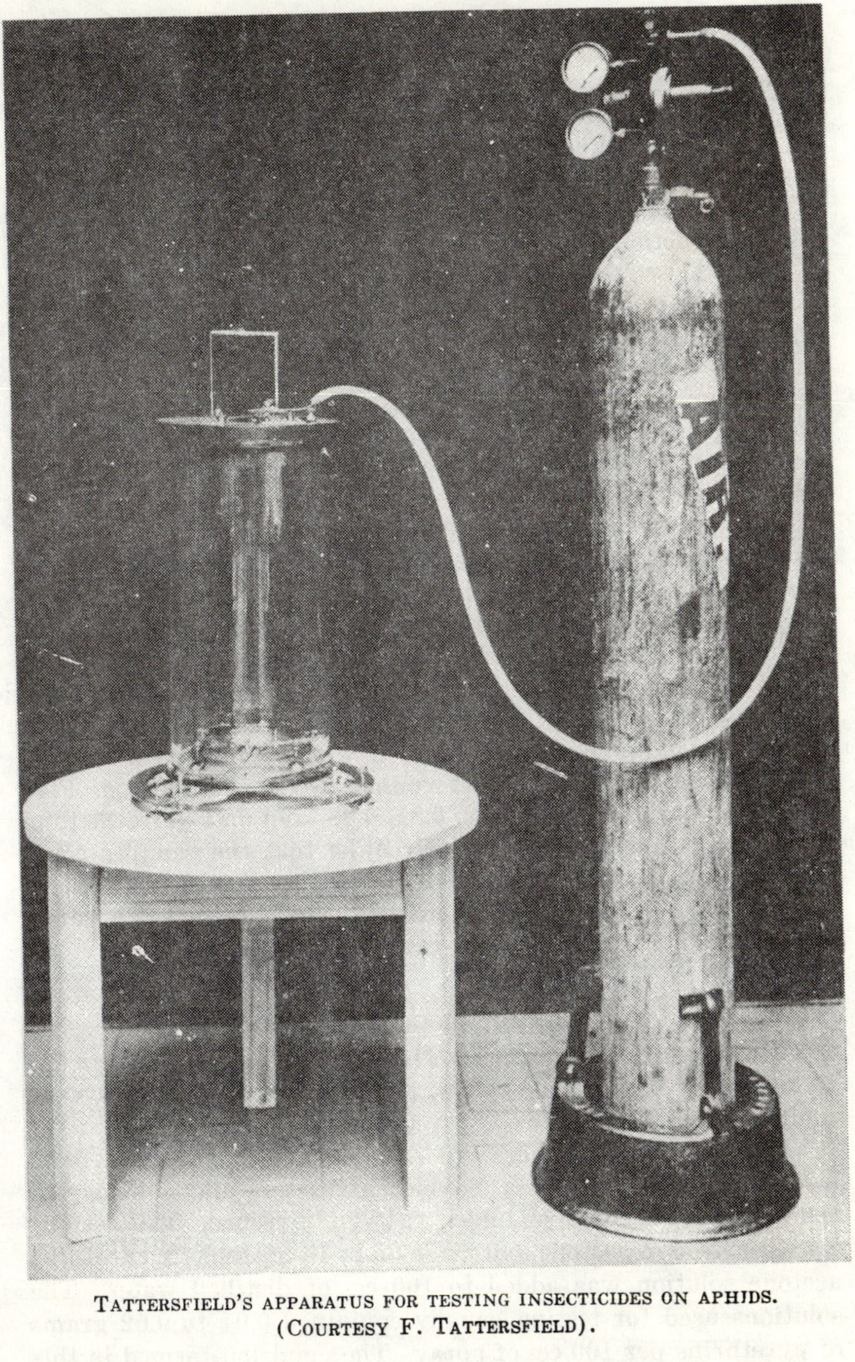

TATTERSFIELD'S APPARATUS FOR TESTING INSECTICIDES ON APHIDS.
(COURTESY F. TATTERSFIELD).

of pyrethrins per 100 cc., whereas McDonnell, *(605)*, using dusts, required 76 to 142 mg. of pyrethrins per 100 g. to obtain the same percentage kills on aphids.

The apparatus employed by Tattersfield *(877, 882)* consists of a glass cylinder 19.5 cm. in diameter by 44 cm. high, used as a spray chamber, and an atomizer so constructed that very fine adjustment of the amount of spray delivered can be obtained. The construction of this equipment is too complicated to be given in detail here; the assembled apparatus is shown on page 89. Tattersfield has recently improved the design of the atomizers employed *(874)*. The test is conducted with 10 adult wingless female *Aphis rumicis* L., which are specially reared for the purpose under carefully controlled conditions. The aphids are placed on a disc of flannel in the glass cylinder, 36 cm. from the sprayer and are sprayed with a suspension of pyrethrins made by diluting an alcoholic extract of pyrethrum with 0.5 per cent saponin solution. Constant pressure of 1055 g. per square cm. is used for spraying. The disc with the aphids is transferred to a petri dish, a few bean leaves are added and a gauze cover is applied. The dish is then placed in a warm, shady place for observation after one to two days.

According to Tattersfield considerable technical experience is necessary to employ the method with precision. Tattersfield *(880)* has compared the kill obtained, with his biological method, with the pyrethrin content determined by his acid method. Five samples containing 0.71, 0.72, 0.81, 0.88, and 0.91 per cent total pyrethrins yielded kills so nearly alike that the samples could not "be separated from each other with any degree of precision"; that is the biological method could not detect a variation of about 25 per cent in pyrethrin content.

Hartzell and Wilcoxon's Procedure

Hartzell and Wilcoxon *(406)* have also used *Aphis rumicis* L., on dwarf nasturtium plants, as test insects. The tests were conducted as follows:

"The sample of flowers (10 g.) was extracted with low boiling petroleum ether in a Soxhlet extractor, and the solvent removed by evaporation under reduced pressure. The residue was dissolved in acetone and made up to 10 cc., and 1.2 cc. of this acetone solution was added to 100 cc. of distilled water. The solutions used for testing usually contained 0.01 to 0.02 grams of pyrethrins per 100 cc. of spray. The emulsion formed in this way was sprayed on the nasturtium plants infested with 200-300 aphids, using an atomizer and compressed air at a pressure of 38 cm. of mercury. During the spraying the plants were slowly rotated by hand. Two plants were used for each test. No spreader was used since the presence of an effective spreader such as soap complicates the interpretation of the results owing to its own toxicity. Immediately after the application the plants were placed on flats containing paper surrounded by a tanglefoot barrier and ruled in squares to facilitate counting. Counts were made after 24 hours under a binocular microscope. Every aphid on the leaves, stems and paper was examined and probed with a needle, when necessary, to decide whether it was alive." Aphids able to move legs or antennae were considered alive; those insects found in the tanglefoot border were also counted as alive. In 26 experiments the average difference between duplicate tests was 6.4 per cent.

Hartzell and Wilcoxon determined the relation between pyrethrin content and per cent mortality in the following manner: "A petroleum ether extract of pyrethrum flowers was

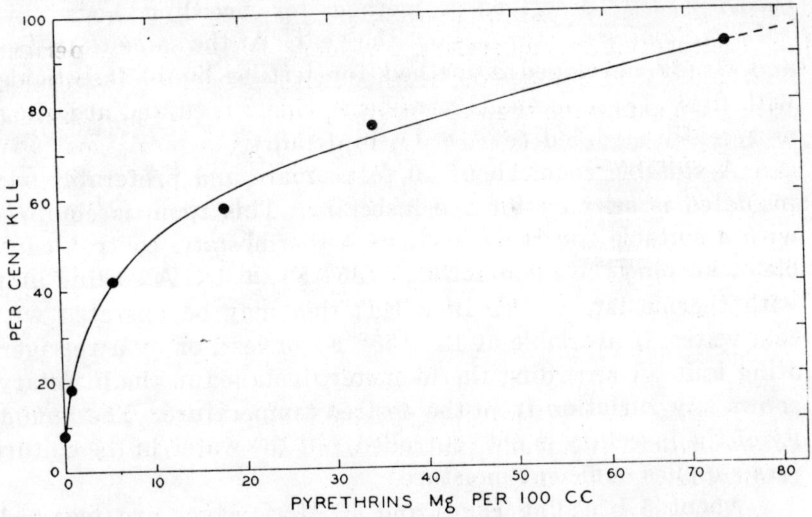

FIG. VIII. RELATION BETWEEN PYRETHRIN CONTENT AND PERCENTAGE KILL, FOR PYRETHRUM SOLUTIONS APPLIED TO *Aphis rumicis* WITHOUT WETTING AGENT. (HARTZELL AND WILCOXON).

partially purified with methanol as described by Staudinger. Analysis by Tattersfield's method showed a pyrethrin content of 64 per cent. Weighed portions of this were dissolved in acetone and a series of emulsions was prepared containing increasing amounts of pyrethrins. Toxicity tests were performed using the above solutions, all sprayings being made the same day." The relation between pyrethrin content and toxicity, so obtained, is shown in Fig. VIII.

Peet-Grady Method

When the kerosene extracts of pyrethrum were introduced commercially as household insecticides, about 1919, methods were sought for testing their efficiency. The most obvious test, and the one generally employed, consisted in releasing a few flies in a small room and then introducing a definite amount of kerosene extract with a sprayer. It was gradually discovered that the vitality of the flies, type of sprayer, temperature and humidity as well, of course, as the concentration of the spray used, were all factors affecting the toxicity of the spray.

One of the principal difficulties in these tests was the lack of a suitable supply of flies during a large part of the year. In 1928 Grady *(369)*, following the procedure recommended by Glaser *(333)*, described a method for breeding house flies *(Musca domestica)* throughout the year. At the same time, Peet and Grady developed a method for testing liquid insecticides with flies. These methods have been widely used and are given, as recently modified *(676, 681)*, in detail.

A suitable room about 10 feet square and preferably well insulated is selected for the insectary. This room is equipped with a suitable source of heat, with thermostatic control, capable of keeping the room at 29.5° (85° F), ± 1°. A cooling unit, with thermostat, is also installed; this may be operated with cold water, if available at 13° (55° F) or less, or by a refrigerating unit. A recording thermometer, installed in the insectary, shows any variation from the desired temperature. The humidity of the insectary is not controlled, but the water in the culture jars supplies sufficient moisture.

About 5 breeding cages and 10 stock cages are required. Their dimensions are: length, 18 inches; breadth, 9 inches; height, 10 inches. The floor is made of a board ½ inch thick to which the frame is attached. The frame is constructed of ½-inch strips. The upright strips are nailed to the floor of the cage and connected by ½-inch crosspieces. The sides, top, and back are made of wire fly screening, (1/16-inch mesh) tacked to the frame. The front is a piece of glass, set in grooves, which serves as a door.

The rearing jars are battery jars 6 inches in diameter by eight inches high. The jars are covered with cheese cloth held in place by elastic bands.

In each of the breeding cages about 200 flies are kept; when the insects die or become soiled or injured they are replaced. Two 200-cc. beakers filled with wet horse manure are placed in each cage for the flies to oviposit on. Every day the beakers

BREEDING CAGES, CULTURE JARS AND OTHER EQUIPMENT FOR REARING FLIES.

containing the eggs and newly hatched larvae are removed and replaced by other beakers of fresh manure. As many cultures as are required are started each day by transferring about 500 eggs and larvae from the 200-cc. beaker to each rearing jar, which is about three-fourths filled with fresh, loosely-packed manure to which 200 cc. of water and 75 cc. of yeast suspension have been added. Every other day 10 cc. more of the yeast suspension are added to each jar; if the amount of yeast is insufficient, or if too many eggs and larvae are placed in a jar, the adult flies will be stunted and weak. The yeast suspension is prepared from commercial, bakers' yeast using 450 g. to 2 liters of water. This suspension should preferably be autoclaved to kill fungi, but this is not essential; the yeast suspenion is stored in a refrigerator.

The culture jars, prepared as above described, are kept in the insectary at 29.5°, the adults emerging in 9 to 11 days. As soon as possible after emergence the adult flies are transferred to stock cages where they are held either for insecticidal tests or for breeding. The stock cages are dated to show the date of emergence of the flies they contain, each day's flies being kept separately. The flies reach the height of activity and resistance when five days old, hence only flies of this age are used for the biological tests which are conducted as follows:

A small room is constructed 6x6x6 feet inside. The frame is conveniently made of 2x4-inch lumber and the walls, floor and ceiling are made of "transite" board, an asbestos board, fitted to the framing so as to present a smooth interior surface. The corners and joints are sealed with sodium silicate cement and the entire inner surface is coated with sodium silicate to render it non-absorbent. In the center of the ceiling a glass window is set with an electric light directly above it.

A tightly-fitting door, large enough to admit a man, is placed in one wall. The adjoining walls are fitted with glass windows in the center, and in each lower corner a 6x6-inch hole is cut, covered with wire gauze and equipped with a tight-fitting hatch. Two half-inch holes are bored in each wall 6 inches from the ceiling; these are closed with corks. The chamber is equipped with a ventilating fan at the top, for changing the air after each test. The opening into the fan is conveniently closed by a sheet iron door, sliding behind a wire gauze cover. An air compressor of some kind is required, which should deliver air at a constant pressure of more than 30 pounds; this is reduced by a suitable control valve to 12½ pounds, which is the pressure used in spraying. The sprayer is a special one, No. 5004, made by the De Vilbiss Co., Toledo, Ohio, on specifications adopted by the

National Association of Insecticide and Disinfectant Manufacturers. The bottle which the manufacturer supplies with the atomizer is graduated so that when filled to the mark no more than 12 cc. of spray can be delivered from the cylinder.

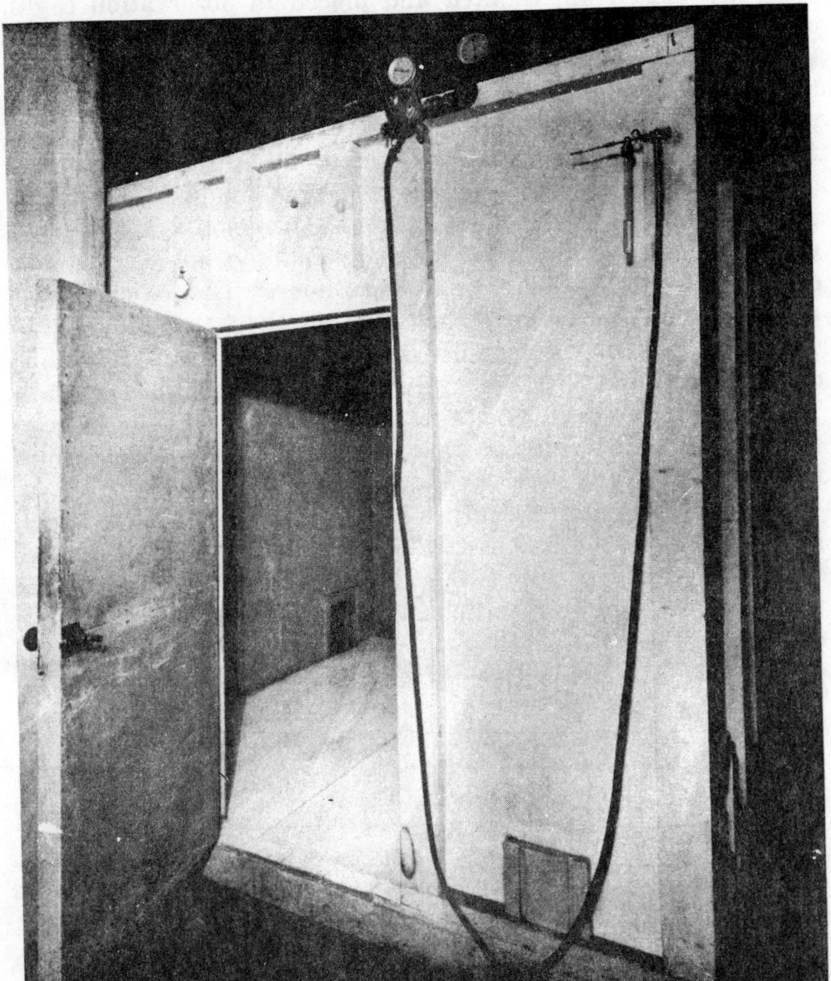

PEET-GRADY TEST CHAMBER.

About 100 five-day old flies are introduced from a stock cage into the chamber and all of the openings are tightly closed. Twelve cc. of insecticide solution are sprayed into the chamber, under 12½ pounds pressure, in about equal amounts through the eight ½-inch holes in the walls. At the end of ten minutes, accurately timed, the screened ports are opened, the ventilating fan is started and the flies clinging to the walls and ceiling are counted. The flies which have dropped to the floor, disabled, are carefully picked up, counted and placed in observation cages, constructed like the stock cages, excepting that they are only 6x6x6 inches in size. A wad of moist cotton gauze is placed in the observation cage to supply water to the flies and the cage is placed in the insectary for 24 hours. Then the number of dead flies in the observation cage is counted. From the total number of flies used and the number disabled at the end of ten minutes the "knockdown" is calculated, while the percentage kill is determined from the total number taken and the number actually dead in the observation cage after 24 hours. The toxicity is not rated on the "knockdown" but only on the "percentage kill."

Between tests the chamber is ventilated for 20 to 30 minutes and the walls and floor are wiped down with an absorbent cloth. Three or four tests by the above method serve to indicate roughly the toxicity of a sample, but for more exact determinations it is generally necessary to run 10 or more tests.

One of the chief difficulties is that the flies quite frequently become infested with mites, which lower their vitality. Such infestations may occur at any time, but are more frequent in the summer. These infestations, which are frequently unnoticed, can be partially avoided by using the following medium, suggested by H. H. Richardson *(715)*, instead of manure:

Wheat bran	1500 g.	Mix thoroughly.
Alfalfa meal	800 g.	
Water	5000 cc.	Mix together and then add to the bran-alfalfa meal mixture, stirring thoroughly.
Yeast suspension (prepared from 453 g. baker's yeast to 2000 cc. water; keep on ice.)	300 cc.	
Diamalt (Fleischmann's Yeast Co.)	25 cc.	

Adoption of Peet-Grady Method as an Official Method

Although a great deal of work has been done on this method, very few of the results have been published. Peet has stated that the average variation between tests is approximately 10 per cent. Weed, however, reported two series of 16 tests each. The first conducted by releasing the flies after the spray was introduced showed 29.6 per cent variation with 52.4 per cent kill compared with 41.4 per cent variation with 73.5 per cent kill when the flies were introduced before spraying. In a series of 101 tests conducted by Mr. Grady, in co-operation with the writer, on 24 samples the maximum variations between the tests were 23, 10, 14, 13, 28, 16, 27, 29, 28, 19, 6, 12, 13, 18, 12, 10, 12, 11, 1, 6, 12, 9, and 12 per cent, or an average of 13.5 per cent. The pyrethrin content of the samples varied from 10 to 150 mg. per 100 cc., and the average kill was 45.3 per cent.

In using the Peet-Grady method for evaluating pyrethrum flowers, the pyrethrin content of the spray should be neither too low nor too high. The percentage of kill is by no means directly proportional to the pyrethrin content. This is especially true when the percentage of kill is above 60 per cent. Serious errors in judging the pyrethrin content of high test flowers can readily be made if the solutions are prepared with one pound of flowers to the gallon. In order to obtain the best results in comparing samples, the pyrethrin content of the spray should be between 50 and 80 mg. per 100 cc. when the kill will fall between 35 and 60 per cent.

In 1931 Peet *(675)*, as chairman of the Standardization Committee of the Insecticide and Disinfectant Manufacturers' Association, reported the results of tests made by four different laboratories on six samples. The samples were prepared by mixing "forty single pints of each of six of the principal insecticides on the market." The odors were masked with pyridine and the samples were identified only by letters. The results obtained by the four laboratories were (Table XXII):

TABLE XXII. RESULTS OBTAINED ON CHECK SAMPLES BY PEET-GRADY METHOD (PEET)

Sample	Average per cent kill Lab. No. 1	Lab. No. 2	Lab. No. 3	Lab. No. 4
A	63.4	60.7	61	39.4
B	70.8	79.1	63	45.2
C	60.5	63.4	58	49.3
D	72.7	72.5	73	52.4
E	67.2	64.0	69	59.5
F	64.8	55.4	64	68.3

On the basis of Peet's report, the National Insecticide and Disinfectant Manufacturers' Association adopted the following standard:

"The members of the Insecticide and Disinfectant Manufacturers' Association agree that a minimum standard for a general household liquid spray insecticide should be 95 per cent down ten minutes after spraying, and at least 60 per cent kill, twenty-four hours after spraying, as determined by the Peet-Grady Method on house flies. In addition, the liquid base should exceed 48.9° (120° F) in flashpoint, as determined by the Tagliabue open cup method, and should not be referred to as kerosene, kerosene petroleum, or petroleum insecticide base in the future, but as a hydrocarbon distillate base. These tests shall be conducted at a temperature of 29.5° (85°F) and 60 to 70 per cent relative humidity. To compare the resistance of flies used in one laboratory to those used in another laboratory, this method will be followed out using ordinary kerosene as originating in the Pennsylvania field. Per cent knock-down and per cent kill by this oil will be set forth in the complete specification. The flies used in this test shall be five days old."

The specifications for the Pennsylvania kerosene to be used for comparing the resistance of flies used by different laboratories were:

A. P. I. Gravity........	49-50
Flash point...........	above 48.9° (120° F)
Initial boiling point....	above 177° (350° F)
End point............	not above 277° (530° F)
Saybolt color.........	30 plus
Odor................	slight
Iodine number, Hanus..	below 1

"Such an oil, when used without additions, by the Peet-Grady method against house flies, should not give more than 20 per cent down and 6 per cent kill. House flies showing these average results shall be considered to have standard resistance to the action of liquid spray insecticides."

The adoption of such a standard by the Association was unfortunate. Examination of Peet's report shows that the differences between the lowest kills and the highest kills found by the different laboratories were as follows:

Laboratory No.	Highest Kill	Lowest Kill	Difference
1	72.7	60.5	12.2
2	79.1	55.4	23.7
3	73.0	58.0	15.0
4	68.2	39.4	28.9

In spite of the wide spread between the strongest and weakest samples shown above, no two of the laboratories agreed as to the relative strengths of the six samples. The relative strengths given to the six samples by the four laboratories were as follows:

Lab. No.	Sample A	Sample B	Sample C	Sample D	Sample E	Sample F
1	5th	2nd	6th	1st	3rd	4th
2	5th	1st	4th	2nd	3rd	6th
3	5th	4th	6th	1st	2nd	3rd
4	6th	5th	4th	3rd	2nd	1st

This means that if a customer had submitted the six samples to any two of these laboratories, he would not have a much better idea of the relative strength of the sprays after the tests than he had before. If the standard of 60 per cent were in force:

Laboratory 2 would have rejected sample F as below standard,

Laboratory 3 would have rejected sample C as below standard,

Laboratory 4 would have rejected samples A, B, C, and D.

that is, the samples on which the standard was based are below standard themselves, a paradoxical condition, to say the least.

The actual relative strengths of the six samples were absolutely unknown. The logical method of preparing these samples would have been to make a strong extract of pyrethrum and then, by simple dilution, prepare samples whose relative strengths would be in the ratio of 100, 90, 80, 70, 60, and 50, respectively. Applying the Peet-Grady method to such a series of samples would establish whether or not such differences can be detected by the method; having thus determined the accuracy of the method, the toxicity of the best commercial products could be determined.

Gothard's Criticism of the Peet-Grady Method

In 1932, Mr. N. J. Gothard succeeded Dr. Peet as chairman of the Standardization Committee. In his report for that year, Gothard expressed the following opinions *(367)* with which the writer, as a member of the committee, was in full agreement:

"Although the Insecticide Committee has set up a definite minimum standard for general household liquid spray insecticides, to be determined by the Peet-Grady method, it is the opinion of the chairman of this committee, and of at least a majority of the members, that there still remains considerable work to be done on the Peet-Grady method before it can actually serve as a standard method by which uniform, comparable results can be obtained by all laboratories.

"The Peet-Grady method was of course originally worked out as a method to be used only in the laboratory of the authors of the method and has since been adopted by numerous others for use in their own laboratories. As long as the results from the method are confined to a single laboratory, the method is very satisfactory and can be relied upon to give consistent, comparable results as long as all the factors involved in the method are carefully controlled by that laboratory. However, when it is attempted to compare results between different laboratories, it becomes apparent that there are certain points about the method which need further investigation before the method can be truly said to be standard.

"The following points appear to be at this time the ones which most need further study.

"1. The temperature at which the test is to be run, and possibly the temperature of the insectary in which the flies are raised.

"2. The effect of variation in droplet size of the spray. This is tied up with the type of atomizer and the question of whether ordinary commercial atomizers as commonly sold actually do throw uniform sprays even though they are of identical make.

"3. The possible increase in uniformity of results to be obtained by the use of paper on the floor of the killing chamber.

"4. The development of a suitable standard for the measurement of the resistance of flies of various origins.

"With regard to the temperature at which the test is to be run, it is now specified that the temperature shall be 85° F. It has been pointed out that this is considerably higher than any normal room temperature, except on the hottest days of summer so that it is difficult to maintain the temperature of the chamber at that point. For instance, let us assume that the surrounding room temperature is 75° F, and that the chamber is brought up to 85° F, and a test made. The chamber then has to be aired out thoroughly by means of an exhaust fan. This means, of course, that air at 75° F is brought into the chamber with the result that after the chamber has been aired thoroughly, the temperature is not 85° F but 75° F, so that before the test can be run again, the chamber must again be brought up to 85° F. To do this properly requires considerable time with the result that comparatively few tests can be run per day. As a matter of fact, the chairman's laboratory has always maintained a temperature of 78° to 80° F in the chamber and I doubt if any laboratories do actually maintain a temperature of 85° F. In view of this, I believe that it is desirable to make a study of the effect of temperature differences in the killing chamber on the percentage of kill obtained. Possibly it will be found desirable to change the present prescribed temperature of 85° F. With regard to the temperature of the insectary, this does not necessarily have to be the same as that of the killing chamber, but it is probably desirable that it should be. Whatever the temperature of the insectary finally decided upon, it is probably necessary that all laboratories follow the standard practice if uniform results are to be obtained since the resistance of flies is closely connected with the temperature at which they are raised.

"With reference to the effect of variation in droplet size of the spray, it is believed that this is a matter of great importance and one to which too little attention has been paid. It was shown in Dr. Peet's original paper that the size of droplet has a marked effect on the test. Uniformity was sought by prescribing the use of a certain specific atomizer operated under a uniform pressure. However, it has been my experience that ordinary commercial atomizers, of the type specified in the method, do not always throw sprays of the same droplet size, so that even though two laboratories are using supposedly identical atomizers, as far as brand and manufacturer are concerned, it is by no means sure that they are getting the same type of spray in their killing chamber. Since these variations in droplet size will affect results markedly, it is of great importance that a study be made of the effect of variation in droplet size and that steps should be taken to insure that each user of the method can obtain an atomizer which will throw exactly the same spray as every other atomizer used.

"With reference to the development of a suitable standard for the measurement of resistance of flies, there is no question in the minds of at least a majority of the committee that this will be necessary before comparable results can be obtained by different laboratories. In spite of every effort to raise strong healthy flies of uniform resistance, different laboratories do not seem to be able to raise flies of uniform resistance. It may be that when the factors heretofore discussed have been satisfactorily settled some, at least, of this apparent difference of resistivity between flies from different laboratories may disappear. However, it is felt that regardless of other factors, there is always liable to be a difference between flies in different laboratories or even between successive lots in one laboratory. In order to obtain comparable results between different laboratories or between results obtained in one laboratory at different periods, it is necessary to know the relative resistance of the flies used. This, it is felt, can only be known by reference to a standard sample whose average killing power has been definitely established. It is obvious, I believe, that the use of any straight run petroleum distillate cannot be satisfactory for this purpose for two reasons. First, it has never been shown that the kill obtained as between strong and weak flies, with a petroleum distillate is proportional to the kill obtained on the same strong and weak flies with a normal insecticide.

"Thus, it is quite conceivable that a normal insecticide might show a decided difference between two lots of flies which would not be detected by a petroleum distillate and in that event, the distillate would be worthless as a measure of resistance. In addition to this, the kill obtained on normal flies with a petroleum distillate is so low that it is in no way comparable with that obtained by the usual insecticides. Thus, in the proposed standard using a highly refined Pennsylvania distillate, it is specified that the insects used should not show more than 20 per cent down and 6 per cent dead. Let us assume that a given laboratory finds that it cannot secure lower than eight per cent kill. Then they must either discard all results obtained or attempt to correct their figures by reference to the standard and this is impossible since the kill is so small that small differences over the standard of 6 per cent are disproportionately large in percentage.

"It seems apparent that a standard should have a killing power in the range of the ordinary insecticides so that results obtained against flies using such a standard may be compared directly with those obtained when testing an unknown sample."

The 1932 report of Mr. Gothard again emphasized the fact that different laboratories, testing the same sample by the Peet-Grady method, frequently obtain entirely different percentage kills. It should be remembered that flies, like pyrethrum, are a natural product and are therefore subject to the variation found in all living organisms.

The writer has suggested benzophenone, which is easily obtained pure, in crystalline form, as a suitable material for use as a standard. Solutions of 5 to 10 per cent benzophenone in kerosene yield kills similar to those obtained with high grade household insecticides by the Peet-Grady method. Comparison of unknown sprays with benzophenone solution would permit the calculation of an "insecticidal coefficient" for the unknown. Nelson *(635)* has shown that a 5.5 per cent solution of benzophenone in deodorized kerosene has essentially the same action, in the Peet-Grady test, as a high grade pyrethrum-oil household spray. The advisability of adopting a standard for household insecticides is generally recognized. That the adoption of the present Peet-Grady method as the standard method of the Insecticide Association was premature can hardly be denied. The reports of the Association's Insecticide Standardization Committee for 1933, 1934 and 1935 have confirmed this opinion.

Nelson *(635)* suggested the following changes in the Peet-Grady method:

1. Adoption of more than one dosage per sample in order to obtain a kill of approximately 50 per cent regardless of sample strength or fly resistance.
2. Adoption of a standard check sample to replace the arbitrary 60 per cent minimum kill.
3. Inclusion of moribund with dead flies in the 24 hour count.
4. Addition of paralysis or knockdown readings during the 10 minute exposure, to show speed of paralytic action.
5. The use of paper on the floor of the test chamber.
6. Open type recovery cages.
7. Adoption of a standard breeding medium to replace manure.
8. A "periscope" type of observation window.

Peet *(680)* as chairman of the Association's Standardization Committee for 1934 reported that only the second and seventh of Nelson's suggestions were generally favored.

Weed *(955)*, chairman of the 1935 committee, reported: "The association must consider these aspects of the Peet-Grady method: (1) that there may be wide differences between laboratories running the same sample; (2) that an undetermined variation is inherent in the method; (3) that accurate placements in the correct relative position of *appreciably* different strength materials are characteristic of most laboratories using the method; (4) that undue selling pressure has resulted in abuse of these results; (5) that through the use of a standard insecticidal material your committee hopes to bring about better agreement; (6) that this committee feels that in view of the above, the results of the Peet-Grady test should not be employed as a sales argument."

Method of O'Kane and Associates

O'Kane, Westgate, Glover and Lowry *(651)* have used a novel method for testing insecticides. Their description of the method follows:

"An automatic air pump stores air under pressure in a large steel tank. An air line leads from this to a Hoke valve where the pressure is reduced to a uniform 20 pounds. From the reducing valve a short rubber hose leads to a De Vilbiss Atomizer, No. 28, which is so constructed as readily to be taken apart for cleaning. The atomizer draws the material under study from a small glass container. Uniform amounts of material are used. The finely-divided spray is discharged at an angle of 45 degrees upon a turntable, 5 inches in diameter, which is made to revolve at a speed of one revolution per second while the spray material is being discharged. Approximately eighteen seconds are required to discharge 10 cc. of the diluted spray materials.

Apparatus for testing spray materials. (O'Kane, Westgate, Glover and Lowry). (Courtesy W. C. O'Kane).

"On the turntable is placed first a disc of filter paper to absorb excess spray, then a disc of wire gauze, slightly turned up at the margin. On the wire gauze the insects are placed. In the case of aphids on leaves the gauze disc is omitted, the filter paper is left in position, and the leaves are pinned to the wooden turntable through the filter paper.

"Adult house flies were confined in twelve-mesh wire cylinders, 4½ inches high and 6 inches in diameter. These were placed on the turntable and sprayed in the usual manner.

"A removable cylinder of kraft paper 15½ inches in diameter and 10½ inches high surrounds the turntable. The top is open so that no volatile products are retained. The paper cylinder is removed and discarded when contaminated. A phonograph motor with multiple spring and automatic adjustable governor revolves the turntable. It can easily be set by a micrometer screw so as to give any number of revolutions desired, from 15 to 150 per minute. The apparatus is shown on page 104.

"In the earlier work, after the insects had been sprayed they were placed in metal pill boxes. Later this plan was discarded. Larger insects are now placed on paper, each lot being covered with a 4-inch or 5½-inch hemisphere of wire gauze. The hemisphere can readily be flamed off when contaminated. Aphids are handled as follows. The detached leaves containing the aphids are placed in petri dishes. Each dish is then covered with a disc of cheese-cloth held taut by metal embroidery rings.

"The cylinders containing house flies were suspended by cords immediately after spraying and the flies were not removed until the end of the experiment."

This method was used by its authors for comparing the toxicity of aqueous solutions of insecticides. It cannot be used for mineral oil extracts of pyrethrum because the oil alone will give 100 per cent kill under the conditions of the test.

Richardson's Method

Richardson *(709)* has recommended the time required to paralyze flies, in preference to the percentage of kill, for evaluating pyrethrum insecticides. His apparatus consists of:

"A rectangular box (2½x2½x4 feet) with a glass top and sides; one side is built so that the glass can be moved back and forth. A copper wire screen cage (1½x2½x1½ feet) which is placed on the inside of the box at one end (C in illustration, page 106). This cage is open on the bottom and its shape is such that it can be moved in and out of the box with ease. A piece of heavy brown wrapping paper is always placed underneath it when it is in the testing box. Actual tests have proven this paper to be of considerable importance since it absorbs the oil droplets, which fall during the test, and thus prevent the flies from becoming entangled in the oil. Significant increases in mortality are produced if it is not used in the test. It is recommended that this paper be changed for each test though little difference has been noted if one paper is used for several tests.

"The spraying device (a No. 29 De Vilbiss atomizer), is located centrally on the outer side of one end (A). It is so placed that its nozzle projects into the testing chamber. The spray container, prepared from a graduated burette, is attached by a cork to the atomizer, and is graduated so that the amount of solution can be read to tenths of a cubic centimeter.

RICHARDSON'S APPARATUS FOR TESTING FLY SPRAYS.
(COURTESY H. H. RICHARDSON).

"A compressed air line is attached to the atomizer. This develops a constant five-pound pressure by use of a Hoke reducing valve. (B).

"An electric fan situated at the end opposite the atomizer, is regulated to give a constant speed. (D). (A 9-inch fan; 300-350 r.p.m.)

"A De Khotinsky temperature regulator maintains a temperature of 28.3° (83° F) in the box. The temperature of the surrounding room is controlled roughly by means of radiators. (E).

"After airing out the test box, the wire screen cage is placed in position with the piece of heavy wrapping paper under it. Forty to sixty house flies of the same age (reared under constant conditions) are introduced into the wire screen cage through an opening in the side. The box is closed tightly and the fan is switched on. Upon opening the air line valve, the spray (1.6 cc.) is atomized into the cage where it is distributed by the air current from the fan.

"At thirty-second intervals, counts are made of the number of paralyzed flies (i.e. lying on their backs). These counts are continued until well over 50 per cent of the flies are down. (All time measurements are made with stop watch.) The test is carried out for eight minutes after which even the weak pyrethrum extracts have paralyzed a majority of the flies. At the conclusion of the test, the box is opened and the screen cage is carefully removed. The flies are counted and then by means of the paper on which they lie, are transferred into a previously prepared cylindrical wire cage (8x2½ inches).

"Following this the testing box and screen cage are thoroughly wiped and aired out and the apparatus is ready for another experiment. Food (diluted milk soaked up by a piece of absorbent cotton) is placed at the disposal of the treated flies. After 24 hours, counts are made of the dead and live flies.

"On one particular day, the fly population appears to be non-differentiated and homogeneous, but it varies to some extent from day to day. This makes it necessary to carry out a complete series of tests each day. The same control solution is used for all series so that tests performed on different days are made comparable by use of the value for the control.

"The speed of paralytic action is taken as the time when 50 per cent of the flies are paralyzed. From the counts made at 30-second intervals and the total number counted at the end of the test, the 50 per cent paralytic point can be easily calculated by interpolation. It has been found that this value varies directly with the strength of the kerosene-pyrethrum extract; the stronger the extract the faster the paralyzing action.

"To test the reproducibility of the results given by this method a series of nine similar experiments was made; all with the same strength of pyrethrum extract, performed under the same conditions with flies of the same age. The results of this series are given as follows (Table XXIII):

TABLE XXIII. REPRODUCIBILITY TESTS (RICHARDSON'S METHOD)

No.	No. of flies	50% paralytic point (sec.)	Dead after 24 hours No.	%
1	47	81	7	15
2	42	75	6	14
3	50	78	9	18
4	39	77	4	10
5	38	78	5	13
6	55	79	11	20
7	43	77	6	14
8	45	80	7	16
9	64	81	8	12
Means		78.40		14.89
Stand. deviation of the observations..		2.00		2.86
Percentage error of the observations..		2.55		19.20
Stand. deviation of the mean.........		0.66		0.99
Percentage error of the mean........		0.84		6.70

"It can be seen from these data that the values for both the speed of paralytic action and the percentage mortality are quite reproducible.

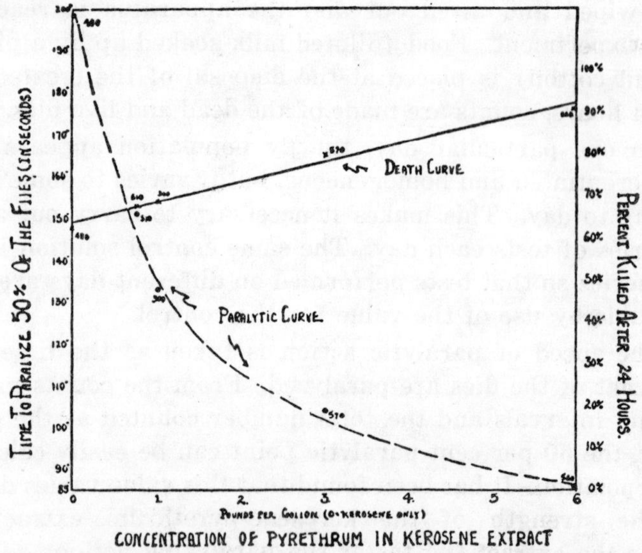

FIG. IX. RELATION OF PYRETHRUM EXTRACT CONCENTRATION TO TIME OF PARALYTIC ACTION AND PER CENT KILL OF FLIES. (RICHARDSON).

"Figure IX shows the results obtained in the study of a series of pyrethrum extracts varying in concentration from kerosene alone to a six-pound to one gallon extract. The number beside each point gives the number of insects used in determining that value. The values represent the averages of a complete series of tests performed on ten different days. Approximately 500 flies were treated in investigating each solution with an average of 50 to 55 flies per test.

"A study of this graph brings out the fact that the speed of paralytic action records very sensitively varying strengths of pyrethrum extract. The death curve, however, changes slowly over the range and can only be used for determining greatly different strengths of extract. For example, a three-fourths pound per gallon extract differs from a one pound per gallon extract in mortality, only 1 per cent; in speed of paralytic action, 18 seconds. Such differences in speed of paralytic action are easily discerned. However, in mortality data a difference of 1 per cent is insignificant."

The Roach Test

The time necessary to paralyze roaches has also been used as an index of the insecticidal value of pyrethrum-oil sprays *(52)*. The American cockroach *(Periplaneta americana)* and the German cockroach *(Blattella germanica)* have been used for this purpose.

"An adult croton-bug is placed in a hemispherical wire cage which is placed on a clean piece of blotting paper. The cage is placed between a small electric light and the atomizer. The roach will take a position away from the light turning its abdomen toward the atomizer. The roach is always sprayed on the abdomen with a constant amount (0.4 cc.) of spray from a constant distance (25 cm.). The amount of spray applied is kept constant by using a special cut-off atomizer of De Vilbiss' make, connected to a compressed air line carrying 12 pounds pressure. The instant the spray has been applied a stop-watch is started and the time interval is taken when the roach lies on its back and is unable to remain on its feet. A perfectly clean cage and a clean piece of blotting paper are used for each roach. Ten such tests are made and from this the mean time interval or 'paralyzing time' is determined." The method is not widely used. The chief source of error is the variability of the age and vitality of the roaches.

Saling *(764)* and Buchmann *(139)* have used the time required to paralyze roaches for evaluating pyrethrum, employing a paraffin oil extract, administered orally or subcutaneously by means of a capillary pipet.

Ueno's Method

Ueno *(922)* has suggested the larvae of mosquitoes commonly found in Japan, as test insects. A composite sample from 10 to 20 lots of highest grade flowers is taken; 100 g. of this sample are extracted with petroleum ether in a Soxhlet apparatus and the solution is evaporated at low temperature and the residue weighed. An accurately weighed portion of the residue (0.1 g.) is dissolved in hot alcohol, cooled, filtered and washed to a volume of 10 cc. This standard solution is kept sealed, in the dark. A similar extract of the sample to be tested is prepared. Two hundred cc. of water are placed in each of two 500-cc. beakers; 1 cc. of standard solution is added to the first beaker and 1 cc. of the unknown solution is placed in the second. The solutions are thoroughly stirred. In each of two 200-cc. beakers are placed 100 cc. of water and 200 of the mosquito larvae. These beakers are simultaneously emptied—one into the beaker containing the standard, the other into the beaker containing the unknown. In a few seconds the larvae will sink to the bottom of the beakers. If all the larvae in one beaker die before all of those in the other, the two solutions are of unequal toxicity. If the two solutions differ in toxicity by 20 per cent, from 3 to 6 larvae will remain alive in the weaker solution at the instant when all are dead in the stronger, while if the difference in toxicity is only 10 per cent only from 2 to 3 larvae will be alive. The strength of the unknown solution is now increased or decreased until all of the larvae in both beakers die at the same time. The test is then repeated 10 times when the result should be the same, otherwise the concentration of the unknown must be changed until agreement is obtained in 10 tests. The relative toxicity of the standard and unknown is then easily calculated, making due allowance for the weight of the petroleum ether extract obtained from the 100 g. samples, since this weight varies with different lots of flowers.

Gersdorff's Method

Gersdorff *(309)* has recommended goldfish *(Carassius auratus)* as test animals for rotenone, the criterion used being the death point. The writer has found that pyrethrins are also extremely toxic to fish. The method, therefore, might possibly be used for comparing the toxicity of acetone extracts of pyrethrum.

According to Gersdorff, "the goldfishes should be as uniform in size as possible, they should not be even slightly out of condition. The tests should be carried out at a constant temperature. Ten or twelve fishes should be used at each concentration; the use of this number lowers the percentage error of the mean to 7 per cent or less with the exception of long survival times (over 5 hours). The most suitable criterion is the death point. The death point is not so easily recognized with precision as is claimed by Pittenger, but may be determined with fair accuracy."

Levy *(552)* has also employed goldfish and other species of small fish for evaluating pyrethrum. Instead of adding the pyrethrum to the water in which the fish are kept the dose is given by intramuscular injection. In evaluating a solution of unknown pyrethrin content, the dilution necessary to cause 50 per cent mortality after 24 hours is determined and the pyrethrin content is calculated from a mortality curve, which Levy has constructed from data obtained in experiments with a partially purified mixture of pyrethrins I and II.

Nelson's Method

One of the greatest weaknesses of the Peet-Grady type of test lies in the fact that every insect does not receive exactly the same dosage of the spray being tested. To overcome this difficulty, Nelson and associates *(636)* administer individual doses to the test insects by means of a capillary pipet, made by drawing out a 4 mm. glass tube. The pipet is calibrated by weighing the amount of mercury that it will contain, from which the volume of the pipet is easily calculated. The German roach and house fly are satisfactory test insects, but the American roach is not satisfactory. The pyrethrum-oil insecticide to be tested is diluted with nine parts of absolute alcohol; for roaches a dose of 5 cubic millimeters is applied, but for flies only 0.75 cubic millimeter is used. If the kill differs considerably from 50 per cent, stronger or weaker dilutions in alsolute alcohol are used. When flies are employed, they are chilled for five minutes at –1° (30° F.) in an electric refrigerator. The insecticide is then applied to the center of the thorax of the quiescent insect, which is transferred to a small screen-covered jar, containing water and powdered milk on a wad of cheese cloth. Thirty flies are placed in each jar and a series of 10 jars is used for each sample of insecticide. The percentage of moribund and dead flies is determined at the end of 24 hours.

The method has also shown some promise for testing aqueous emulsions of oils and pyrethrum-oil sprays. Advantages claimed are: greater accuracy because of controlled individual dosage; careful separation of flies for size and age; small amount and low cost of equipment. Disadvantages are: only relative toxicity is measured; speed of paralysis is not indicated; field conditions are not reproduced; considerable manipulative skill is required.

CAMPBELL'S METHOD

Campbell, Sullivan and Jones *(152)* have adapted the principle of Tattersfield's method (page 90) to the testing of oil sprays containing pyrethrum and derris. House flies, reared in much the same way as for the Peet-Grady method, are used as test insects.

The apparatus (Fig. X) consists of a copper tray on which is placed a glass cylinder (17.5 inches high and 8.25 inches inside diameter). On top of the cylinder is a bell jar of the same diameter. On the top of the bell jar a paint spray gun is mounted so as to deliver a misty spray upon the flies, which are confined, at the bottom of the glass cylinder, in a 6 inch Petri dish, the bottom of which is covered with filter-paper and the top of which is screened with 14 mesh wire cloth, held in place by a celluloid band.

The test is conducted as follows: the flies are chilled at –1° (30° F.) for 15 minutes; after mixing, 50 are counted into the Petri dish, which is then covered with the wire screen and returned to the rearing room, where the flies become normal in about 10 minutes. The Petri dish containing the flies is centered on the copper tray upon which the cylinder and bell jar are placed. Five cc. of the liquid to be tested are measured into a 15 cc. graduated centrifuge tube into which the inlet tube of the spray gun is inserted. The spray gun valve is opened for 10 seconds, spraying the liquid into the bell jar and cylinder. The flies are exposed to the mist for 2 minutes after which they are released from the dish in a 9.5 inch cubical cage, having a sliding glass front, sheet rubber back and top and sides covered with perforated cellophane; the Petri dish is removed from the cage.

The pressure at the start of spraying is 20 pounds per square inch, falling to 7 pounds during spraying. The volume of liquid sprayed into the cylinder is controlled by adjusting the depth of the spray gun inlet tube in the centrifuge tube containing the liquid.

The sprayed flies are kept for observation at the same temperature as the rearing room (83° F.); they are supplied with food and moisture by a wad of milk-soaked cotton. Counts of the dead and moribund flies are made at the end of 24, 48 and 72 hours.

Campbell *(484)* has recently modified the method just described. The apparatus is shown in Figure X. "A" shows the equipment framed by the floor, walls, and ceiling of the constant-temperature room in which it is placed. Six glass cylinders (17 inches high and 8¾ inches outside diameter) stand around the center of a table, "B", which can be revolved on a pivot, "D", fixed on a smaller stationary table top below the turntable. Each cylinder is covered with a circular piece of plate glass having a ¾ inch hole bored through the center. The spray from the gun enters the cylinder through this hole, and air and a little mist escape through it. The hole is not stoppered after spraying.

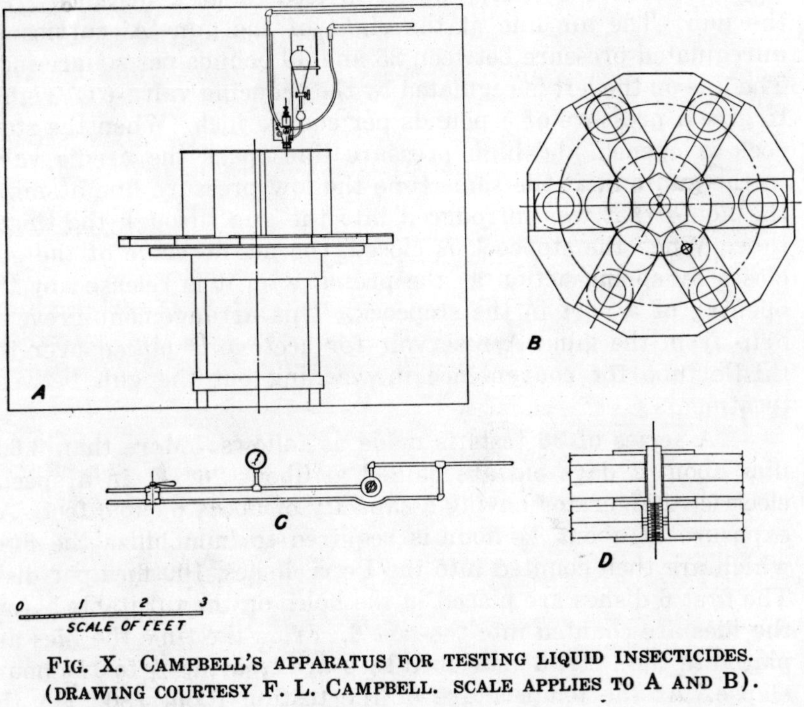

FIG. X. CAMPBELL'S APPARATUS FOR TESTING LIQUID INSECTICIDES. (DRAWING COURTESY F. L. CAMPBELL. SCALE APPLIES TO A AND B).

"The turntable is made of hard maple, in two layers glued and screwed together, with the grain in one layer running at right angles to that in the other to prevent warping. Holes (6 inches in diameter) to contain Petri dishes are cut in the upper layer before the two are fastened together. These holes are shown in 'B' by the small solid circles near the periphery of the table. Guides and stops are provided on the turntable so that the cylinders can easily be brought under the nozzle of the spray gun and can be pulled quickly from their position, represented by the solid circles, where they are filled with mist, to their position, represented by the dot-and-dash circles, where the mist settles on the screened Petri dishes, each containing 100 flies. The holes for the dishes are deep enough to permit the cylinders to slide over them without touching the 14 mesh wire-screen covers of the dishes. The bottom of each dish is covered with a circle of filter paper, and the screen is held over the dish by a celluloid collar band.

"The spray gun is shown in 'A', and its position over a cylinder indicated by the small heavy circle in 'B'. The gun is supplied by two air lines, both of which pass through a three-way stopcock ('C' left), which is located at a distance from the gun. The air line at the right of the gun 'A' carries an unregulated pressure between 25 and 30 pounds per square inch. The line on the left is regulated by the reducing valve ('C', right) to give a pressure of 5 pounds per square inch. When the stopcock is opened, the high pressure line opens the needle valve in the gun and at the same time the low pressure line atomizes the liquid that was introduced into the gun through the thistle tube. When the stopcock is closed, the needle valve of the gun closes by spring action as the pressure on it is released by the opening of a port in the stopcock. This arrangement prevents drip from the gun. A reservoir for acetone is placed over the thistle tube for convenience in washing out the gun between treatments.

"A series of 36 tests is made as follows: More than 3,600 flies about 2 days old are chilled at about 26° F. in a special electric refrigerator having a capacity of about 6 cubic feet. An exposure of about ¾ hour is required to immobilize the flies, which are then counted into the Petri dishes, 100 flies per dish. The first 6 dishes are placed in the holes on the turntable before the flies are counted into the last 6. From the time the flies are placed in the Petri dishes until they are treated, ¾ to 1½ hours elapse. At the temperature of the testing room (80° F.) the flies resume their normal activity in a few minutes and are ready to be treated. The liquids to be tested are measured out in advance, 5 cc. to a test tube.

"With cylinder No. 1 in position under the spray gun the operator pours the contents of the first test tube into the thistle tube leading to the gun, sprays the entire 5 cc. into the cylinder in 4 seconds, and immediately pulls the cylinder over the dish of flies. He then runs acetone into the gun and sprays it through into a cloth to clean the gun. He turns the table to the next ratcheted position, bringing cylinder No. 2 into place under the gun, pours the contents of the second test tube into the gun, and is ready to spray again 30 seconds after the first spraying. He repeats the operations just described and thus goes around the table in 2½ minutes. Each cylinder is allowed to stand over the flies for 10 minutes. It is then pushed back to its original position, and the flies are transferred to 9½ inch cubical cages, which have already been described. The cages are placed in an adjacent constant-temperature room, and the cylinders and the table are wiped off. The second set of 6 dishes of flies is brought in and the cycle is repeated. The 36 treatments are completed in about 1½ hours. Finally, food (milk-soaked cotton) is placed in each of the cages."

Other Methods

Goetze *(360)* has described two unusual methods for determining the toxicity of insecticides. In the first method a mixture of the insecticide with honey and water is fed to bees by means of a capillary pipet. Corn weevils *(Calandra granaria)* are employed in the second method. Twenty of the insects are treated with powdered pyrethrum or an acetone-water extract of pyrethrum and are then transferred to large test tubes which are closed with cotton. The weevils are shaken to the bottom of the tubes, which are carefully laid on a support provided with a paper chart divided in millimeters. Unaffected or slightly affected weevils begin to migrate toward the cotton plug. The more seriously affected insects move more feebly. At the end of half an hour the lengths of the rows of weevils in the various tubes are measured; the shorter the row the stronger the insecticide.

Biological tests employing frogs have been suggested for evaluating pyrethrum *(177)*. The action of pyrethrins on isolated rabbit intestine has also been used as the basis of a physiological assay method *(693)*.

Three methods for evaluating pyrethrum powder have been described by Kemper *(518)*.

It is evident that the Peet-Grady method more nearly approximates the actual conditions under which household insecticides are used than any of the other biological methods. A further discussion of certain sources of error, likely to be encountered in using the Peet-Grady method, will be found in Chapter XII.

CHAPTER VI

CORRELATION OF CHEMICAL ASSAYS AND BIOLOGICAL TESTS

In quantitative chemical analysis the ability of the analyst to obtain concordant results is largely determined by his skill in avoiding errors in following the procedure. In conducting biological tests, not only must the operator avoid errors in manipulation, but the very materials with which he works exhibit a variability of their own, which is beyond his control. The chemist is accustomed to report his analyses in duplicate. The biological investigator is obliged to make a large number of tests to avoid errors due to variations in the test insects or animals. In spite of the fact that biological methods are inherently less accurate than chemical methods, the value of any chemical method for assaying pyrethrum must be established by comparing the chemical analyses with toxicity tests on insects. Such comparisons should be made on different test insects, by different methods, and by different investigators, and once the value of a chemical method is thus established it may quite properly be considered a more accurate index of toxicity than the biological tests with which the comparisons were made.

Staudinger and Ruzicka found that the toxicity of pyrethrum is due to the pyrethrins alone. No other toxic constituent was found by them, nor by any other investigator in the twelve years since their work was published.

Ratio in Which the Two Pyrethrins Naturally Occur

Staudinger and Harder *(840)* have shown that the two pyrethrins do not occur in pyrethrum flowers in equal or definite proportions. This conclusion has been confirmed by Tattersfield and Hobson *(878)*, Martin and Tattersfield *(590)*, Hartzell and Wilcoxon *(406)*, and others. The amounts of pyrethrins I and II in the flowers examined by these workers varied, as shown in the analyses presented in Table XXIV.

The analyses of Staudinger and Harder are not included in Table XXIV because of the inaccuracy of their method. Their results would indicate an even greater variation in the ratio of pyrethrin I to pyrethrin II.

TABLE XXIV. PROPORTIONS IN WHICH PYRETHRINS OCCUR IN PYRETHRUM FLOWERS

Analysts	Source	Pyrethrin I %	Pyrethrin II %	Ratio I:II
Martin and Tattersfield	England	1.32	0.86	1:0.65
Gnadinger and Corl....	Minnesota	0.45	0.31	1:0.69
Martin and Tattersfield	England	0.51	0.44	1:0.86
Gnadinger and Corl....	Michigan	0.45	0.44	1:0.97
Gnadinger and Corl....	Colorado	0.48	0.48	1:1.00
Hartzell and Wilcoxon..	....	0.28	0.39	1:1.40
Martin and Tattersfield	Japan	0.38	0.60	1:1.58
Hartzell and Wilcoxon..	....	0.38	0.63	1:1.66
Tattersfield and Hobson	England	0.29	0.50	1:1.72
Gnadinger and Corl....	Yugoslavia	0.35	0.63	1:1.80
Gnadinger and Corl....	Japan	0.31	0.59	1:1.90
Gnadinger and Corl....	Cregon	0.49	0.96	1:1.95
Jary..................	England	0.57	1.13	1:1.98
Martin and Tattersfield	England	0.21	0.42	1:2.00
Martin and Tattersfield	England	0.36	0.87	1:2.42

Ripert *(745)* states that the two pyrethrins are present in equal amounts in the flowers. This is contrary to the findings of other investigators.

RELATIVE TOXICITIES OF PYRETHRINS I AND II

It was shown in an earlier chapter that the chemical methods for assaying pyrethrum are of two kinds: those which determine pyrethrin I and pyrethrin II separately, and those which determine the total pyrethrin content. If the two pyrethrins are equally toxic to insects, the ratio in which they are present in the flowers is of little importance and a determination of the total pyrethrin content would indicate the toxic value of a given sample. If, however, there is a considerable difference in the toxicity of the two pyrethrins, it would be necessary to determine the proportion of each present in a sample in order to establish its insecticidal value. Therefore, before investigating the relation between the pyrethrin content, determined chemically, and the toxicity to insects, the relative toxicity of the two pyrethrins must first be established. This problem has been studied by Staudinger and Ruzicka, Tattersfield, Hobson and Gimingham, Gnadinger and Corl, and Wilcoxon and Hartzell.

STAUDINGER AND RUZICKA ON RELATIVE TOXICITY

Staudinger and Ruzicka *(842, 850)* isolated pyrethrin I direct from the flowers but they were not able to obtain pyrethrin II in this way. The pyrethrins which they used in determining the relative toxicity were partially synthesized. The pyrethrin I was prepared by treating pyrethrolone, in the cold, with the chloride of chrysanthemum monocarboxylic acid in benzene solution, in the presence of quinoline. The pyrethrin II was similarly prepared from pyrethrolone and the chloride of chrysanthemum dicarboxylic acid monomethyl ester. The two partially synthesized pyrethrins did not yield semicarbazones of sharp melting points, although the composition of the pyrethrins, determined by ultimate analysis, agreed with the theoretical. Staudinger and Ruzicka attributed this to the possible presence of stereoisomers or to changes in the highly unsaturated side chains.

The relative toxicity of the partially synthesized pyrethrins was determined on roaches. The pyrethrins were dissolved in ether, which was then mixed with flour, with gentle warming on a water bath, until the ether was evaporated. Five to six roaches were placed in a flask and dusted with a spoonful of the powder. The tests were made at four concentrations of pyrethrins, 1:500, 1:5000, 1:10,000 and 1:25,000. Staudinger and Ruzicka concluded* "pyrethrin I is somewhat more active than pyrethrin II, it kills roaches in a dilution of 1:10,000 in 10 to 20 minutes, pyrethrin II only in 20 to 40 minutes . . . If one esterifies pyrethrolone with chrysanthemum monocarboxylic acid or with chrysanthemum dicarboxylic acid methyl ester, there are formed esters which are strongly active and which, at a dilution of 1:500, or also 1:10,000, possess the same activity as insect powder, or rather, a correspondingly diluted insect powder . . . The two synthetic pyrethrins are strongly active insect poisons and it is true that the pure pyrethrin I, so obtained, is somewhat more strongly active than the pyrethrin II. The pyrethrin I has approximately the activity of insect powder."

*"Pyrethrin I ist etwas wirksamer als Pyrethrin II, es tötet Schaben in einer Verdünnung 1:10,000 in 10 bis 20 Minuten, Pyrethrin II erst in 20 bis 40 Minuten. . . . Verestert man dagegen Pyrethrolon mit Chrysanthemum-monocarbonsäuer resp. mit Chrysanthemum-dicarbonmethylestersäuer, so entstehen Ester, die stark wirksam sind und die nach Verdünnung 1:500, aber auch 1:10,000 die gleiche Wirkung wie das Insektenpulver resp. ein entsprechend verdünntes Insektenpulver besitzen. . . . Die beiden synthetischen Pyrethrine sind stark wirksame Insektengifte und zwar ist das reine Pyrethrin I, das so gewonnen wird, etwas stärker wirksam als das Pyrethrin II. Das Pyrethrin I hat ungefahr die Wirksamkeit des Insektenpulvers."

TATTERSFIELD'S WORK ON TOXICITY AND PYRETHRIN CONTENT

Tattersfield, Hobson and Gimingham *(880)* did not isolate pyrethrin I and pyrethrin II direct from the flowers, but partially synthesized the two pyrethrins from pyrethrolone and the chlorides of the two chrysanthemum acids, by the same method Staudinger and Ruzicka had used. The pyrethrin I boiled at 145° (0.05 mm.), but there was no proof of the purity of the pyrethrin II. The toxicity of the pyrethrins was determined on aphids, using Tattersfield's method, described on page 88. The synthetic pyrethrins were dissolved in alcohol which was then diluted with 0.5 per cent aqueous saponin solution for spraying.

From the results obtained (Table XXV) Tattersfield, Hobson and Gimingham concluded that pyrethrin I is about ten times as toxic as pyrethrin II to *Aphis rumicis*. At the same time they pointed out that "In view of the method of synthesizing pyrethrin II, there is the possibility that a certain amount of an isomer may be present in the product, which may result in a loss of toxicity."

TABLE XXV. TOXICITIES OF PYRETHRINS I AND II TO *Aphis rumicis* (TATTERSFIELD, HOBSON AND GIMINGHAM)

Pyrethrin	No. of tests	Pyrethrins in spray g. per 100 cc.	N %	S %	M %	D %	M and D %	Marks*
I	2	0.05	..	..	..	100	100	100
I	2	0.025	..	..	..	100	100	100
I	2	0.01	..	..	..	100	100	100
I	3	0.005	..	..	..	100	100	100
I	3	0.0025	..	..	27	73	100	87
I	2	0.001	5	10	70	15	85	53
I	2	0.0005	55	25	10	10	20	21
I	1	0.00025	70	30	..	...	...	8
II	1	0.05	..	..	50	50	100	75
II	3	0.025	..	7	50	43	93	70
II	3	0.01	10	6	67	17	84	52
II	4	0.005	30	23	27	20	47	39
II	3	0.0025	54	20	..	26	26	32

*N, not affected; S, slightly affected; M, moribund; D, dead.
Marks = ¼S + ½M + D.

Tattersfield, Hobson and Gimingham *(880)* also determined the pyrethrins in twelve samples of pyrethrum and tested the toxicity of the samples to aphids. They showed that the toxicity was correlated with the pyrethrin I content and advanced the theory that a determination of pyrethrin I is sufficient to indicate the insecticidal value of a given sample, regardless of the pyrethrin II content. If the analyses of these samples are closely examined, it will be seen (Table XXVI) that the ratio of pyrethrin I to pyrethrin II is, in every case, substantially 1:1. Hence it is obvious that a determination of either the pyrethrin I content, or the pyrethrin II content, or the total pyrethrin content would indicate the relative toxicity of these particular samples, which were all from the same seed stock.

It cannot be fairly inferred from these tests that a determination of pyrethrin I alone is sufficient for the evaluation of all samples of pyrethrum, since the ratio of pyrethrin I to pyrethrin II in other samples varies (page 118).

TABLE XXVI. RATIO OF PYRETHRIN I TO PYRETHRIN II IN TATTERSFIELD'S SAMPLES

No.	Pyrethrin I %	Pyrethrin II %	Ratio, I:II
1	0.36	0.36	1:1.00
2	0.39	0.33	1:0.85
3	0.45	0.46	1:1.02
4	0.28	0.31	1:1.11
5	0.41	0.47	1:1.15
6	0.39	0.42	1:1.08
7	0.46	0.42	1:0.91
8	0.38	0.33	1:0.87
9	0.59	0.53	1:0.90
10	0.37	0.34	1:0.92
11	0.41	0.47	1:1.15
12	0.38	0.33	1:0.87

Martin and Tattersfield *(590)* compared the toxicities of four samples of pyrethrum with their pyrethrin contents, determined chemically, as shown in Table XXVII. "Each sample was extracted with absolute alcohol in the cold, the 10 per cent extract was then diluted with saponin solution of 0.5 per cent concentration, and the dilutions were used for spraying." The insects used were *Aphis rumicis,* and the trials were carried out in the way and with the apparatus described on page 88.

"Each test was given a number, and the examination carried out and the result expressed in ignorance of the actual concentration to which it referred." The test insects were put into four categories: N, not affected; S, slightly affected; M, moribund; D, dead. "The data . . . indicate that the toxicities run in the same order as the pyrethrin contents, namely:

Sw.>L.M.>No. 82>No. 42."

TABLE XXVII. PYRETHRIN CONTENT OF FLOWERS AND TOXICITY TO *Aphis rumicis* (MARTIN AND TATTERSFIELD)

No.	Pyrethrin content of flowers: Tattersfield acid method I %	Tattersfield acid method II %	Tattersfield acid method Total %	Gnadinger-Corl method Total %	Ferricyanide method Total %	Flowers in spray %	Pyrethrins in spray I %	Pyrethrins in spray II %	Pyrethrins in spray Total %	Moribund and dead aphids %	Marks or ratings*
42	0.20	0.22	0.42	0.40	0.40	1.00	0.00200	0.00220	0.00420	100	85
	0.20	0.24	0.44			0.75	0.00150	0.00170	0.00320	100	80
						0.50	0.00100	0.00110	0.00210	70	63
						0.35	0.00070	0.00077	0.00147	50	41
						0.20	0.00040	0.00044	0.00084	15	15
82	0.38	0.60	0.98	0.92	1.00	0.50	0.00190	0.00300	0.00490	100	100
	0.39	0.60	0.99			0.35	0.00133	0.00210	0.00343	100	98
						0.20	0.00076	0.00120	0.00196	100	93
						0.10	0.00038	0.00060	0.00098	32	24
L.M.	0.67	0.87	1.54	1.51	1.58	0.200	0.00140	0.00180	0.00320	100	100
	0.74	0.84	1.58	1.47		0.100	0.00070	0.00090	0.00160	100	93
	0.66	0.99	1.65			0.075	0.00053	0.00067	0.00120	85	74
						0.050	0.00035	0.00045	0.00080	30	38
						0.025	0.00017	0.00023	0.00040	30	31
Sw.	1.13	1.02	2.15	1.98	2.05	0.350	0.00395	0.00357	0.00752	100	100
						0.200	0.00226	0.00204	0.00430	100	98
						0.100	0.00113	0.00102	0.00215	100	95
						0.075	0.00085	0.00076	0.00161	100	88
						0.050	0.00056	0.00051	0.00107	85	77
						0.025	0.00028	0.00025	0.00053	35	28

*Marks or ratings = ¼ of slightly affected aphids, plus ½ of moribund, plus the dead.

Careful study will show, however, that the data do not indicate that the pyrethrin I content alone is sufficient for determining the insecticidal value of the samples.

Tattersfield and Hobson *(878)* compared the pyrethrin content of 8 samples of flowers with their toxicity to *Aphis rumicis*. The biological tests were made in 1927 but the pyrethrin contents were not determined until 1928, hence these results are of doubtful value.

Gnadinger and Corl on Toxicity and Pyrethrin Content

Gnadinger and Corl *(343)* did not employ partially synthesized pyrethrins as Staudinger and Ruzicka, and Tattersfield, Hobson and Gimingham had done. Instead they isolated the two pyrethrins in nearly pure condition direct from the flowers. The toxicity of the pyrethrins was determined on roaches. The pyrethrins were dissolved in alcohol and then diluted with water to the desired strength. The amount of alcohol in the solution applied to the roaches was less than 0.5 per cent. The dilute solutions were freshly prepared before each experiment and the pyrethrins remained uniformly distributed without settling or floating. The results of the experiments, given in Table XXVIII, indicated that pyrethrin I was slightly more toxic than Pyrethrin II to roaches *(Blatta germanica)*, thus agreeing with the conclusion of Staudinger and Ruzicka.

Table XXVIII. Toxicity of the Pyrethrins to Roaches (Gnadinger and Corl)

Composition of solution tested				Roaches dead
Pyrethrin		Water	Ratio	in 24 hours
	mg.	cc.	pyrethrin-water	%
I	33	1000	1:30,300	100
I	20	1500	1:75,000	100
I	12.5	1000	1:80,000	100
I	20	2000	1:100,000	50, Balance disabled
I	20	3000	1:150,000	None dead; all disabled
II	33	1000	1:30,000	100
II	20	1500	1:75,000	100
II	12.5	1000	1:80,000	50, Balance disabled
II	20	2000	1:100,000	50, Balance partly disabled
II	20	3000	1:150,000	None dead; all recovered

In view of the contradictory nature of the results thus far obtained by different investigators, Gnadinger and Corl *(346)* undertook a more comprehensive study of the problem. About 500 grams of the mixed, crystalline semicarbazones of pyrethrins I and II were isolated from Japanese pyrethrum by the method of Staudinger and Ruzicka. This material was repeatedly crystallized from 90 per cent alcohol, 60 per cent alcohol and a mixture of one part of benzene and three parts of petroleum ether, finally yielding 40 g. of pyrethrin I semicarbazone melting at 115°-117° and 30 g. of pyrethrin II semicarbazone melting at 54°-58°.

Analysis of the pyrethrin I semicarbazone showed:

Volatile acids calculated as pyrethrin I semicarbazone......96.94%
Non-volatile acids calculated as pyrethrin II semicarbazone. 8.10%

Analysis of the pyrethrin II semicarbazone showed:

Volatile acids calculated as pyrethrin I semicarbazone......16.94%

The pyrethrin I semicarbazone was 97 per cent pure, while the pyrethrin II semicarbazone contained 17 per cent of semicarbazone of pyrethrin I. The total acidity found in the analysis of pyrethrin I semicarbazone was 5 per cent higher than the theoretical amount. This was probably due to the formation of acid compounds by the action of the sodium hydroxide on pyrethrolone semicarbazone. Staudinger and Ruzicka found that pyrethrolone yielded dehydropyrethrolone, a compound having weak acid properties, on prolonged digestion with alcoholic sodium hydroxide.

The semicarbazones were converted into the pyrethrins by digestion with oxalic acid solution. The crude pyrethrins so obtained were extracted with petroleum ether and the petroleum ether solution was washed once with 10 per cent potassium carbonate solution, three times with 1 per cent sodium hydroxide, three times with water, three times with 3 per cent potassium permanganate, three times with water, once with 1 per cent sodium hydroxide and three times with water. It was then filtered and distilled in vacuum to constant weight at a maximum temperature of 40°.

The monocarboxylic acid in the pyrethrin I was determined by saponification and distillation with steam, and was calculated to pyrethrin I. The percentage of pyrethrin I was also determined from the copper reducing power.

Analysis of pyrethrin I:

Pyrethrin I calculated from volatile acid..................79.7%
Pyrethrin I calculated from copper reducing power........79.5%

Anal. Calcd. for $C_{21}H_{30}O_3$: C, 76.31; H, 9.16. Found: C, 76.51, 76.20; H, 9.06, 8.89.

Similarly the amounts of monocarboxylic acid and dicarboxylic acid in the pyrethrin II were determined and calculated to pyrethrin I and II, respectively.

Analysis of pyrethrin II:

Pyrethrin I calculated from volatile acid................23.3%
Pyrethrin II calculated from non-volatile acid............84.3%

Anal. Calcd. for $C_{22}H_{30}O_5$: C, 70.54; H, 8.08. Found: C, 72.62, 72.00; H, 8.17, 8.09.

The pyrethrin I was 80 per cent pure and contained little pyrethrin II. The ultimate analysis, however, showed practically the theoretical percentage composition for pyrethrin I. The pyrethrin II contained about 20 per cent pyrethrin I and 80 per cent pyrethrin II. As in the analysis of the pyrethrin I semicarbazone, the total acidity was about 7 per cent higher than the theoretical, probably because of the presence of acid decomposition products of pyrethrolone. Ultimate analysis confirmed that the material was a mixture of about 20 per cent pyrethrin I and 80 per cent pyrethrin II.

Solutions of pyrethrin I and pyrethrin II were prepared by dissolving the accurately-weighed pyrethrins in a highly refined mineral oil. These solutions were made within twelve hours from the time the pyrethrins were isolated and in the interval the latter had been kept in vacuum flasks. These precautions were taken to insure that the pyrethrins were not oxidized by exposure to the air. The mineral oil used was water white, nearly odorless and tasteless; specific gravity 0.785 at 15.6°; distilling range, 180°-240°. The pyrethrin solutions were perfectly clear and were colorless. Solutions were prepared containing 150, 125, 100, 75, 50, 25, and 10 mg. of the isolated pyrethrins per 100 cc. The actual amounts of pyrethrins in these solutions were calculated from the analyses given above, which showed that the pyrethrin I was 80 per cent pure and the pyrethrin II consisted of 77 per cent pyrethrin II and 23 per cent of pyrethrin I.

Extracts were prepared from four samples of pyrethrum flowers, which had been assayed by the method of Gnadinger and Corl, using the same oil employed for making the pyrethrin solutions. The ground flowers were macerated for seventeen days with frequent shaking. The extracts were then filtered and kept in well-filled bottles in a dark closet. The composition of these extracts is fully described in the accompanying table (XXIX).

TABLE XXIX. EXTRACTS OF ASSAYED FLOWERS *(Pyrethrum cinerariaefolium)* (GNADINGER AND CORL)

No.	Description	Pyrethrins 6|3|29	Pyrethrins 1|28|30	Composition of oil extracts: Flowers g.	Oil cc.	Pyrethrins mg. per 100 cc.
6	Dalmatian open, 1926 crop	0.60	0.60	63.5	497.5	75
		.56				
7	Dalmatian half open, 1925 crop	.38	.33	69.5	497.3	50
		.38				
13	Dalmatian open, 1926 crop	.43	.39	59.5	497.6	50
82	Japanese, 1929 crop		.97	104.2	996.0	100
			.97			
			.93			

From the extract of Sample 6, dilutions were prepared, containing 60 and 40 mg. of pyrethrins per 100 cc.; from Sample 82, solutions containing 80, 60, 40, 20 and 10 mg. per 100 cc. were made. All of the solutions were kept in the dark when not in use.

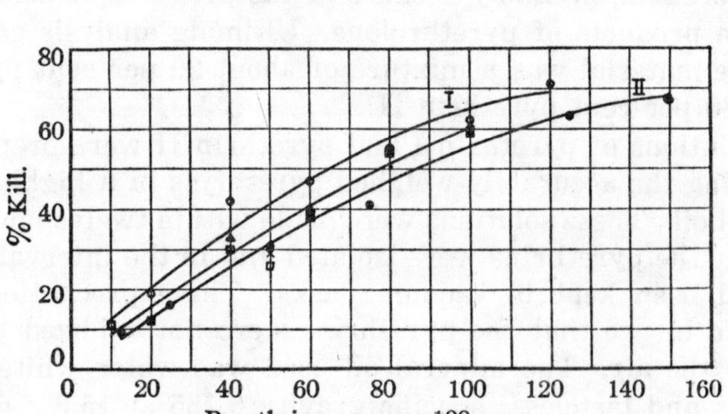

FIG. XI. TOXICITY OF PYRETHRIN I, PYRETHRIN II AND PYRETHRUM FLOWERS TO FLIES. ○, PYRETHRIN I; ●, PYRETHRIN II; ■, FLOWERS NO. 82; ▲, FLOWERS NO. 6; □, FLOWERS NO. 13; ×, FLOWERS NO. 7. (GNADINGER AND CORL).

The biological tests with these solutions were made on flies, by the Peet-Grady method. The tests were made by Mr. Grady, one of the authors of the method, through the courtesy of his employers, and were observed by the writer. The results are reported in Table XXX.

The temperature throughout the experiments was 25.6° and the humidity 43-50 per cent. The same atomizer was used in all experiments. The results of every experiment were recorded with the exception of five which were obviously incorrect. The controls were not sprayed but were kept in the same kind of cage and under the same conditions as the disabled flies.

The values for pyrethrin I and pyrethrin II and pyrethrum flowers, after correcting for the kill due to the oil alone, are plotted in Fig. XI; the curve of the pyrethrin I solution is based on 100 per cent pure pyrethrin I. The material from which the pyrethrin II curve was prepared consisted of 23 per cent pyrethrin I and 77 per cent pyrethrin II.

Inspection of these biological tests indicates at once that the difference in toxicity between pyrethrin I and pyrethrin II cannot be great. Calculations from the kill (Fig. XI) at concentrations of 120, 100, 80, 60, and 40 mg. per 100 cc. show that pyrethrin II is at least 77 per cent as toxic as pyrethrin I. The kill obtained with the extract of flowers No. 82 is further proof that pyrethrin II is nearly as toxic as pyrethrin I. Sample 82

TABLE XXX. TOXICITY OF PYRETHRINS I AND II AND PYRETHRUM FLOWERS TO FLIES *(M. domestica)* (GNADINGER AND CORL)

No. of tests	Description of soln.	Pyrethrin content mg. per 100 cc. I	II	Total	Av. no. flies	Disabled in 10 min. Av. %	Condition of disabled flies in 24 hrs. Recovered Av. %	Dead Max. %	Min. %	Av. %
7	Oil only	0	0	0	97	19	15	6	1	4
7	Pyrethrin I	120		120	97	95	20	83	68	75
4	Pyrethrin II	34	116	150	97	99	28	76	66	71
4	Pyrethrin I	100		100	96	95	29	71	57	66
4	Pyrethrin II	29	96	125	99	98	31	73	60	67
6	Flowers No. 82	39	61	100	99	96	33	69	53	63
6	Pyrethrin I	80		80	97	96	33	66	50	63
5	Pyrethrin II	23	77	100	94	97	33	71	57	64
5	Flowers No. 82	31	49	80	95	96	38	67	52	58
3	Pyrethrin I	60		60	98	92	41	52	50	51
3	Pyrethrin II	17	58	75	96	96	51	48	40	45
6	Flowers No. 82	23	37	60	96	94	52	52	33	42
4	Flowers No. 6			60	95	92	48	46	40	44
5	Pyrethrin I	40		40	94	90	44	52	40	46
4	Pyrethrin II	11	39	50	103	93	58	42	39	35
5	Flowers No. 82	16	24	40	93	91	57	44	26	34
4	Flowers No. 6			40	105	82	45	42	30	37
4	Flowers No. 7			50	96	81	47	40	30	34
4	Flowers No. 13			50	95	78	48	38	26	30
3	Pyrethrin I	20		20	98	69	46	29	18	23
3	Pyrethrin II	6	19	25	96	83	63	20	19	20
3	Flowers No. 82	8	12	20	91	76	60	21	13	16
3	Pyrethrin I	10		10	91	50	35	20	8	15
3	Pyrethrin II	3	10	13	101	64	51	17	8	13
3	Flowers No. 82	4	6	10	93	65	50	22	10	15
7	Controls (not sprayed)	0	0	0	92			0	0	0

was analyzed by three different laboratories, both by Tattersfield's method and the method of Gnadinger and Corl. These analyses are given below:

Laboratory	Tattersfield method, pyrethrins I, %	II, %	Total, %	Gnadinger and Corl method Total pyrethrins, %	
A	0.39, 0.38	0.60, 0.60	0.99, 0.98	0.92	
B				0.95,	0.95
C				0.97,	0.97
				0.97,	0.92

The ratio of pyrethrin I to pyrethrin II in Sample 82 is 1:1.5 but the kill is only slightly less than with pure pyrethrin I and is somewhat greater than with pyrethrin II. The toxicity of Sample 6 is also between that of pyrethrin I and pyrethrin II, while the toxicity of Samples 7 and 13 is about the same as that of pyrethrin II. Therefore the toxicities of oil extracts of pyrethrum, whose total pyrethrin content has been determined by the copper reduction method, fall between the toxicity of pyrethrin I solution and the toxicity of pyrethrin II solution of the same pyrethrin content, made by dissolving the pyrethrins in the same oil. Furthermore, it will be seen from the foregoing that when extracts of pyrethrum flowers, of widely different pyrethrin content, are diluted to the same pyrethrin content, as determined by the copper reduction method, the toxicities are substantially the same.

This work shows that the determination of the total pyrethrin content of pyrethrum flowers by the copper reduction method of Gnadinger and Corl is an accurate index of their toxicity.

In addition to the above experiments seven samples of pyrethrum ranging in pyrethrin content from .39 to 1.62 per

TABLE XXXI. RELATION BETWEEN PYRETHRIN CONTENT OF FLOWERS AND TOXICITY TO FLIES, PEET-GRADY METHOD (GNADINGER AND CORL)

Sample	Pyrethrin content of flowers* %	Flowers per 100 cc. of oil g.	Pyrethrin content of oil mg. per 100 cc.	No. of tests	Kill Max. %	Kill Min. %	Kill Av. %
C	1.62	10.0	162	8	88	72	80
A	1.06	10.0	106	8	70	54	61
B	1.01	10.0	101	8	72	55	62
D	0.72	10.0	72	9	68	46	55
C	1.62	4.4	72	10	68	46	59
14	0.62	8.4	53	8	55	36	45
5	0.40	8.4	34	10	45	22	31
16	0.39	8.4	33	9	45	20	31
Oil only	...	...	0	5	9	5	6

*Gnadinger and Corl method.

cent were carefully assayed. Oil extracts of these samples were then prepared using the same oil for all samples. The biological tests on these seven samples were made by the Peet-Grady method. The results are tabulated on page 128 (Table XXXI).

Sample C had the highest pyrethrin content, 1.62 per cent; when used in the proportion of 10 g. per 100 cc. it gave 80 per cent kill. Sample A containing 1.06 per cent pyrethrins killed 61 per cent, while Sample B, of nearly the same pyrethrin content as A had almost the same kill. Sample D, with a lower pyrethrin content, gave a lower kill. All of these solutions were made with 10 g. of flowers per 100 cc. and the kills are in the same order as the pyrethrin contents. Part of Sample C solution was diluted with oil so that the flower content was 4.4 g. per 100 cc.; at this concentration the solution would have the same pyrethrin content as the solution made from Sample D, if the chemical analyses were correct, and the kills obtained with the two solutions should be substantially the same. This was found to be the case. The pyrethrin content of the seven samples, determined by the Gnadinger-Corl method, was, in every case, an accurate index of the toxicity to flies, determined by the Peet-Grady method. This confirmed the conclusion reached in the earlier work with Mr. Grady.

Wilcoxon and Hartzell on Relative Toxicity of Pyrethrins

Hartzell and Wilcoxon *(406)* also compared the toxicity of pyrethrum to aphids (using their method, page 90) with the pyrethrin content determined by Tattersfield's acid method and the Gnadinger-Corl method, with the following results (Table XXXII):

Hartzell and Wilcoxon state: "Inspection of these results shows that in every case a positive difference in per cent kill is associated with a positive difference in the analytical figures. There is little correspondence, however, between the actual magnitude of the differences in per cent kill and toxicity for each pair. There are several reasons for this. From the curve previously shown (page 91) it may be seen that a given difference in mortality represents a varying difference in concentration depending on the portion of the curve under consideration. In addition to this the figures for per cent kill are subject to a considerable error of random sampling. From our experience it appears that whenever two samples show a significant differ-

TABLE XXXII. RELATION BETWEEN PYRETHRIN CONTENT AND TOXICITY TO APHIDS (HARTZELL AND WILCOXON)

Sample	Kill		Pyrethrin content of flowers Acid method	Gnadinger method	Difference between pairs
	%	Diff.	%	%	%
1	37		1.01	1.03	
1A	27	10	0.79	0.81	0.22
2	55		0.96	...	
2A	33	22	0.78	...	0.18
3	55		0.96	...	
3A	33	22	0.70	...	0.26
4	68		0.67	...	
4A	37	31	0.58	...	0.09
5	73		0.75	0.73	
5A	52	21	0.52	0.59	0.23
6	41		0.99	...	
6A	10	31	0.08	...	0.91
7	53		0.98	0.82	
7A	28	25	0.67	0.64	0.31

ence in toxicity by the biological test, the analytical results will confirm this fact."

Recently, Wilcoxon and Hartzell *(968)* isolated the pyrethrins by their physical method (page 35) and effected a partial separation of pyrethrins I and II. In the process of separation the petroleum ether layers were richer in pyrethrin I and the methanol layers were richer in pyrethrin II. Samples containing widely different ratios of pyrethrin I to pyrethrin II were prepared and made up to the same total pyrethrin content in acetone. The only proof of the purity of the pyrethrins in these solutions was the determination of the pyrethrin content by Tattersfield's acid method, which, under these conditions, might include altered pyrethrins. The toxicities of these solutions were determined on *Aphis rumicis* by the Hartzell-Wilcoxon procedure (page 90). The samples were compared in pairs, both members of a pair being run at the same time. These experiments are summarized in Table XXXIII, and indicate that pyrethrin I is much more toxic to aphids than pyrethrin II.

Wilcoxon and Hartzell also compared the toxicities of their isolated pyrethrins with the toxicities of assayed pyrethrum flowers; the following is their description of the experiments: "Sam-

TABLE XXXIII. COMPARATIVE TOXICITY TO *Aphis rumicis* OF PYRETHRUM EXTRACTS, VARYING IN THE RATIO OF PYRETHRIN I TO II (WILCOXON AND HARTZELL)

Pyrethrin I %	Pyrethrin II %	Total pyrethrins %	Mortality %	
23.1	28.5	51.6	55.5	. .
9.5	42.1	51.6	25.6	. .
21.6	24.4	46.0	62.5	53.0
8.3	41.6	49.9	32.0	26.0
43.6	57.1	100.7	22.8	23.5
14.7	76.5	91.2	9.5	10.3
50.6	44.4	95.0	51.1	49.4
2.8	92.6	95.4	17.0	9.1

ples of analyzed flowers were extracted with petroleum ether, the solvent was removed at low temperature and the residue dissolved in acetone. Portions of a purified preparation, the analytical results indicating 100.7 per cent total pyrethrins, were dissolved in acetone, and from these spray solutions were made up so that the sprays from the flowers and from the purified extract contained the same content of pyrethrin I. Toxicity experiments were then performed using *Aphis rumicis* on nasturtium in the usual manner." The results are shown in Table XXXIV. Wilcoxon and Hartzell comment: "In every case the flowers gave a higher kill than the extract. The average difference, 13.5 per cent, while significant, is not very great, and it does not appear that any serious error is likely to arise if the flowers are rated on the content of pyrethrin I alone. Several possibilities suggest themselves as the cause of the higher toxicity of the flowers. It may be due to the toxicity of pyrethrin II, of which the content

TABLE XXXIV. COMPARATIVE TOXICITY TO *Aphis rumicis* OF PURIFIED PREPARATIONS AND CRUDE EXTRACTS FROM THE FLOWERS, BOTH SAMPLES HAVING THE SAME CONTENT OF PYRETHRIN I (WILCOXON AND HARTZELL)

Experiment	Ratio of II to I in extract	Mortality with extract %	Ratio of II to I in flowers	Mortality with flowers %
1	1.31	64.4	2.76	72.8
		57.5		72.7
2	1.24	43.5	1.88	55.3
		42.0		53.3
3	2.06	51.5	1.68	68.6
		50.8		68.2

was somewhat higher in the flowers than in the extracts in two of the three cases, or there may be another substance in the flowers which has some slight toxicity in addition to the pyrethrins. Another possibility is that the impurities in the flowers serve as wetting agents or emulsifiers, thus making the spray more efficient when made from the flowers."

It will be recalled that Hartzell and Wilcoxon constructed a curve (Fig. VIII, page 91) showing the relation between pyrethrin content and toxicity to *Aphis rumicis* determined by their method. If the results in Table XXXIV are compared with Fig. VIII, we find that the differences between the kills obtained with the solutions of purified pyrethrins and solutions of the assayed flowers, adjusted to the same pyrethrin I content, are not small, as Wilcoxon and Hartzell state, but, on the contrary, are very large. In fact, the kills obtained with solutions made from the assayed flowers correspond to pyrethrin contents which are double the pyrethrin contents indicated for the purified pyrethrin solutions; this is shown in Table XXXV. It is true that Fig. VIII is based on total pyrethrin content, but if pyrethrin II is as inert as Wilcoxon and Hartzell state, the curve should, nevertheless show the relationship between toxicity and pyrethrin I content

TABLE XXXV. RE-CALCULATION OF DATA OF WILCOXON AND HARTZELL, FROM TABLE XXXIV AND FIG. VIII

Exp.*	Solutions having same pyrethrin I content prepared from			
	Purified pyrethrins		Assayed flowers	
	Mortality to aphids* %	Indicated pyrethrin content** mg. per 100 cc.	Mortality to aphids* %	Indicated pyrethrin content** mg. per 100 cc.
1	60.9	18	72.7	34
2	42.7	6	54.3	12
3	51.1	10	68.4	25

*Table XXXIV.
**Fig. VIII.

These wide differences in results may be due in part to the causes suggested by Wilcoxon and Hartzell, or to the unreliability of the biological method employed.

The biological test used by Hartzell and Wilcoxon differs in several important respects from Tattersfield's method employing the same test insect, *Aphis rumicis.* Tattersfield removes the aphids from the host plant and makes a considerable number of tests, using 10 adult wingless female aphids for each experiment results of this kind lend themselves more readily to statistical treatment. Hartzell and Wilcoxon spray nasturtium plants infested with about 200 aphids, in all stages of development

Only 2 plants are used for each solution tested. Shepard and Richardson *(800)* have criticized this kind of test because "Evaporation of spray liquids from aphid-infested leaves is very uneven, many aphids remaining immersed in the spray much longer than others. Young and old members of the usual aphid colony react differently to toxic materials." When no spreader is used, as in the Hartzell and Wilcoxon test, the spray does not spread in a continuous film over the nasturtium foliage, but collects in small globules, which are not uniformly distributed. Some of these globules enclose aphids, other aphids are scarcely wet by the liquid; the kill obtained is thus largely a matter of chance. In tests on deteriorated pyrethrum, Hartzell and Wilcoxon, using their own method, were able to find little correspondence between the actual magnitude of the differences in pyrethrin content and toxicity.

Ripert and Gaudin *(748)* isolated pyrethrin I and pyrethrin II from the flowers by the physical method described on page 36. The pyrethrin I was 96 per cent pure, the pyrethrin II, 98.9 per cent pure. The toxicities of the pyrethrins were compared, by the method of Gaudin and Carron *(305)*, using fish as test animals. Ripert and Gaudin found fish unsuitable because the pyrethrins acted differently on them, pyrethrin II causing excitation, but pyrethrin I causing slow paralysis without excitation. They therefore measured the relative toxicity of the two pyrethrins on frogs, following the method of Trevan. By this method they found that in order to obtain 50 per cent of deaths, it was necessary to inject 0.75 mg. of pyrethrin II per kilogram of animal, or, 0.80 mg. of pyrethrin I, or, 0.66 mg. of a mixture of equal parts of pyrethrins I and II. They concluded that pyrethrin II is slightly more toxic to frogs than pyrethrin I and that the mixture of pyrethrins is more toxic than either pyrethrin alone.

Ripert and Gaudin *(748)* have confirmed their original conclusion that pyrethrin II is more toxic than pyrethrin I to frogs, fish and mice, but have found that pyrethrin I is more toxic than pyrethrin II to flies.

McDonnell's Experiments

McDonnell and his associates *(605)* were unable to find any correlation between pyrethrin content and toxicity. Their tests on aphids were made by an obsolete dusting method and the chemical analyses were made 2 years after the biological tests were begun; in that time the powdered flowers probably deteriorated about 50 per cent. The writer assayed some of the samples used by McDonnell and also tested them biologically by the Peet-Grady method. The results are reproduced in Table XXXVI; for convenience, the kills obtained by McDonnell on aphids are included in the last column.

TABLE XXXVI. COMPARISON OF PYRETHRIN CONTENT, TOXICITY TO APHIDS (MCDONNELL), AND TOXICITY TO FLIES (GNADINGER)

Sample	Pyrethrin* content of flowers, %	Flowers per 100 cc. of oil g.	Pyrethrin content of oil mg. per 100 cc.	Average kill flies** %	aphids*** %
14	0.62	8.4	53	45	56
5	0.40	8.4	34	31	58
16	0.39	8.4	33	31	65

*Gnadinger-Corl method.
**Peet-Grady method.
***McDonnell's tests on aphids.

Samples 5 and 16 had almost the same pyrethrin content and gave the same kill by the Peet-Grady method, but the kills obtained on aphids by McDonnell were different. Sample 14 contained about 50 per cent more pyrethrins than Samples 5 and 16 and showed a correspondingly higher kill by the Peet-Grady method. McDonnell, however, obtained a lower kill with Sample 14 than with 5 or 16. There is little doubt but that McDonnell's dusting tests on aphids were not a true index of the toxicity of the samples of pyrethrum which he used.

RICHARDSON'S WORK ON TOXICITY AND PYRETHRIN I CONTENT

Richardson *(711)* compared the toxicity of 4 samples of pyrethrum, determined on flies, by his method (page 105), with the pyrethrin I content determined by Tattersfield's short acid method. Richardson concluded that the pyrethrin I content can be used to indicate the relative value of different samples of pyrethrum. Unfortunately he made 3 serious errors in his work:

1. He determined only pyrethrin I in his 4 samples and failed to determine pyrethrin II. It is quite possible that the ratio of pyrethrin I to pyrethrin II was approximately the same in this limited number of samples, in which case the toxicity would be proportional to either the pyrethrin I or pyrethrin II content.
2. He prepared his extracts by an inefficient method of percolation *(716)* which made it necessary to apply a cor-

rection for the "extractive efficiency" of kerosene. This correction was an approximation and in view of the fact that practically complete extraction of the pyrethrins by kerosene is easily attained, when proper methods of extraction are employed, it is difficult to understand why correction for "extractive efficiency" is necessary.

3. He assumed that pyrethrins I and II are present in the flowers in approximately equal quantities, which is not the case (page 118).

Because of these errors Richardson's conclusion cannot be considered proved.

Other Investigations

One of the laboratories using the roach test (page 109) has reported that the pyrethrin content, determined by the copper reduction method, is an index of the toxicity to roaches. On the other hand, Goetze *(360)* failed to get complete concordance between pyrethrin content and toxicity determined by his methods. Profft and Körting *(703)* found only a very general relationship between the pyrethrin content and toxicity to grain weevils.

Summarizing the data thus far presented:

1. Staudinger and Ruzicka found pyrethrin I only slightly more toxic to roaches than pyrethrin II.
2. Gnadinger and Corl concluded that pyrethrin II is about 80 per cent as toxic to roaches and house flies as pyrethrin I.
3. Tattersfield, Hobson and Gimingham found pyrethrin I about 10 times as toxic to aphids as pyrethrin II.
4. Wilcoxon and Hartzell also reported pyrethrin I far more toxic to aphids than pyrethrin II.
5. Staudinger and Ruzicka, and Tattersfield, Hobson and Gimingham employed partially synthesized pyrethrins in determining the relative toxicity of pyrethrins I and II. Gnadinger and Corl used pure pyrethrin I and nearly pure pyrethrin II isolated directly from the flowers by the semicarbazone method. Wilcoxon and Hartzell employed solutions containing different ratios of pyrethrins I and II, purifying the pyrethrins by repeated extractions with petroleum ether and 80 per cent methanol. They did not prove the purity of their pyrethrins, which were much less toxic than extracts made from assayed flowers, adjusted to the same pyrethrin I content. On the other hand, the isolated pyrethrins of Staudinger and Ruzicka and Gnadinger and Corl had about the same toxicity as pyrethrum flowers correspondingly diluted. Neither Tatters-

field, Hobson and Gimingham nor Wilcoxon and Hartzell mentions any precautions taken to prevent decomposition of the isolated pyrethrins.

6. Martin and Tattersfield, Gnadinger and Corl, and Hartzell and Wilcoxon showed correlation between total pyrethrin content, determined chemically, and toxicity to insects. Martin and Tattersfield found that the acid method, ferricyanide method and copper reduction method indicate the same relative toxicity as determined by Tattersfield's method on aphids. Hartzell and Wilcoxon showed that the acid and copper reduction methods indicate a positive difference in pyrethrin content for each positive difference in toxicity to aphids. Gnadinger and Corl found that the pyrethrin content, determined by their copper reduction method, indicated the toxicity to flies, determined by the Peet-Grady method.
7. McDonnell was unable to find any correlation between pyrethrin content and toxicity of powdered pyrethrum to aphids, as determined by dusting experiments.
8. Tattersfield, Hobson and Gimingham, Wilcoxon and Hartzell and Richardson attempted to prove that a determination of pyrethrin I is sufficient to indicate the toxicity of a sample, but critical examination of their data does not bear out their conclusions.
9. Ripert and Gaudin found pyrethrin II more toxic than pyrethrin I to frogs, fish and mice, but with flies the reverse was true.
10. It has been suggested that pyrethrin II might have comparatively little toxicity when used in aqueous suspensions against aphids and yet be nearly as toxic as pyrethrin I when used against roaches or in oil solutions against flies. This would explain the discrepancies between the results of the different investigators. A somewhat similar condition is known to exist between the pyrethrins and rotenone, when applied to different insects (Chapter XIV).

CHAPTER VII

COMPARATIVE VALUE OF COMMERCIAL GRADES OF PYRETHRUM

THE OPEN AND CLOSED FLOWER CONTROVERSY

In the early literature the statement that immature flowers are more toxic than mature flowers is found repeatedly. This idea was generally accepted by the trade and it had become so firmly fixed that, in 1929, closed flowers commanded a premium

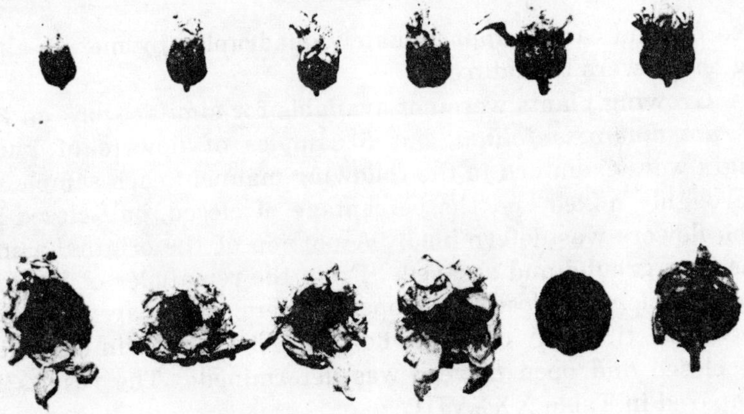

TYPICAL CLOSED AND OPEN DALMATIAN FLOWERS.

of 5 to 6 cents a pound over half closed flowers, which, in turn, cost 2 to 3 cents a pound more than open flowers. In that year Gnadinger and Corl *(344)* collected and assayed *Pyrethrum roseum,* throughout the flowering season, from plants grown in Minnesota. These analyses are given in Table XXXVII.

Before the buds were formed, the roots, leaves and stems did not contain any pyrethrins. As the buds formed and developed, their pyrethrin content gradually increased. The last and most mature sample contained 4 times the percentage of pyrethrins found in the unexpanded buds. The average weight of the open flowers was 2 to 3 times that of the closed flowers, and the pyrethrin content of the former was about double that of the closed flowers. Therefore the yield of pyrethrins from a given

TABLE XXXVII. ANALYSES OF *Pyrethrum roseum* AT DIFFERENT STAGES OF MATURITY (GNADINGER AND CORL)

No.	Collected	Description	Av. wt. g.	Diam. mm.	Pyrethrins dry-basis %
1	May 7	Roots	....	...	None
2	May 7	Entire plant except roots; buds very small	....	...	None
3	May 23	Unexpanded buds; ray florets not showing	0.023	3-7	0.22
4	May 30	Closed flowers; ray florets visible	0.052	4-9	0.40
5	June 5	Half-open; disk florets visible	0.086	5-11	0.54
6	June 11	Closed flowers; same description as No. 4	0.057	4-9	0.56
7	June 11	Open flowers; rays completely expanded	0.106	5-12	0.67
8	June 21	Open flowers; disk florets expanded	0.154	7-15	0.78
9	June 28	Open flowers; completely expanded	0.096	6-12	0.87

area of land can be approximately quadrupled by merely allowing the flowers to mature.

Growing plants were not available for similar work on *Pyrethrum cinerariaefolium*, but 5 samples of flowers of known origin were examined in the following manner: each sample was thoroughly mixed and the percentage of closed, half-closed and open flowers was determined. A portion of the original sample was then ground and assayed. From the remainder of the original sample, the closed and open flowers were carefully separated, selecting only complete flowers. The pyrethrin content of the closed and open flowers was determined. The results are compared in Table XXXVIII.

TABLE XXXVIII. PYRETHRIN CONTENT OF OPEN AND CLOSED *Pyrethrum cinerariaefolium* (GNADINGER AND CORL)

No.	Kind	Description	Av. weight g.	Diam. mm.	Pyrethrins %
20	Japanese	Original sample	....	...	0.96
20	Japanese	Closed flowers	0.109	6-9	0.80
20	Japanese	Open flowers	0.208	9-15	0.94
25	Japanese	Original sample	....	...	0.84
25	Japanese	Closed flowers	0.101	4-9	0.76
25	Japanese	Open flowers	0.238	11-14	0.97
23	Dalmatian	Original sample	....	...	0.53
23	Dalmatian	Closed flowers	0.064	4-8	0.61
23	Dalmatian	Open flowers	0.143	8-12	0.81
24	Dalmatian	Original sample	....	...	0.52
24	Dalmatian	Closed flowers	0.074	5-8	0.48
24	Dalmatian	Open flowers	0.178	10-14	0.71
26	American	Original sample	....	...	0.85
26	American	Closed flowers	0.086	6-8	0.64
26	American	Open flowers	0.135	9-15	1.03

The open flowers contained from 18 to 61 per cent more pyrethrins than the closed flowers from the same lots.

The conclusion of Gnadinger and Corl, that the pyrethrin content of the flowers increases as they mature, was confirmed in 1931 by Tattersfield *(871)* in an elaborate study of the development of pyrethrum. Fryer, Tattersfield and Gimingham *(287)* had earlier concluded, from biological tests, that the correct time

FLOWER HEADS OF *P. cinerariaefolium.* AT THE RIGHT, ABOUT THE PROPER STAGE FOR HARVESTING.

to harvest the flowers is when the majority is fully open. Tattersfield summarized his experiments as follows *(871)*:

"Pyrethrum flowers from plants grown upon a bed at Harpenden were examined. The plants were divided into blocks and randomized, the flowers being harvested from a dozen plants each week over a period of 8½ weeks; the flower heads ranged from the small bud stage in the first week to the overblown stage in the last week. The yield in number and weight of heads per plant, the diameters of the receptacles and the content of pyrethrins I and II were determined. There was a considerable amount of variation in all these factors in the flowers from different plants.

"A statistical analysis showed that there was no significant variation in the numbers of the flowers with time, but that position of the plant in the bed had a significant effect. The time of harvesting had a significant effect on the content of pyrethrins whether taken separately or together and whether expressed in percentages, parts per flower head or parts per plant. There was on the different dates a significant correlation between the contents of pyrethrin I and II expressed in parts per flower head or plant.

"The data in this experiment indicate that for the material examined, there is a quantitative development of the active principles in the flower heads from the small bud stage up to the time of maturity of the flowers, which more than keeps pace on the whole with the increase in weight of the flowers. Thus the content of pyrethrins, both relatively and absolutely, rises to a maximum at the maturity of the flowers. The mean percentage content of pyrethrins fell after pollination and the fading of the flowers; this corresponds with the rapid increase in weight of the heads on the formation of seed. There would thus appear to be a loss, which might be serious, both in percentage content of active principles and in yield of flowers if harvested before being fully open. There seems to be no useful purpose served in leaving the flowers to the overblown condition."

Distribution of Pyrethrins in the Flower Head

It was well known that certain parts of the flower head contain more pyrethrins than others, but the distribution of the pyrethrins in the different flower parts was not determined chemically until 1929. Gnadinger and Corl *(344)* carefully separated a sample of uncompressed Japanese flowers into the principal parts, using for this purpose flowers from which none of the parts was missing. The analyses of these parts are given in Table XXXIX.

Table XXXIX. Distribution of Pyrethrins in Japanese Pyrethrum (Gnadinger and Corl)

	Composition of flowers %	Pyrethrins %	Percentage of total pyrethrins
Achenes	34.2	2.27	92.4
Receptacles	11.3	0.26	3.5
Involucral scales	11.5	0.15	2.0
Disk florets	25.8	Trace	...
Ray florets	17.2	Trace	...
Original sample	...	0.84	...
Stems	...	0.15	...

The achenes contained 92.4 per cent of the total pyrethrins;

the receptacles and scales contained 3.5 and 2.0 per cent respectively, while the disc and ray florets contained only traces. Analysis of the original sample showed 0.84 per cent pyrethrins, while calculation from the analysis of the parts gave 0.82 per cent pyrethrins. The flowers used in this experiment were fully matured; it is extremely difficult to separate the parts from immature flowers. The achenes from mature flowers contain a high proportion of fixed oil; a sample of ripe achenes containing 0.70 per cent pyrethrins also contained 12.9 per cent of petroleum ether-soluble extractive.

Martin and Tattersfield *(591, 592)* found the following distribution of pyrethrins in fully open flowers, confirming the results of Gnadinger and Corl:

	Composition of flowers %	Total pyrethrins %
Petals	25.2	0.18
Receptacles and scales	20.4	0.27
Disk florets	31.4	0.48
Achenes	23.0	4.54

Martin and Tattersfield also found a fourfold increase in the weight of the achenes during ripening with a proportional reduction in percentage of pyrethrins (see page 140).

SECTION OF DRIED, MATURE FLOWER HEAD, *P cinerariaefolium.*

Comparative Value of Japanese and Dalmatian Pyrethrum

The relative value of Japanese and Dalmatian flowers has long been a debated question; it was quite generally conceded that the latter were superior. In 1930 a writer stated in the Bul-

Table XL. Pyrethrin Content of Pyrethrum Flowers (Gnadinger and Corl, 1929)

Dalmatian *P. cinerariaefolium* Grade	Crop year	Pyrethrins %	Japanese *P. cinerariaefolium** Crop year	Pyrethrins %
Closed	1926	0.38	1928	0.58
Closed	1925	0.39	1925	0.62
Closed	1926	0.39	1926	0.62
Closed	1928	0.40	1926	0.64
Closed	1928	0.40	1926	0.68
Closed	1926	0.41	1926	0.71
Closed	1926	0.42	1926	0.74
Closed	1925	0.44	1928	0.74
Closed	1925	0.45	1928	0.80
Closed	1926	0.45	1928	0.81
Closed	1928	0.52	1928	0.84
Closed	1925	0.53	1928	0.86
Closed	1928	0.53	1926	0.87
Closed	1925	0.57	1928	0.87
Half-closed	1925	0.38	1928	0.92
Half-closed	1926	0.38	1928	0.96
Half-closed	1926	0.38	1928	0.99
Half-closed	1928	0.53	1928	1.10
Half-closed	1927	0.57	1928	1.17
Open	1925	0.39	1928	1.20
Open	1925	0.40	1928	1.21
Open	1926	0.40	American *P. roseum*	
Open	1927	0.43	1929	0.25
Open	1926	0.43	1929	0.56
Open	1925	0.47	1929	0.73
Open	1928	0.51	1929	0.79
Open	1926	0.58	1929	0.82

*Japanese flowers not graded.

Table XLI. Summary of 1929 Analyses of Pyrethrum (Gnadinger and Corl)

Samples analyzed	Species	Description	Pyrethrin content Max. %	Min. %	Av. %
14	*Cinerariaefolium*	Dalmatian closed	0.57	0.38	0.448
5	*Cinerariaefolium*	Dalmatian half-closed	0.57	0.38	0.448
8	*Cinerariaefolium*	Dalmatian open	0.58	0.39	0.451
27	*Cinerariaefolium*	Dalmatian, all grades	0.58	0.38	0.449
21	*Cinerariaefolium*	Japanese	1.21	0.58	0.853
48	*Cinerariaefolium*	All sources	1.21	0.38	0.630
5	*Roseum*	American	0.82	0.25	0.630

letin of the (British) Imperial Institute: "Owing to its superior quality, the Dalmatian product has recently gained an increasing demand in the United Kingdom, France, Italy, Germany, and even the United States."

In 1929 Gnadinger and Corl *(345)* assayed 53 samples of pyrethrum from different sources. Twenty-eight commercial samples of whole flowers were collected from dealers in the United States and from agents in Japan and Europe. Twenty samples of powdered flowers, which had been kept in airtight containers for 2 years, were kindly supplied by Dr. C. C. McDonnell, U. S. Department of Agriculture. Five samples of *Pyrethrum roseum* were collected in Minnesota and Iowa. Samples of *Pyrethrum carneum* could not be obtained. The whole flowers were ground to 40 mesh, taking care to avoid heating during grinding, and the pyrethrin contents were determined by the copper reduction method. The analyses are given in detail and also in summarized form, Tables XL and XLI.

The pyrethrin content of the Dalmatian flowers ranged from 0.38 to 0.58 per cent. The Japanese flowers contained from 0.58 to 1.21 per cent of pyrethrins and averaged twice the pyrethrin content of the Dalmatian flowers.

At the time these analyses were made (1929) it was generally believed that pyrethrum does not deteriorate appreciably if properly stored. Since then Gnadinger and Corl have shown that there is a marked deterioration when ground flowers are stored for one year. These earlier analyses consequently do not represent the pyrethrin content of freshly harvested flowers, since they were made in the first half of 1929, when the 1928 crop was 6 to 11 months old and before the 1929 foreign crops were available. Since then a large number of analyses have been made of both Japanese and Dalmatian flowers. For convenience the pyrethrin content of the flowers from the two sources will be considered separately.

Study of Dalmatian Pyrethrum

According to experts of the Yugoslav Government, the best flowers are produced in two districts of the Province of Primorska (map, page 6):

District of Split; including the municipalities of Drvenik and Šolta, on the island of the same names, and the municipality of Trogir on the mainland.

District of Supetar; including the municipalities of Supetar, Milna and Nerežišće on the island of Brač.

A second grade, said to be nearly as good as flowers from Primorska, is obtained from the District of Dubrovnik, Province of Zetska. The flowers from the other producing districts are considered inferior. It should be remembered that these opinions are based on the color, size and general appearance of the flowers and not on the pyrethrin content.

In 1931 the Kingdom of Yugoslavia undertook a complete reorganization of the Dalmatian pyrethrum industry. A comprehensive educational campaign was begun, to acquaint growers with proper methods of cultivating, harvesting, storing and marketing. The writer was invited to cooperate in one phase of this work, with the ultimate object of increasing the pyrethrin content of the commercial Dalmatian product *(351)*.

In February, 1932, a number of samples of the 1931 crop was submitted by the Yugoslav Government. Since Dalmatian flowers are harvested in May and June, these samples were 8 to 9 months old when assayed. The analyses are given in Table XLII.

TABLE XLII. ANALYSES OF 1931 CROP DALMATIAN PYRETHRUM EIGHT TO NINE MONTHS AFTER HARVESTING (GNADINGER AND CORL)

Province and locality	Dalmatian classification Grade	Quality	Pyrethrins %
Primorska			
Trogir	Half-closed	Best	0.63
Trogir	Closed	Best	0.71
Trogir	Mixed	Best	0.76
Trogir	Mixed	Best	0.79
Gornje selo, Šolta	Half-closed	Best	0.72
Komiža, Vis	Closed	Best	0.64
Milna, Brač	Half-closed	Best	0.34
Supetar, Brač	Half-closed	Best	0.64
Bobovišće, Brač	Closed	Best	0.70
Selca, Hvar	Open	Slightly weak	0.58
Rudine-Starigrad, Hvar	Mixed	Slightly weak	0.72
Blato, Korčula	Closed	Inferior	0.44
Zetska			
Rose	Half-closed	Inferior	0.70
Kutovo	Open	Inferior	0.39
Ston	Open	Inferior	0.42
Ston	Open	Inferior	0.44
Mljet	Open	Inferior	0.45
Mljet	Open	Inferior	0.53
Konavlje	Open	Inferior	0.52
Savska			
Vrbnik, Krk	Mixed	Inferior	0.43

The average pyrethrin content of the 20 samples from the 1931 crop (0.58 per cent) was only slightly higher than the average for the 27 Dalmatian samples (0.45 per cent) reported in 1929. Moreover, eight lots of Japanese flowers, assayed when 8 to 9 months old, contained 0.76 to 1.04 per cent pyrethrins, averaging 0.87 per cent. Either the Dalmatian flowers had contained a lower percentage of pyrethrins than the Japanese flowers, when harvested, or the pyrethrins in the Dalmatian flowers had decomposed at a more rapid rate. Since it was now known that storage plays an important part in the deterioration of pyrethrum, it was decided to postpone further work until the 1932 crop should become available.

Samples of the new crop were collected by a special expert of the Yugoslav Government and were forwarded from Beograd to Minneapolis, where they were promptly assayed. Some of these samples were from plants grown on the experimental plots of the Agricultural College at Split. These plants were originally obtained from different districts. The analyses of the samples from the Agricultural College were made within 60 days of

TABLE XLIII. ANALYSES OF FLOWERS GROWN AT THE AGRICULTURAL COLLEGE AT SPLIT, YUGOSLAVIA (GNADINGER AND CORL)

No.	Description	Crop	Pyrethrins %
1	Very immature "buttons," collected 5|17|32	1932	0.44
2	Half-closed, from same plants as No. 1, collected 5|17	1932	0.68
3	Open, from same plants as No. 1, collected 5|17|32	1932	0.81
3a	Stems, from same plants as No. 1	1932	0.10
4	Half-closed, color dark	1928	0.25
5	Open, color dark, mature, some flower parts missing	1928	0.23
6	Closed, color good, very immature	1931	0.29
19	Flowers with stems, color good, slightly immature (stems 49.1 per cent)	1932	0.56
20	Open, color fair, mature	1932	0.88
20a	Ray florets from No. 20	1932	Trace
20b	Disk florets from No. 20	1932	Trace
20c	Achenes from No. 20	1932	1.00
20d	Receptacles and scales from No. 20	1932	0.12
7	Flowers from plants originally from several provinces	1932	0.98
8	Plants originally from Trogir, color good, mature	1932	0.84
9	Plants originally from Mljet, color good, sl. immature	1932	1.06
14	Plants originally from Mljet, color good, sl. immature	1932	0.84
10	Plants originally from Hvar, color fair, sl. immature	1932	0.91
11	Originally from Hvar and Trogir, color good, sl.diseased	1932	0.94
15	Originally from Hvar and Trogir, color good, mature	1932	1.01
12	Plants originally fromDubrovnik, color good, mature	1932	1.03
13	Plants originally from Slano, color good, mature	1932	1.03

TABLE XLIV. ANALYSES OF COMMERCIAL LOTS 1932 CROP DALMATIAN PYRETHRUM (GNADINGER AND CORL)

No.	Province and locality	Dalmatian grade	Age when assayed months	Av. weight g.	Stems %	Moisture %	Pyrethrins %	Remarks*
	Primorska:							
17	Marina near Trogir........	......	2	0.125	trace	9.0	0.83	Immature, "shade dried."
34	Marina near Trogir........	......	2½	0.117	0.8	9.9	0.82	Immature.
33	Trogir..................	......	2½	0.140	0.3	7.6	0.98	Slightly immature.
46	Marina near Trogir........	f.a.q.†	3½	0.098	trace	8.8	0.91	Slightly immature.
48	Seget near Trogir..........	f.a.q.	3½	0.093	trace	9.2	0.83	Mature but very small.
47	Seget near Trogir..........	f.a.q.	3½	0.127	trace	8.2	0.94	Mature.
49	Trogir..................	½ closed	3½	0.122	trace	7.6	0.83	Slightly immature.
50	Trogir..................	¾ open	3½	0.130	trace	8.8	0.86	Slightly immature.
51	Trogir..................	mixed	3½	0.105	0.3	8.2	0.91	Very slightly immature.
71	Labin near Trogir.........	......	4½	0.114	0.3	7.0	0.70	Open; "primitively grown."
18	Milna, Brač (island)......	......	2	0.135	trace	7.7	0.88	Slightly immature; "shade dried."
16	Supetar, Brač............	......	2	0.192	trace	8.3	1.01	Mature; "shade dried."
27	Supetar, Brač............	......	2½	0.170	0.4	8.3	0.93	Mature.
28	Nerežišće, Brač...........	......	2½	0.185	trace	7.7	0.96	Mature.
29	Milna, Brač..............	......	2½	0.195	trace	6.7	1.02	Mature; "sun dried."
30	Milna, Brač..............	......	2½	0.185	trace	6.7	0.96	Mature; "shade dried and pressed."
31	Milna, Brač..............	......	2½	0.195	trace	8.1	0.98	Mature; "shade dried."
42	Pučišće, Brač............	f.a.q.	3½	0.082	trace	7.5	0.70	Immature, closed.
41	Bol, Brač................	mixed	3½	0.123	trace	8.9	0.82	Slightly immature.
70	Sutivan, Brač............	......	4½	0.096	0.7	6.0	0.67	Mixed open and closed; "best quality."
54	Vis, Vis (island)..........	f.a.q.	3½	0.128	trace	7.5	0.90	Half closed.
66	Vis, Vis.................	¾ open	4½	0.108	1.8	5.8	0.57	Small; immature; "medium quality."
67	Vis, Vis.................	......	4½	0.102	1.6	7.1	0.60	Small; immature; "superior quality."
63	Komiža, Vis..............	mixed	4½	0.135	0.8	7.5	0.80	Mixed open and closed; "superior quality."
64	Pošpilje, Vis.............	......	4½	0.125	0.3	7.3	0.83	Half closed; "good quality."
65	Oklјučina, Vis............	......	4½	9.122	0.5	7.0	0.94	Half closed; "very good quality."
61	Vrboska, Hvar (island)....	......	4½	0.125	0.6	9.2	0.51	Open; "medium quality."
62	Vrboska, Hvar............	open	4½	0.154	0.3	8.7	0.67	Open; "inferior quality."
60	Bogomlje, Hvar...........	open	4½	0.141	0.4	8.3	0.70	Open; "poor quality."
58	Hvar, Hvar...............	open	4½	0.125	1.1	5.9	0.79	Open; "medium quality."
59	Hvar, Hvar...............	open	4½	0.164	trace	7.5	1.03	Open; "poor quality."
39	Korčula, Korčula (island)..	½ closed	3½	0.085	0.8	8.0	0.77	Very immature.

No.	Province and locality	Dalmatian grade	Age when assayed months	Av.weight g.	Stems %	Moisture %	Pyrethrins %	Remarks*
40	Korčula, Korčula	closed	3½	0.090	0.5	9.2	0.80	Very immature.
38	Korčula, Korčula	open	3½	0.112	trace	11.2	0.98	Slightly immature.
77	Korčula, Korčula	......	4½	0.089	2.6	8.7	0.74	Half closed; "medium quality, wild grown."
32	Šolta, Šolta (island)	......	2½	0.140	0.8	8.0	0.93	Slightly immature.
44	Šolta, Šolta	f.a.q.†	3½	0.093	trace	8.0	0.90	Slightly immature.
53	Maslinica, Šolta	......	3½	0.112	0.3	7.0	0.95	Slightly immature.
26	Gradac, near Makarska....	......	2½	0.157	0.3	8.1	0.93	Slightly immature.
24	Tučepi, near Makarska.....	......	2½	0.200	0.8	8.5	0.94	Mature.
25	Vojici, near Makarska......	......	2½	0.195	0.7	7.0	0.94	Mature.
43	Omis, near Split..........	¾ open	3½	0.122	trace	7.9	0.70	Immature.
72	Šestanovac, near Split	......	4½	0.123	0.4	7.4	0.61	Mature; "wild grown."
45	Drvenik, Drvenik (island) ..	f.a.q.	3½	0.147	trace	7.4	1.00	Mature.
35	Primosten	......	2½	0.142	0.3	11.9	0.96	Slightly immature.
68	Sali-Preko, near Zara......	closed	4½	0.071	3.8	6.3	0.88	Closed.
69	Šibenik..................	mixed	4½	0.094	1.1	7.5	0.65	Mixed; "inferior quality."
73	Metkovič	......	4½	0.122	0.3	6.7	0.56	Open; "medium quality."
	Zetska:							
37	Trebinje	open	3½	0.162	trace	8.9	0.94	Mature.
36	Trebinje	......	3½	0.122	trace	11.0	0.99	Mature, but small.
76	Rasovac, near Trebinje.....	mixed	4½	0.151	0.3	7.3	1.03	Open; "good quality, wild grown."
75	Konavlje, near Dubrovnik..	......	4½	0.141	0.8	7.6	0.61	Open; "very poor quality."
74	Aleksic, near Dubrovnik....	......	4½	0.179	trace	7.5	0.83	Open; "very poor quality."
21	Neum	......	2½	0.130	0.9	8.6	0.88	Slightly immature.
22	Mljet (island)	......	2½	0.195	0.4	9.0	0.73	Mature.
23	Ston	......	2½	0.120	0.4	7.7	0.81	Immature.
56	Ston	......	3½	0.164	1.3	9.8	0.09	"Said to be P. caucasicum, not P. cinerariaefolium."
	Savska:							
55	Aleksandrova, Krk (island)	f.a.q.	3½	0.114	0.4	9.0	0.94	Half closed.
57	Krk, Krk................	½ closed	4½	0.110	1.8	8.3	0.89	Half closed.

* Phrases in quotation marks are Dalmatian comments.
† Fair average quality of the season.

harvesting and are reported separately (Table XLIII) from the analyses of the commercial samples (Table XLIV).

The analyses of Samples 1, 2, and 3 confirm the fact that the pyrethrin content increases as the flowers mature. The stems from these flowers contained 0.10 per cent pyrethrins, which was expected, as was also the low pyrethrin content of Sample 19, containing 49.1 per cent stems. Samples 4, 5, and 6 had decomposed in storage. Samples 20, 20a, 20b, 20c, and 20d showed again that the pyrethrins in mature flowers occur chiefly in the achenes. The pyrethrin content of Samples 7 to 15 was much greater than previously found in Dalmatian pyrethrum, averaging 0.96 per cent, or about the same as new crop Japanese pyrethrum.

The most important factors affecting the pyrethrin content of the commercial samples (Table XLIV) are the degree of maturity of the flowers, the length of time between harvesting and assaying, and the locality where grown.

The average weight per flower head varied from 0.071 to 0.200 g. The degree of maturity cannot always be judged from the average weight because some lots of small, mature flowers have a lower average weight than large half-closed flowers. Nevertheless, a correlation is indicated between the average weight and the pyrethrin content (Table XLV).

TABLE XLV. RELATION OF PYRETHRIN CONTENT TO AVERAGE WEIGHT OF FLOWER HEADS (GNADINGER AND CORL)

Group	Samples	Av. age mos.	Pyrethrin content Min. %	Pyrethrin content Max. %	Pyrethrin content Av. %	Av. weight per flower head g.
A	17	4.20	0.51	0.79	0.66	0.119
B	15	3.43	0.80	0.89	0.84	0.120
C	21	3.07	0.90	0.98	0.94	0.143
D	5	3.40	1.00	1.03	1.02	0.170

The average age of group A is greater than that of group B, which probably accounts for the considerable difference in the pyrethrin content of these groups, which have nearly the same average weight.

If the commercial samples are grouped according to their maturity, determined by physical examination, the relation between degree of maturity, average weight and pyrethrin content is shown (Table XLVI).

The average age of the three groups is substantially the same. Tattersfield *(871)* has shown that there is a marked increase in weight per flower head after pollination, which is not

TABLE XLVI. RELATION BETWEEN MATURITY, AVERAGE WEIGHT AND PYRETHRIN CONTENT (GNADINGER AND CORL)

Classification	Samples	Av. age mos.	Av. weight per head g.	Av. pyrethrin content %
Immature	10	3.45	0.102	0.75
Slightly immature	23	3.50	0.120	0.87
Fully mature, overblown	25	3.56	0.156	0.85

accompanied by further synthesis of pyrethrins. The percentage of pyrethrins in the flowers drops, therefore, after pollination, but the weight of pyrethrins per flower head remains unchanged. In Table XLVI the average weight of the mature flowers is greater than that of the slightly immature flowers, but the percentage of pyrethrins is slightly less.

The fact that all the samples in Table XLIV were not collected at the same stage of maturity nor assayed when the same age, makes it difficult to draw definite conclusions regarding the merits of the different producing districts. The analyses are summarized by Provinces in Table XLVII.

TABLE XLVII. ANALYSES OF COMMERCIAL SAMPLES SUMMARIZED BY PROVINCES (GNADINGER AND CORL)

Province and locality	Samples	Av. age mos.	Av. weight g.	Av. pyrethrin content %
Primorska				
Makarska	3	2.50	0.184	0.94
Šolta	3	3.26	0.115	0.93
Brač	10	2.80	0.156	0.89
Trogir	10	3.25	0.117	0.86
Korčula	4	3.75	0.094	0.82
Vis	6	4.33	0.120	0.77
Hvar	5	4.50	0.142	0.74
Split	2	4.00	0.123	0.66
Miscellaneous	5	3.90	0.115	0.81
Av. for Primorska	48	3.30	0.130	0.83
Zetska				
Trebinje	3	3.83	0.145	0.99
Dubrovnik	2	4.50	0.160	0.77
Miscellaneous	3	2.50	0.148	0.81
Av. for Zetska	8	3.50	0.150	0.85
Savska				
Krk	2	4.00	0.112	0.92
Av. for all provinces	58	3.45	0.132	0.82

The average pyrethrin content of the samples from Makarska, Šolta, Brač, Trogir, Trebinje and Krk is quite satisfactory. Twenty-six of the 58 commercial lots contained 0.90 per cent or more of pyrethrins.

It has been clearly shown that certain Provinces of Yugoslavia can produce pyrethrum equal to the best Japanese product. On the other hand, much of the Dalmatian crop has been of low pyrethrin content because of improper methods of cultivating, harvesting, storing, packing and marketing.

In 1933 the Yugoslav Government established a fully equipped laboratory for research on pyrethrum at the Experiment Station at Split. In March, 1934, regulations for the control of the marketing of pyrethrum were promulgated by the Government at Belgrade. These regulations provided for the sampling, assaying, grading and marking of pyrethrum flowers and powder. Three grades of pyrethrum flowers were established:

"Dalmatian Pyrethrum": containing not less than 0.8 per cent pyrethrins, not more than 3 per cent stems, not more than 1 per cent sand.

"Pyrethrum": containing not less than 0.5 per cent pyrethrins; not more than 5 per cent stems, not more than 2 per cent sand.

"Pyrethrum Second Class": any product of lower quality than "Pyrethrum."

The regulations provided for the issuance of certificates by the Experiment Station showing the crop year, marks and numbers of the lot, gross weight of lot, pyrethrin content and official grade. Exportation of flowers was prohibited unless accompanied by a certificate not more than 6 months old.

The Experiment Station at Split is also investigating the problems involved in cultivating, harvesting and storing pyrethrum.

Pyrethrin Content of Japanese Flowers

It has been mentioned (page 14) that Japanese flowers are classified as 1st, 2d, 3rd, 4th and 5th grade; flowers lower than 5th grade are not considered merchantable. This classification

Grade	Relative prices cents per lb.	Proportion of crop %
1	18.0	17
2	17.7	
3	17.1	58
4	15.9	25
5	14.7	

is based on the color, maturity and general appearance of the flowers. The differences in the prices of the 5 grades and the distribution of the crop among them, for the Hokkaido crop, for the year 1927 *(38)* are given on the preceding page.

These price differentials and the distribution of the crop between the grades naturally vary somewhat from year to year; the price differentials are usually set at the beginning of the season. In 1932 there were practically no 1st grade flowers and very few of the 2d and 3rd grades; the bulk of the crop was 4th and 5th grades. American importers are not offered the various

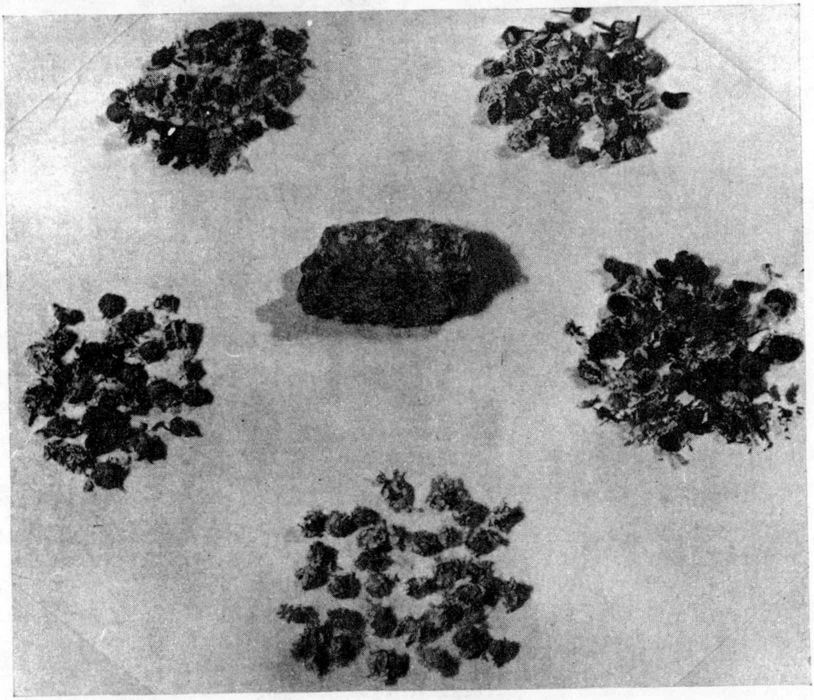

THE FIVE GRADES OF JAPANESE FLOWERS WITH SECTION OF COMPRESSED BALE. (SEE PAGE 153).

grades separately, but they seem to be used for domestic trading only. Flowers sold to this country are generally "fair average quality of the season" (f.a.q.), but a grade of somewhat better appearance can be obtained for a slight premium. The Japanese are loath to export 1st, 2d or 3rd grade flowers unmixed, even for a considerable premium. The f. a. q. grade is a mixture of grades, in such proportion as the exporter has to offer; it con-

TABLE XLVIII. EXAMINATION OF 1932 CROP GRADED JAPANESE PYRETHRUM (GNADINGER)

Where grown	Grade	Description	Av. wt. g.	Stems %	Debris* %	Moisture %	Pyrethrins %
Okayama	1	Mostly half-closed; color bright, none badly discolored....	0.139	0.9	Trace	10.4	0.94
Okayama	2	Darker than 1; badly discolored flowers, 1.5%.............	0.129	0.5	2.8	7.8	0.89
Okayama	3	Darker than 2; badly discolored flowers, 3.5%.............	0.133	0.8	4.6	10.1	0.98
Okayama	4	Darker than 3; badly discolored flowers, 5.8%.............	0.131	1.7	5.5	7.6	0.89
Okayama	5	Considerably darker than 4; badly discolored flowers, 11.5%	0.133	1.1	24.9	8.7	0.75
Hiroshima	1	Mostly half-closed; color bright, none badly discolored.....	0.131	0.3	Trace	7.0	0.89
Hiroshima	2	More mature than 1; very little darker than 1; none discolored	0.120	1.1	0.6	7.5	0.93
Hiroshima	3	Mostly open; slightly darker than 2; none badly discolored..	0.139	0.8	2.0	8.4	1.02
Hiroshima	4	All open; darker than 3; none badly discolored...........	0.135	0.7	9.7	8.2	1.06
Hiroshima	5	All open; much darker than 4; badly discolored, 18.0%.....	0.120	1.2	15.5	7.5	0.79
Hokkaido	3	Compressed....................................	....	...	...	...	1.06

* Detached parts of flowers.

tains no 1st or 2d grade flowers and usually consists of about 50 per cent of 4th grade and 50 per cent of 5th grade. Moreover, if the carry-over from the preceding year is large, 20 to 30 per cent of old crop flowers may be mixed with the new crop. The hydraulic pressing of Japanese flowers makes it extremely difficult to determine the relative amounts of the different grades present.

As might be expected, there is no relation between the grading of Japanese pyrethrum and the pyrethrin content, except in the case of the 5th grade, which is always more or less discolored or damaged and usually low in pyrethrins. Flowers from Hokkaido are considered the best; those from Hiroshima are considered next in quality.

Recently the Imperial Industrial Research Institute of Hokkaido has begun to make analyses and issue certificates of analysis for export shipments of pyrethrum. Official Inspection Certificates, showing the pyrethrin content, are also issued by the Government Farm Products Inspection Bureau, Otaru.

The different grades of Japanese flowers are shown on page 151, together with a small section from a bale, showing the degree of compression. Grade 1 is at the bottom of the picture; grade 2 at the left; grade 3, top left; grade 4, top right; grade 5, right.

The analyses presented in Table XLVIII were made on graded samples of the 1932 crop from Okayama and Hiroshima. The flowers were assayed within 2 months of harvesting.

During the last 7 years the writer has assayed a large number of shipments of Japanese flowers. The analyses in Table

TABLE XLIX. ANALYSES OF JAPANESE PYRETHRUM (GNADINGER AND CORL)

1929 crop Pyrethrins %	1930 crop Pyrethrins %	1931 crop Pyrethrins %	1932 crop Pyrethrins %	1933 crop Pyrethrins %	1934 crop Pyrethrins %	1935 crop Pyrethrins %
0.97	0.97	1.04	0.90	0.90	0.94	1.00
1.08	1.04	1.00	1.06	0.90	0.97	0.98
1.10	1.06	1.05	0.93	0.90	0.96	0.95
1.03	1.00	1.03	0.87	1.04	0.95	0.90
0.97	1.06	0.98	0.92	0.99	0.94	0.91
0.93	1.03	0.96	1.05	0.93	0.90	1.02
0.90	1.03	1.06	1.01	0.99	0.87	0.92
0.89	0.96	1.03	0.96	0.90	0.92	0.93
0.95	1.05	0.95	0.94	0.90	0.89	0.93
...	1.05	0.97	0.98	1.01	0.92	0.98
...	...	1.12	1.06	0.94	0.94	0.98
...	...	0.94	0.96	0.87	0.91	0.91
...	...	0.93	1.01	0.95	0.99	0.85
Av. 0.98	1.03	1.00	0.97	0.94	0.93	0.94

XLIX represent lots examined within 6 months of harvesting, for the years 1929, 1930, 1931, 1932, 1933, 1934 and 1935. These analyses represent several million pounds of flowers.

There is little variation in the pyrethrin content from 1929 to 1935. It should be emphasized that these samples do not represent the entire crop, but only that part reaching the market within 6 months of harvesting. The effect of storage on the last part of the crop to reach the market is discussed in Chapter VIII.

Pyrethrin Content of Kenya Flowers

Tattersfield was among the first to analyze Kenya flowers; he reported a maximum of 1.44% pyrethrins. He found that the ratio of pyrethrin I to pyrethrin II in Kenya flowers was about 1:1. The analyses, by the copper reduction method, of a number of commercial lots of Kenya flowers are given below (analysts, Gnadinger and Corl):

Crop	Pyrethrins %
1933	0.90
1934	1.49
1934	1.19
1935	1.29
1935	1.47
1935	1.48
1935	1.47
1935	1.32
Av.	1.33

Flowers from Other Sources

Several analyses of flowers from minor sources of supply have been reported; some of these analyses are presented in Table L.

Table L. Pyrethrin Content of Flowers from Minor Sources of Supply.

Where grown	Pyrethrins %	Analysts
Bulgaria	0.36	Bojtschinow
Bulgaria	0.83-1.17	Gnadinger and Corl
Bulgaria	0.65-1.35	Nikolov
California	1.10	Gnadinger and Corl
China	1.20	China Chemical Works
Cyprus	0.37-0.75	Gnadinger and Corl
France	0.45-1.41	Ripert
Russia	0.24	Gnadinger and Corl
Spain	0.57	Gnadinger and Corl
Switzerland	0.34-0.56	Staudinger and Harder

CHAPTER VIII

EFFECT OF STORAGE, LIGHT AND HEAT ON PYRETHRUM PRODUCTS

Pyrethrum flowers are harvested in May, June and July in the countries which produce them commercially. Shipments from the new crop begin to arrive in the United States in July and August. September marks the end of the insecticide season in this country. It is obvious, therefore, that the pyrethrum crop of a given year cannot be used against insects until the following year; frequently flowers are carried over for a second year before reaching the ultimate user. It has been recognized that long periods of storage may result in decomposition of the pyrethrins, but until 1929 no method was available which was sufficiently accurate to determine the extent of the loss.

Abbott (*1*) concluded that whole flowers can be stored in sealed glass jars for 5½ years without loss of toxicity, and that whole and ground flowers are not damaged by exposure to weather for 3 months. He also found that heating for 18 hours at 130° to 140° destroyed the activity. These results were obtained by dusting experiments on aphids and roaches.

EFFECT OF SHORT EXPOSURES TO SUNLIGHT, ULTRAVIOLET LIGHT AND HEAT

Hartzell and Wilcoxon (*406*) exposed ground pyrethrum in air to sunlight, ultraviolet light and heat. The pyrethrin content

TABLE LI. EFFECT OF SUNLIGHT, ULTRAVIOLET LIGHT AND HEAT ON GROUND PYRETHRUM (HARTZELL AND WILCOXON)

Sample	Kill, aphids		Pyrethrin content: Tattersfield acid method			Gnadinger method	Difference between pairs by acid method
			I	II	Total	Total	
	%	diff.	%	%	%	%	%
Control	37		0.38	0.63	1.01	1.03	
Exposed to sunlight 3 days	27	10	0.24	0.55	0.79	0.81	0.22
Control	55		0.32	0.64	0.96	...	
Exposed to u. v. light 3 hours	33	22	0.23	0.55	0.78	...	0.18
Control	55		0.32	0.64	0.96	...	
Exposed to u. v. light 6 hours	33	22	0.23	0.47	0.70	...	0.26
Control	68		0.28	0.39	0.67	...	
Heated 10 hrs. at 100°	37	31	0.23	0.35	0.58	...	0.09
Control	73		0.26	0.49	0.75	0.73	
Heated 40 hrs. at 100°	52	21	0.18	0.34	0.52	0.59	0.23

of the exposed flowers was determined chemically and the toxicity to aphids was determined by their method (page 90). Their results (Table LI) showed that sunlight, ultraviolet light and heat caused a loss of pyrethrins, determined chemically, and a loss of toxicity to aphids. It is interesting to note that in the flowers, pyrethrins I and II decomposed to about the same extent. In the writer's experiments with isolated pyrethrins, however, pyrethrin II appeared to decompose more rapidly than pyrethrin I.

Tattersfield's Experiments with Pyrethrum Dusts and Antioxidants

Tattersfield *(873)* prepared dusts by extracting pyrethrum with petroleum ether which was then incorporated with talc or kieselguhr and the petroleum ether was evaporated. The resulting powders were stored in the dark. Such a dust, as rich in pyrethrins as a high test pyrethrum powder, was distributed in thin layers in petri dishes, which were exposed to air in the dark and in white, red and blue light for 3 days. Then known weights of the dusts were extracted with sufficient alcohol to yield a 10 per cent extract and the toxicity of these extracts to aphids was compared with that of an unexposed control. The toxicity tests were made by the method described on page 88.

Tattersfield also exposed similar dusts in flasks filled with oxygen, carbon dioxide, nitrogen and air, the first 3 in the sunlight, the last in the dark. The effect of the exposure on the toxicity to aphids was determined as above. The effectiveness of various antioxidants for preventing decomposition of the pyrethrins was also investigated. Generally 2.5 per cent of the antioxidant was incorporated in the dust before exposure.

Tattersfield gives the following summary of his results *(873)*:

"Pyrethrum powders and dusts, prepared by grinding or by the incorporation of extracts of pyrethrum on absorbent earths, such as talc and kieselguhr, lose their insecticidal activity on exposure to light and air. The loss is more rapid in the case of artificially prepared dusts than with ground flower heads.

"Both light and air play an important part in the process of inactivation, as samples of kieselguhr-pyrethrum or talc-pyrethrum dusts stored in closed vessels in the dark or exposed to the air in the dark are relatively stable; also samples exposed to light in an atmosphere of carbon dioxide, nitrogen or in vacuum lose little of their toxicity under the same conditions of illumination; samples exposed in oxygen, however, rapidly lose their activity. Both wet and dry oxygen were effective in destroying the activity of the dusts, but apparently at different rates.

"The incorporation of antioxidants with talc-pyrethrum and kieselguhr-pyrethrum dusts retards loss of activity due to exposure to light and air. Such compounds as pyrocatechol, resorcinol, hydroquinone and pyrogallol confer a large measure of protection against loss of toxicity. Phenol and phloroglucinol were not effective. Tannic acid exerted a considerable measure of protection. The protection was greater in the case of artificially prepared dusts than with ground pyrethrum flowers, although it seems also to be exerted to some extent in the latter case. There is no conclusive evidence that antioxidants, naturally occurring in pyrethrum, play any part in stabilizing the pyrethrins against inactivation. The greater part of the protection would appear to be due to particle size or to cellular inclusion."

It should be remembered that Tattersfield's samples were exposed for only a few days.

In a later paper Tattersfield and Martin *(881)* confirmed the stabilizing effect of tannic acid and hydroquinone on pyrethrum dusts. They also found that ground pyrethrum loses its pyrethrin content, when exposed to sunlight, in an atmosphere of nitrogen, but the loss is much less rapid than in air. Whole flowers lose pyrethrins less rapidly than ground flowers. Pyrethrin II decomposes at a slower rate than pyrethrin I. They concluded that loss of activity can be followed fairly closely by determining pyrethrin I and applying a small correction.

Effect of Prolonged Storage under Commercial Conditions

The effect of storage on pyrethrum has also been investigated by Gnadinger and Corl *(347)*. Both ground and powdered Dalmatian and Japanese *P. cinerariaefolium* were used for their experiments. Three hundred kilograms of Japanese flowers were carefully ground and mixed and packed in different types of containers. All of the containers were stored at 26° to 30°, not exposed to direct sunlight. The pyrethrin content of the flowers was determined at intervals; all analyses were calculated to a moisture-free basis and are reported in Table LII.

At the time the flowers were ground, a kerosene extract was prepared for biological testing. This extract contained the active principles from 6.4 g. of moisture-free flowers in 100 cc., corre-

TABLE LII. EFFECT OF STORAGE ON PYRETHRIN CONTENT OF GROUND FLOWERS (GNADINGER AND CORL)

No.		Pyrethrins (moisture-free basis)						Total loss in pyrethrin content
	Days in storage	0 %	30 %	60 %	90 %	182 %	365 %	%
		500-gram containers						
1	Open metal trays...	1.17	1.03	1.03	0.94	0.88	0.66	43.6
2	Unlined burlap bags	1.17	1.06	1.00	0.97	0.88	0.71	39.3
3	Vacuum coffee cans.	1.16	...	...	...	...	0.72	37.9
4	Friction top tin cans	1.17	1.01	1.02	1.01	0.89	0.78	33.3
		Commercial packages						
5	Fiber drum	1.17	...	...	...	...	0.72	38.4
6	Slack barrel	1.18	...	...	...	...	0.78	33.3
7	Unlined burlap bag	1.16	...	...	...	...	0.79	32.5
8	Metal drum.........	1.17	...	...	...	...	0.82	30.0

sponding to 75 mg. of pyrethrins per 100 cc. It was prepared by macerating the flowers with kerosene for 2 days (correcting for the volume of kerosene-soluble material in the flowers). The mixture was then shaken for 8 hours, again macerated for 2 days, and filtered. This kerosene extract was tested on flies by the Peet-Grady method. Part of the extract was stored with the samples of flowers. A portion of the kerosene was reserved for the preparation of extracts at the end of the storage period; at that time extracts were made from the flowers which had been stored 1 year in open trays and friction-top tin cans. These extracts also contained the active principles from 6.4 g. of moisture-free flowers, corresponding to 42 and 50 mg. of pyrethrins in 100 cc., respectively. The results of the biological tests are given in Table LIII.

TABLE LIII. EFFECT OF STORAGE ON TOXICITY OF GROUND PYRETHRUM TO FLIES (GNADINGER AND CORL)

Extract	Description*	Pyrethrins mg. per 100 cc.	Date tested	Tests	Av. kill %	Loss in toxicity %	Loss in pyrethrins %
1	Made from freshly ground flowers at beginning of storage	75	1/22/31	25	42	..	..
2	Same as 1, stored in tin, in room with flowers for 1 year	..	1/25/32	10	44	0	..
3	Made from flowers stored for 1 year in open trays, sample 1 Table LII	42	1/25/32	10	34	22.7	43.6
4	Made from flowers stored 1 year in friction-top cans, sample 4 Table LII	50	1/25/32	10	35	20.4	33.3

*All extracts contained the active principles from 6.4 g. of moisture-free flowers in 100 cc.

It is apparent from Table LII that a loss of more than 30 per cent in pyrethrin content occurs when freshly ground flowers are stored for 1 year, regardless of the kinds of containers used. The loss is greatest in open trays, but for other types of containers, including commercial packages, it is practically the same. Storage in partial vacuum does not prevent decomposition of the pyrethrins. Reference to Table LIII shows that losses of 43.6 and 33.3 per cent of the pyrethrin content, as determined chemically, are accompanied by losses of toxicity of 22.7 and 20.4 per cent, determined on flies. The toxicity of pyrethrin solutions is not directly proportional to their pyrethrin content; the results obtained by Grady, in collaboration with the writer (page 126), indicate that reduction of the pyrethrin content from 75 to 42 and 50 mg. per 100 cc. should result in losses of about 35 and 25 per cent, respectively. Considering the inaccuracies inherent in biological methods, the losses in toxicity correspond well to the losses in pyrethrin content.

Some months after the foregoing experiments, several of the samples reported in Table LII were again assayed, with the results shown in Table LIV.

TABLE LIV. ADDITIONAL ANALYSES OF SAMPLES REPORTED IN TABLE LII.

No.	Days in storage	Pyrethrins, dry basis 0	365	642	1137	Additional loss of pyrethrins	Total loss of pyrethrins
		%	%	%	%	%	%
1	Open trays	1.17	0.66	0.51	...	22.7	56.4
2	Vacuum cans...	1.16	0.72	0.70	0.63	12.5	45.7
4	Friction top cans	1.17	0.78	...	0.71	7.7	39.3
6	Slack barrel ...	1.18	0.78	0.71	...	9.1	39.8
7	Burlap bag	1.16	0.79	0.61	...	22.8	47.4

From Table LIV it appears that the loss in the tight containers gradually reaches a limit, possibly because of exhaustion of oxygen, but in the open type containers the loss continues.

Gnadinger and Corl also analyzed, at different intervals, a number of samples stored in friction-top cans. Some of these samples had been ground 2 to 3 years before the first analysis was made. The second analysis indicates, therefore, that the decomposition continues for more than 2 years, but apparently at a reduced rate, Table LV.

No storage experiments were made with whole flowers because of the difficulty of sampling them accurately. However, in Table LVI the analyses of a large number of shipments of Japanese flowers are given. These analyses are grouped by crop years and by months; the figure for each month represents the

TABLE LV. EFFECT OF STORAGE ON MISCELLANEOUS SAMPLES (GNADINGER AND CORL)

Description	Crop	First assay Date	First assay Pyrethrins %	Second assay Date	Second assay Pyrethrins %	Storage mos.	Loss of pyrethrins %
Dalmatian, po.	1926	June, 1929	0.43	Feb., 1930	0.39	8	9.3
Dalmatian, po.	1925	June, 1929	0.38	Feb., 1930	0.33	8	13.2
Dalmatian, po.	1925	June, 1929	0.40	Dec., 1930	0.30	18	25.0
Dalmatian, po.	1926	June, 1929	0.45	Jan., 1932	0.30	30	33.3
Japanese, grd.	1930	Oct., 1930	1.07	Jan., 1932	0.83	15	22.4
Japanese, grd.	1930	Oct., 1930	1.00	Jan., 1932	0.71	15	29.0
Japanese, po.	1926	June, 1929	0.61	Dec., 1930	0.41	18	33.3
Japanese, po.	1926	June, 1929	0.64	Dec., 1931	0.51	30	20.3
Japanese, po.	1926	June, 1929	0.68	Dec., 1931	0.52	30	23.5
Japanese grd.	1928	June, 1929	1.06	Dec., 1931	0.75	30	29.2
American, grd.	1928	May, 1929	0.90	Dec., 1931	0.64	31	28.9

average of all shipments analyzed in that month. There is a noticeable decline in the pyrethrin content each year, as the age of the whole flowers increases.

This decline in pyrethrin content has been confirmed by Hoyer *(448)*.

EFFECT OF COLD STORAGE ON PYRETHRUM

Gnadinger, Evans and Corl *(354)* have investigated the loss of pyrethrins that occurs in the first 30 days after harvesting and have also determined the effect of cold storage on pyrethrum. These preliminary experiments are summarized in Table LVII.

TABLE LVII. EFFECT OF COLD STORAGE ON GROUND PYRETHRUM (GNADINGER, EVANS AND CORL)

Sample	Days in storage	Temperature of storage °C	Pyrethrins %	Loss of pyrethrins %
1	0	....	1.14	...
1	30	20 to 35	1.07	6.1
1	180	20 to 35	1.00	12.3
1	180	–2 to –5	1.18	none
2	0	....	1.12	...
2	30	20 to 35	1.02	8.9
2	180	20 to 35	0.90	19.6
2	180	–2 to –5	1.11	0.9
2	330	–2 to –5	0.92	17.8

The 2 samples were assayed within 10 days of harvesting. The ground flowers were then stored in air-tight tin cans under the conditions indicated in Table LVII. The samples stored at room temperature showed an appreciable loss after 30 days and a still further loss after 180 days. The samples stored at $-2°$ to

TABLE LVI. ANALYSES OF JAPANESE FLOWERS FROM SEVEN CROPS (GNADINGER AND CORL)

1929 crop Assayed	Pyrethrins %	1930 crop Assayed	Pyrethrins %	1931 crop Assayed	Pyrethrins %	1932 crop Assayed	Pyrethrins %	1933 crop Assayed	Pyrethrins %	1934 crop Assayed	Pyrethrins %	1935 crop Assayed	Pyrethrins %
Sept., 1929	0.97	Sep., 1930	1.01	Aug., 1931	1.03	July, 1932	0.90	Aug., 1933	0.90	Aug., 1934	0.95	Aug., 1935	0.99
Oct., 1929	1.08	Oct., 1930	1.03	Sep., 1931	1.04	Aug., 1932	1.00	Sept., 1933	1.01	Sept., 1934	0.92	Sept., 1935	0.93
Nov., 1929	1.10	Nov., 1930	1.06	Oct., 1931	0.98	Sep., 1932	0.90	Oct., 1933	0.96	Oct., 1934	0.90	Oct., 1935	0.91
Dec., 1929	1.00	Dec., 1930	1.04	Nov., 1931	1.02	Oct., 1932	1.03	Nov., 1933	0.90	Nov., 1934	0.94	Nov., 1935	0.98
Jan., 1930	0.91	Jan., 1931	0.96	Dec., 1931	0.96	Nov., 1932	0.96	Dec., 1933	0.97	Dec., 1934	0.96	Dec., 1935	0.93
Feb., 1930	0.95	Feb., 1931	1.05	Jan., 1932	1.03	Dec., 1932	0.99	Jan., 1934	0.88	Jan., 1935	0.97	Jan., 1936	0.95
Mar., 1930	0.99	Apr., 1931	0.96	Feb., 1932	0.94	Jan., 1933	1.03	Feb., 1934	0.90	Feb., 1935	0.90	Feb., 1936	0.86
Apr., 1930	0.84	June, 1931	0.82	Mar., 1932	0.86	Feb., 1933	0.96	Mar., 1934	0.84	Mar., 1935	0.87	Mar., 1936	0.88
July, 1930	0.82	July, 1931	0.79	Apr., 1932	0.77	Mar., 1933	0.84	Apr., 1934	0.85	Apr., 1935	0.84	Apr., 1936	0.79
Aug., 1930	0.73	Aug., 1931	0.74	June, 1932	0.74	Apr., 1933	0.85	May, 1934	0.84	May, 1935	0.86		
Sep., 1930	0.68			July, 1932	0.69	May, 1933	0.78	June, 1934	0.74				
Nov., 1930	0.61			Aug., 1932	0.72	June, 1933	0.70	July, 1934	0.67				
						July, 1933	0.64						

- 5° did not lose pyrethrins during the first 180 days, but there was a noticeable loss after 330 days.

Relation Between Maturity of Flowers and Decomposition in Storage

It has been observed that some lots of flowers lose pyrethrins more rapidly than others; in some samples the loss was only 3 per cent in 6 months; in others, 20 per cent of the pyrethrins decomposed in 60 days. The loss of pyrethrins during storage is of considerable economic importance, hence it seemed advisable to determine, if possible, the reason for the difference in keeping quality of various lots. As one phase of this problem, the relation between the maturity of the flowers and their decomposition in storage has been investigated by Gnadinger, Evans and Corl *(355)*.

The flowers used in this experiment were grown at Fort Collins, Colo., and were selected from plants of uniform growth and flower yield. The samples, which represented three stages of maturity, were all picked on the same day and dried at the same time and in the same way. The stages of maturity and the number of samples collected at each stage were:

Stage 1. Bud stage. Corollas of marginal florets erect, first row of disk florets open. Three samples, average dry weight 70 g.

Stage 2. Nearly mature stage. Two-thirds of disk florets fully open. Five samples, average dry weight 64 g.

Stage 3. Overblown stage. Collected five days after all disk florets fully open. Five samples, average dry weight 88 g.

The samples of a given stage were not composited. Each sample was thoroughly mixed and divided into two equal parts. One part was ground and assayed by the copper reduction method, 10 days after picking; the moisture content was also determined. The other part was placed in a cheese cloth bag and suspended from a hook in the ceiling of a large room. The bags were woven closely enough to prevent sifting. The temperature of the room varied from 22° to 32° C. All samples were stored under the same conditions; after approximately 60 days in storage, they were ground and assayed. The analyses, before and after storage, are compared in Table LVIII.

The immature flowers (Stage 1) showed no loss of pyrethrins during 60 days' storage. The nearly mature flowers (Stage 2)

TABLE LVIII. MATURITY AND KEEPING QUALITY OF FLOWERS

Stage	Sample	Avg. weight per flower g.	Days in storage	Moisture %	Pyrethrins, moisture-free basis Before storage %	Pyrethrins, moisture-free basis After storage %	Pyrethrins lost during storage %
1	A	0.101	0	6.5	0.96	...	....
1	A	0.101	60	9.1	...	0.98	none
1	B	0.101	0	9.3	0.97	...	....
1	B	0.101	63	9.1	...	0.98	none
1	C	0.115	0	11.5	0.99	...	....
1	C	0.115	63	8.5	...	0.97	2.0
1	Avg.	....	..	...	0.97	0.98	none
2	A	0.166	0	9.6	1.12	...	....
2	A	0.166	60	8.4	...	1.09	2.7
2	B	0.184	0	8.2	1.09	...	....
2	B	0.184	60	9.0	...	1.12	none
2	C	0.180	0	6.9	1.05	...	...
2	C	0.180	63	8.7	...	1.01	3.8
2	D	0.158	0	7.3	1.08	...	....
2	D	0.158	63	8.4	...	1.09	none
2	E	0.160	0	7.2	1.10	...	....
2	E	0.160	64	9.2	...	1.07	2.7
2	Avg.	....	..	...	1.09	1.08	1.1
3	A	0.242	0	6.6	0.95	...	....
3	A	0.242	60	8.5	...	0.83	12.6
3	B	0.224	0	6.2	1.00	...	....
3	B	0.224	60	9.5	...	0.86	14.0
3	C	0.221	0	5.8	1.02	...	....
3	C	0.221	63	9.5	...	0.86	15.7
3	D	0.240	0	7.4	0.99	...	....
3	D	0.240	63	7.8	...	0.87	12.1
3	E	0.230	0	5.8	0.95	...	....
3	E	0.230	64	9.0	...	0.85	10.5
3	Avg.	....	..	...	0.98	0.85	13.0

showed an average loss of 1.1 per cent, which is within the limits of error of the assay method. The over mature flowers (Stage 3) showed a definite average loss of 13.0 per cent pyrethrins. The moisture content of the different samples was rather uniform after storage.

It would seem, from this experiment, that the selection of the proper time for harvesting pyrethrum is even more important and difficult than hitherto supposed; if the flowers are picked when immature, they will not have attained their maximum pyrethrin content; if picked when too mature, they will lose pyrethrins rapidly during storage.

Effect of Baling on Loss of Pyrethrins During Storage

The practice of compressing Japanese pyrethrum in bales originated before the pyrethrins were identified and long before it was known that pyrethrum flowers deteriorate in storage. Baling was probably resorted to in order to reduce ocean freight charges, rather than to preserve the flowers.

If baling inhibits loss of pyrethrins during storage, the flowers should be compressed immediately after drying. At present, the flowers are stored in uncompressed bags until the time for shipment arrives; they are then compressed into bales and loaded on the vessel. Under these conditions, flowers may not be compressed until eight or ten months after harvesting. If flowers were baled at harvest time, the difficulty of sampling, at the time of shipping, would be greatly increased; moreover, it would be necessary to dry the flowers very thoroughly after harvesting, to prevent "heating" of the bales, with consequent loss of pyrethrins.

Japanese flowers are compressed at a pressure of about 6000 lbs. per square inch. In Yugoslavia a pressure of about 2000 lbs. is used. Kenya flowers are somewhat less compressed than Japanese.

Several investigators have surmised that the baling of flowers prevents decomposition of pyrethrins during storage, but, heretofore, no experimental evidence on the effects of baling has been presented.

Gnadinger, Evans and Corl *(355)* have investigated the effect of baling, on flowers grown at Avon, Colorado. These flowers were harvested July 17, 1935, and were dried until July 26th. On August 3rd, the lot of flowers was thoroughly mixed and divided into six equal parts. A sample was taken from each of the six parts for assay and the remainder of each of the six parts was divided into portions of 2.20 lbs. each. The subdivisions of the six parts, each weighing 2.20 lbs., were then treated in the following manner:

Part 1. Stored in white cotton seed bags, no pressure.
Part 2. Baled under pressure of 1000 lbs. per sq. in.
Part 3. Baled under pressure of 5000 lbs. per sq. in.
Part 4. Baled under pressure of 10,000 lbs. per sq. in.
Part 5. Baled under pressure of 16,000 lbs. per sq. in.
Part 6. Sprayed with a solution of Antioxidant No. 5 (page 170) and baled under pressure of 10,000 lbs. per sq. in.

The bales were compressed with a Carver hydraulic press, using a specially constructed box to hold the flowers. The press and bales made from Parts 1 to 5 are shown below; the bag and each of the bales weighed 2.20 lbs. net. The plates which formed the tops and bottoms of the bales were of stainless steel. The bags and bales were suspended from the ceiling, to permit free circulating of air during storage at room temperature.

The analyses of the samples, at the time of baling, after 93 days in storage and after 263 days in storage are given in Table LIX.

The change in pyrethrin content at the end of 93 days in

HYDRAULIC PRESS AND BALES USED IN STUDY OF BALING PRESSURES.

TABLE LIX. EFFECT OF COMPRESSION ON LOSS OF PYRETHRINS IN STORAGE

Part	Gauge pressure lbs.-sq. in.	Pressure on flowers lbs.-sq. in.	Vol. of 2.2 lb. bale cu. in.	Weight per cu. in. oz.	Days in storage	Moisture in grd. flowers %	Pyrethrins (dry basis) %	Pyrethrins lost %
1	0	0	485	.072	0	4.6	1.41	....
1	0	0	485	.072	93	7.5	1.38	....
1	0	0	485	.072	263	6.0	1.17	17.0
2	1,000	55	126	.279	0	6.2	1.45	....
2	1,000	55	126	.279	93	6.9	1.33	....
2	1,000	55	126	.279	263	6.5	1.12	22.8
3	5,000	278	90	.391	0	5.8	1.44	....
3	5,000	278	90	.391	93	7.5	1.38	....
3	5,000	278	90	.391	263	5.1	1.21	16.0
4	10,000	555	72	.488	0	5.5	1.44	....
4	10,000	555	72	.488	93	7.1	1.42	....
4	10,000	555	72	.488	263	6.4	1.22	15.3
5	16,000	889	61	.577	0	5.8	1.47	....
5	16,000	889	61	.577	93	6.9	1.52	....
5	16,000	889	61	.577	263	5.1	1.27	13.6
6	10,000	555	72	.488	0	5.0	1.48	....
6	10,000	555	72	.488	93	8.2	1.43	....
6	10,000	555	72	.488	263	7.0	1.29	12.8

storage was not greater than the error of the copper reduction method, except in the case of Part 2.

After storage for 263 days, the loss of pyrethrins was greatest in the bales which had been made under 1000 pounds gauge pressure. This was probably because the pressure was sufficient to break up the flowers, but not great enough to exclude air. The pyrethrin loss in the uncompressed flowers was only slightly greater than in highly compressed flowers. The antioxidant did not prevent decomposition of the pyrethrins, probably because it was applied only on the surface of the flowers.

EXPERIMENTS WITH ISOLATED PYRETHRINS

The work thus far described was done with whole or ground flowers or with extracts of pyrethrum. Gnadinger and Corl isolated a mixture of pyrethrins I and II by a modification of the method of Staudinger and Ruzicka; the mixture consisted of 66.8 per cent pyrethrin I and 33.2 per cent pyrethrin II. The pyrethrins were dissolved in petroleum ether so that 0.888 g. of total pyrethrins was contained in 100 cc. of solution. Measured portions of this solution were evaporated in beakers and the residues were exposed to the air at room temperature, in the sunlight, and at 90° to 95° in the dark. Similar portions were measured into U-tubes and the solvent was completely removed in a rapid current of nitrogen. The tubes were sealed while filled with nitrogen and were exposed to sunlight and heat. When the pyrethrins are decomposed by air, light or heat they become insoluble in petroleum ether; the pure pyrethrins are completely soluble in petroleum ether. The amount of decomposition, in the different experiments, was determined from the amount of petroleum ether-insoluble material present after exposure. The results of the experiments are summarized below:

No.	Description	Temperature °C	Decomposed pyrethrins %
1	Exposed to air and sunlight for 3 days....	10-25	97.1
2	Exposed to air in the dark for 8 hours...	90-95	65.3
3	Exposed to nitrogen in the dark, 8 hours..	90-95	0.6
4	Exposed to nitrogen and sunlight 12 days	10-25	4.2

The petroleum ether-soluble portion of No. 4, after exposure for 12 days, weighed 0.4275 g.; this was diluted with kerosene to a volume of 700 cc. The toxicity of this solution was compared, by the Peet-Grady method, with that of a similar solution containing 0.444 g. of the unexposed pyrethrins in 700 cc. The exposed pyrethrins averaged 44 per cent kill in 10 tests; the unexposed pyrethrins averaged 46 per cent kill in 10 tests. The pyrethrins exposed to sunlight in nitrogen for 12 days did not suffer a greater loss of toxicity than would be expected from the formation of 4.2 per cent of petroleum ether-insoluble material.

Effect of Antioxidants on Concentrated Pyrethrum Extracts

Ripert *(745)* states that certain antioxidants occur naturally in pyrethrum. He also attacks the conclusions of Gnadinger and Corl and of Tattersfield and Martin, who found that pyrethrum deteriorates in storage, or, on exposure to light. The losses in pyrethrin content found by these investigators are, according to Ripert, not due to decomposition of pyrethrins, but are caused by the unsaturated acids present in the flowers. These unsaturated acids, Ripert states, become oxidized and coat the pyrethrin particles, thus rendering them insoluble in petroleum ether. In proof of this hypothesis, Ripert says that if ether or ethylene dichloride be used instead of petroleum ether, much higher pyrethrin contents will be found. He completely overlooks the fact that altered or oxidized pyrethrins are much more soluble in ether and ethylene dichloride than in petroleum ether, hence the former solvents give higher results. Ripert's theory does not explain the decomposition of pyrethrins in solution.

Mills and Fayerweather *(618)* have patented the use of tertiary alkyl-substituted *o*-dihydroxybenzenes as stabilizers for pyrethrum.

Hughes *(458)* has patented a process for preserving pyrethrum, which comprises spraying the freshly harvested and dried flowers with cymene, kerosene or other petroleum oil and packing in containers to prevent evaporation of the solvent.

Voorhees *(936)* employs amino-anthraquinone compounds for preparing light-stable pyrethrum extract, using, for example, 0.02 per cent of 1,4-toluido-anthraquinone.

Yates *(983)* has found that pyrethrins in highly refined mineral oil deteriorate more rapidly than in less refined oils. Oil containing less than 2 per cent of sulfonatable compounds had a satisfactory odor, but when combined with pyrethrum, an antioxidant or stabilizer was necessary to prevent deterioration. Yates found thymol especially effective as a stabilizer. His Canadian patent claims most of the common antioxidants as stabilizers for mineral oil extracts of pyrethrum.

In the course of experiments on the development of highly concentrated extracts, Gnadinger, Corl and Clark *(353)* found that properly made kerosene extracts of comparatively low concentration (2.5 per cent pyrethrins) were quite stable, (Table LX), but extracts of high concentration (10 to 15 per cent pyrethrins) were unstable, during prolonged storage. Slight changes in the highly concentrated extracts could be detected readily by chemical assay methods, but the detection of such changes in household pyrethrum sprays by these methods was extremely difficult.

TABLE LX. EFFECT OF STORAGE ON KEROSENE-PYRETHRUM EXTRACTS OF LOW CONCENTRATION

Manufacturer	Days in Storage 35° C.	Pyrethrin content (Seil method) I %	II %	Total %
A	0	1.04	1.45	2.49
A	120	1.07	1.38	2.45
A	275	1.07	1.36	2.43
B	0	0.93	1.41	2.34
B	120	1.00	1.39	2.39
C	0	1.00	1.28	2.28
C	120	0.90	1.19	2.09

By using decalin (decahydronaphthalene)* as solvent, in preparing the concentrates, it was possible to avoid interference by antioxidants occurring naturally in kerosene or other mineral oils. At the high pyrethrin concentration used, neither decalin nor kerosene interferred with the use of the copper reduction method for assaying the extracts. This was proved in the following manner:

30 g. of ground pyrethrum were extracted with petroleum ether (b.p. 20°-60°) in a Soxhlet extractor. The petroleum ether extract was filtered into a 250 cc. volumetric flask and made to the mark; 100 cc. of the extract were measured into each of two 400-cc. beakers; to one beaker 1 cc. of decalin was added. The

TABLE LXI. EFFECT OF DECALIN ON DETERMINATION OF PYRETHRINS

Age of Flowers mos.	Pyrethrin content With decalin %	Without decalin %
38	0.58	0.58
35	0.67	0.66
1	0.76	0.77
1	0.84	0.84
6	1.21	1.23

* Patent applied for.

petroleum ether was then evaporated and the analyses by the copper reduction method were completed in the usual way.

The pyrethrin contents found were almost identical, as shown in Table LXI.

A similar experiment, using 1 cc. of kerosene instead of decalin showed:

With kerosene	0.98%
Without kerosene	0.96%

The procedure finally adopted for assaying the decalin or kerosene extracts of pyrethrum, containing 10 to 15 per cent pyrethrins, was that described on page 228.

Having developed a method for preparing and accurately assaying highly concentrated pyrethrum extract, a number of experiments were made to determine the effect of antioxidants on these extracts during storage.

The effect of storage, at 35°, on highly concentrated pyrethrum extract, without antioxidants, is shown in Table LXII. The analyses were made by the copper reduction method.

TABLE LXII. EFFECT OF STORAGE AT 35° ON CONCENTRATED EXTRACTS

Solvent used	Original pyrethrin content %	Days in storage	Pyrethrin content after storage %	Loss of pyrethrins %
Decalin	14.1	60	12.4	12.0
Decalin	14.1	120	11.3	19.8
Decalin	14.4	84	10.8	25.0
Decalin	13.5	120	9.9	26.7
Decalin	13.4	97	11.0	17.9
Kerosene	11.5	93	9.8	14.8

A concentrated extract was prepared from freshly-made oleoresin of pyrethrum using decalin as solvent. Antioxidants were added immediately to different portions of this extract, of identical pyrethrin content, and the resulting solutions were stored in screw-cap, air-tight tin cans, at 35° C. A few of the antioxidants were of unknown composition, being sold under trade names. The concentrated extract was assayed at the beginning and end of the storage period; the extracts containing antioxidants were analyzed after 60 to 120 days in storage. The efficiency of the various compounds is shown in Table LXIII.

Antioxidant No. 5 was effective for 60 days, but not for 120 days. None of the other antioxidants was sufficiently effective under the conditions of storage.

A second decalin-pyrethrum extract was prepared. Part of

TABLE LXIII. EFFECT OF ANTIOXIDANTS ON DECALIN-PYRETHRUM EXTRACT STORED AT 35°

Antioxidant Name	Amount used* %	Days in storage	Pyrethrins present %	Pyrethrins decomposed %
None, original extract	..	0	14.1	...
Antioxidant No. 5	0.1	60	14.3	none
Antioxidant No. 5	0.1	120	12.3	12.8
Alpha-naphthylamine	0.5	81	13.0	7.8
Hydroquinone	0.1	81	12.8	9.2
Thymol	0.5	82	12.6	10.6
Eugenol	0.5	83	12.3	12.8
Para-aminophenol	0.1	97	12.3	12.8
Monobenzyl para-aminophenol	0.1	85	12.1	14.2
Dibenzyl para-aminophenol	0.1	85	12.1	14.2
Alpha-naphthol	0.5	82	11.9	15.6
Beta-naphthol	0.5	83	11.9	15.6
Diphenylamine	0.5	85	11.9	15.6
Steam distilled pine oil	2.0	85	11.9	15.6
Phenyl-alpha-naphthylamine	0.5	85	11.7	17.0
2-Amino-5-hydroxytoluene	0.1	97	11.6	17.7
Para-hydroxyphenyl-morpholine	0.1	84	11.4	19.1
Flectol H	0.5	84	10.8	23.4
Lecithin	0.2	97	10.8	23.4
Flectol, White	0.1	84	10.4	26.2
Turpentine	2.0	85	10.0	29.1
None, original extract	..	85	10.8	23.4

* Because of the difficulty of determining solubility in the concentrate it is not certain that all of the antioxidant dissolved, in every case.

this extract was stored, without the addition of any antioxidant, in screw-top tin cans and in amber glass bottles at 35° and at 6°. To a second part of this extract, 0.1 per cent of Antioxidant No. 5 was added; this solution was stored, in cans and bottles, at 35° and 6°, in air, under partial vacuum (45 mm.) and in nitrogen. The analyses of the different solutions are summarized in Table LXIV.

The extract which contained no antioxidant lost pyrethrins to the same extent in tin and amber glass containers. The loss when stored at 6° was about half the loss when stored at 35°. The extract containing 0.1 per cent Antioxidant No. 5 also lost at the same rate in amber glass and tin containers but the losses were much smaller than in the case of the extract without antioxidant. The loss at 6° was again much lower than at 35°. Storage in partial vacuum and in nitrogen did not prevent decomposition of pyrethrins.

TABLE LXIV. EFFECT OF TEMPERATURE AND TYPE OF CONTAINER ON ACTION OF ANTIOXIDANT IN DECALIN-PYRETHRUM EXTRACT

Antioxidant Name	Antioxidant Amount used %	Days in storage	Temp. of storage ° C.	Container	Pyrethrins per cent %	Pyrethrins decomposed %
None, original solution	..	0	..	......	13.4	...
None, original solution	..	62	35	tin can	10.9	18.6
None, original solution	..	62	35	amber glass	10.9	18.6
None, original solution	..	120	35	tin can	9.9	26.1
None, original solution	..	120	6	tin can	11.6	13.4
Antioxidant No. 5	0.1	62	35	tin can	12.8	4.4
Antioxidant No. 5	0.1	62	35	amber glass	13.0	3.0
Antioxidant No. 5	0.1	97	35	tin can	12.2	8.9
Antioxidant No. 5	0.1	99	6	tin can	12.9	3.7
Antioxidant No. 5	0.1	98	35	partial vacuum	11.8	11.9
Antioxidant No. 5	0.1	98	35	nitrogen filled	12.0	10.4
Antioxidant No. 5	0.1	120	35	tin can	11.5	14.1

A concentrated kerosene extract was next prepared. Portions of this extract, to which antioxidant was added, were stored in tin cans at 35°, 27°, room temperature and 6°. After about three months in storage the samples were again assayed, with the results shown in Table LXV.

TABLE LXV. EFFECT OF STORAGE ON KEROSENE EXTRACT OF PYRETHRUM

Antioxidant Name	Antioxidant Amount %	Temperature of storage ° C.	Days in storage	Pyrethrin content Originally present %	Pyrethrin content After storage %	Loss of pyrethrins %
No. 5	0.1	35	93	11.5	10.3	10.4
No. 5	0.1	27	95	11.5	10.8	6.1
No. 5	0.1	Room temp.	95	11.5	11.3	1.7
No. 5	0.1	6	94	11.5	11.6	0.0

The combined effect of the antioxidants (present in kerosene and added) and storage at 6° completely prevented decomposition. There was still some decomposition at higher temperatures, however.

In a final experiment, different percentages of Antioxidant No. 5 were added to freshly prepared concentrated extract containing 11.0 per cent pyrethrins. Samples were assayed and then stored in air tight cans at 35° C. for 236 days, when they were again assayed. The analyses, given below, show that increasing the amount of antioxidant decreased the loss of pyrethrins during storage.

Antioxidant No. 5 %	Pyrethrin content Before storage %	Pyrethrin content After storage %	Loss %
1.00	11.0	10.6	3.6
0.50	11.0	8.5	22.7
0.25	10.9	7.5	31.2

EFFECT OF STORAGE ON PYRETHRUM HOUSEHOLD SPRAYS

It has been shown that pure pyrethrins are readily decomposed when exposed to air or when heated, and that ground pyrethrum flowers, during storage, lose a considerable part of their pyrethrin content. The experiment with potassium permanganate (page 80) proved that the pyrethrins are not readily oxidized in petroleum ether solution, hence it might be inferred that the pyrethrins in oil solution are more stable than in the pure state. Gnadinger and Corl have investigated the effect of storage on pyrethrum extract. An extract was prepared from assayed flowers by percolation with kerosene; it contained 75 mg. of pyrethrins per 100 cc. Portions of this extract were packed in tin cans and in flint, amber and blue glass bottles. Three packages of each kind were stored out of doors in an eastern exposure for 13 months. The temperature varied from −29° to 43°. Three of the tin cans were stored indoors, unexposed to direct sunlight, at a temperature of 26° to 35°. Part of the extract was tested for toxicity to flies by the Peet-Grady method at the beginning of the storage in December, 1930. The final toxicity tests were made in January, 1932. The tests are described in Table LXVI.

TABLE LXVI. EFFECT OF STORAGE ON PYRETHRUM HOUSEHOLD INSECTICIDES

Kind of container	Where stored	Average kill: Original extract	Average kill: Same extract after storage for 13 months	Loss in toxicity
		%	%	%
Tin cans	Indoors, no direct sunlight	42	44	None
Tin cans	Outdoors, direct sunlight	42	44	None
Amber glass	Outdoors, direct sunlight	42	53	None
Flint glass	Outdoors, direct sunlight	42	21	50
Blue glass	Outdoors, direct sunlight	42	4	90

No loss of toxicity occurred in the samples stored in tin. No reason could be found for the high toxicity of the samples stored in amber glass. Flint and blue glass were unsatisfactory containers for pyrethrum extract. The walls of the blue glass bottles were much thinner than those of the flint glass bottles. This probably accounts for the greater loss of toxicity in the extract stored in blue glass. A storage experiment, conducted with a concentrated pyrethrum extract made by the ethylene dichloride process is described on page 197.

Hoyer *(445)* obtained anomalous results in experiments on the effect of sunlight on four household sprays. The sprays contained about 100 mg. of pyrethrins per 100 cc. and were made with "odorless" and "regular" oil bases. The pyrethrin content was determined by the Seil method; toxicity was determined by the Peet-Grady method. At the end of four days' exposure, the toxicity of the "odorless" sprays increased noticeably, but that of the "regular" spray decreased. The pyrethrin I content of two of the "odorless" sprays decreased, as did that of the "regular" spray, but the pyrethrin I content of the other "odorless" spray increased. The pyrethrin II content of all the "odorless" sprays increased; that of the "regular" spray remained the same. At the end of 120 days' exposure, the "odorless" sprays had lost from 73 to 83 per cent of their toxicity; the "regular" spray had lost 68 per cent. The pyrethrin I and pyrethrin II content of two of the "odorless" sprays had greatly decreased; the pyrethrin I and pyrethrin II content of the third "odorless" spray had greatly increased; the pyrethrin I content of the "regular" spray had decreased, but the pyrethrin II content had increased. Hoyer does not state whether or not the sprays were filtered before assaying.

SUMMARY

Ground pyrethrum loses a part of its activity, through decomposition of the pyrethrins, when exposed in thin layers to air and sunlight for a few days. When stored for several months in the dark, even in tins under partial vacuum, there is a pronounced loss of pyrethrins. Antioxidants apparently retard decomposition to some extent. Ground flowers stored in air-tight tins at room temperature for 6 months lose an appreciable amount of pyrethrins; similar samples stored at $-2°$ to $-5°$ showed no loss of pyrethrins in 6 months, but there was a noticeable loss in 11 months. Whole flowers also lose pyrethrins by decomposition in storage, but apparently at a slower rate than ground flowers.

The keeping quality of whole pyrethrum flowers is apparently influenced by the stage of maturity at which they are picked. Over-mature flowers decompose more rapidly than immature or nearly mature flowers.

Pure pyrethrins decompose rather rapidly on exposure to air and sunlight or ultraviolet light. They are also rapidly decomposed when heated at 100° in air. When exposed to sunlight in nitrogen they do not decompose appreciably in 12 days.

Baling whole flowers does not prevent decomposition of pyrethrins during storage. There was little difference in the pyrethrin loss of uncompressed flowers and flowers baled under pressures of 5000 to 16,000 pounds per square inch.

Pyrethrins apparently decompose more rapidly in some oils than in others. This is probably due to the removal of natural antioxidants from the oils during refining. Antioxidants added to concentrated pyrethrum extracts prevent decomposition to a marked extent. Storage at 0° to 6° C. also inhibits decomposition. Sunlight accelerates the decomposition of oil extracts of pyrethrum.

Certain antioxidants are quite effective in preventing decomposition of pyrethrins in highly concentrated extracts (8 to 15 per cent pyrethrins). Moderately concentrated extracts (2 to 3 per cent pyrethrins) can be made which do not require antioxidants to prevent decomposition in storage.

The effect of moisture on the pyrethrins has not been quantitatively studied, although moisture is known to facilitate decomposition. The effect of light and heat on mineral oil solutions of the pyrethrins is also discussed in Chapter XI.

CHAPTER IX

ADULTERATION OF PYRETHRUM

Before methods for determining the percentage of pyrethrins in flowers were available, the chemical analysis of pyrethrum was limited to an examination for added foreign materials. That no accurate biological tests existed at that time is best shown by the erroneous opinions then held, which have since been disproved by chemical assay methods and improved biological tests.

The pyrethrin content is of primary importance to the user of pyrethrum; whether a given lot contains added foreign matter is of secondary importance. For example, a shipment of unadulterated pyrethrum containing 0.2 per cent pyrethrins can legally be imported and sold in this country, whereas a lot containing 0.9 per cent pyrethrins and 6 per cent of added stems would be adulterated under the present standards of the Food and Drug Administration. That the addition of stems and other foreign matter should be prohibited is, of course, evident; that a minimum standard of pyrethrin content should be adopted is equally obvious. Whether a low pyrethrin content is due to decomposition in storage or to added foreign material is of little importance to the user; the effect is the same in either case.

Types of Adulteration

Few products have been so extensively adulterated as powdered pyrethrum. The adulteration of the whole flowers or coarsely ground flowers is comparatively rare. Among the adulterants which have been used are the following *(608)*: lead chromate, barium chromate, potassium chromate, turmeric, fustic, saffron and yellow ochre—all used to give color. Different species of flowers used as adulterants include: *Pyrethrum corymbosum, P. indicum, P. parthenium* (feverfew), *Chrysanthemum caucasicum, C. coronarium,* (crown daisy), *C. frutescens* (Marguerite), *C. leucanthemum* (ox-eye daisy), *C. segetum* (marigold), *Anthemis nobilis* (Roman chamomile), *A. arvensis* (corn chamomile), *A. cotula* (mayweed), *Calendula officinalis* (calendula), *Tanacetum vulgare* (tansy), *Matricaria chamomilla* (German chamomile) and others. Miscellaneous adulterants which have been found in pyrethrum powder include almond shells, aloes, arsenic, borax, brick dust, euphorbium, fleabane, hellebore, jalap, nux vomica, pepper, quassia, sand, sawdust, senna, soapbark, starch, sumac, rice hulls and stems and leaves of pyrethrum *(608)*.

Most of these adulterants were found in pyrethrum in European countries. The passage of the Federal Insecticide Act of 1910 put an end to the cruder forms of adulteration, so that in later years the most common adulterants in the United States have been powdered pyrethrum stems and daisy flowers. In 1926 the Department of Agriculture stated that probably more than 90 per cent of the adulterated pyrethrum found, contained ground stems, which are considered inert.

Mention should also be made of the substitution of inferior grades for presumably better grades. For example, until about 1930 purchasers preferred powder made from "all closed Dalmatian" flowers and paid a premium of 5 to 6 cents a pound for this grade. The unscrupulous seller frequently supplied powdered Japanese flowers and thus unwittingly did the buyer a service, for it has since been proved that the Japanese flowers were considerably richer in pyrethrins than the immature "all closed" grade. At the present time Dalmatian flowers are the poorer variety and are probably being substituted for Japanese flowers.

Former Methods of Detecting Adulteration

The United States Department of Agriculture considers the terms "Pyrethrum," "Pyrethrum powder" and "Insect powder" synonymous, when applied to insecticides, and permits the use of only the three species of pyrethrum mentioned on page 1, in insecticides so labelled. The regulations of the Department do not permit more than 5 per cent of naturally adherent stems, nor more than 2 per cent of acid-insoluble ash in shipments of pyrethrum *(37, 39, 41)*.

Before the pyrethrins were identified, an enormous amount of work was done for the purpose of detecting the addition of powdered stems and daisies to pyrethrum powder. Although the methods developed for this purpose have been rendered obsolete by the chemical assay methods, which determine the pyrethrin content, a description of the earlier methods may be of interest.

The Department of Agriculture determined stems in powdered pyrethrum by means of the following formula *(608)*: $X = \frac{100\,(a-c)}{a-b}$, in which X is the percentage of stems in the sample; a is the average percentage of nitrogen in flowers;

b is the average percentage of nitrogen in stems; *c* is the nitrogen content of the sample. The value of *a* for mixtures made of open flowers and stems differs from the value of *a* for mixtures of closed flowers and stems. In the case of open flowers and stems, *a* is 1.267, *b* is 0.765 and $X = \frac{100\ (1.267 - c)}{1.267 - 0.765}$. In mixtures of closed flowers and stems, *a* is 1.784, *b* is 0.765 and the formula becomes $X = \frac{100\ (1.784 - c)}{1.784 - 0.765}$. In using this method, therefore, the analyst was first obliged to determine whether the sample was made from open flowers and stems or from closed flowers and stems. The Department suggested 4 ways for deciding this question: 1, examination by a microscopist; 2, by calculating the lowest cost from the market prices of closed and open flowers and stems; 3, from the intensity of the green color of the ether extract; 4, from the crude fiber content. A similar method was based on the phosphorus pentoxide content of flowers and stems instead of the nitrogen content.

The addition of daisy flowers (*C. leucanthemum*) to pyrethrum was easily detected, in the case of whole flowers, by the following differences *(754)*:

Pyrethrum: achenes 5-ribbed; possess a small toothed crown; golden yellow; ray floret achenes more curved than disk floret achenes.

Daisy: achenes usually 10-ribbed; ribs prominent, white, alternating with black stripes; knob-like projection on broad end of achene; achene much smaller than that of pyrethrum.

In powdered samples microscopic examination served to detect the addition of daisy flowers. According to Roark and Keenan *(754)*, the powdered achenes of pyrethrum are strikingly different from those of the daisy, under the microspoe. The achene tissues cf pyrethrum are characterized by numerous crystals, but those of the daisy exhibit no such crystals, containing instead a notable amount of brownish-red material; the color of this material is intensified on heating with 50 per cent aqueous chloral hydrate solution. Another diagnostic character is the tissue from the white ribs of the daisy achene. The other flower parts of the daisy, when powdered, are not readily distinguished from those of pyrethrum.

That chemical analysis of the type in use in 1919 failed to distinguish daisy flowers from pyrethrum is shown by the work of Roark and Keenan *(754)*, reproduced in Table LXVII. Gnadin-

TABLE LXVII. COMPARISON OF ANALYSES OF PYRETHRUM AND DAISIES (ROARK AND KEENAN)

Sample	Nitrogen	Crude fiber	Pentosans	Ether extract	Petroleum ether extract	Phosphorus pentoxide
	%	%	%	%	%	%
Pyrethrum, closed	1.784	22.09	16.66	6.87	4.11	0.691
Pyrethrum, open	1.267	31.02	21.11	5.81	4.06	0.532
Pyrethrum, stems	0.765	40.66	18.21	3.22	1.97	0.234
Daisy flowers	1.860	22.56	12.45	5.03	3.58	0.840

ger and Corl have since shown that daisies contain no pyrethrins. The analyses in Table LXVII are also of interest because they indicate the kind of chemical examination ordinarily made on pyrethrum in 1919. Needless to say, there is no direct connection between pyrethrin content and such determinations as nitrogen, crude fiber, pentosans, ether extract, phosphorus pentoxide and manganese.

Microscopic examination served to detect starch and certain other adulterants. The following is the procedure suggested by the Department of Agriculture *(608)*. "The powder to be analyzed microscopically should be thoroughly mixed. This is best done by spreading the sample on a sheet of white paper and mixing the powder with a spatula. Flattening the powder out upon the paper often reveals the presence of unground pieces of material which can be transferred to a microscope slide and examined. After mixing, a composite sample is taken from various parts of the sample. A small amount of the powder is placed upon a microscope slide, a drop or two of distilled water added and the cover glass adjusted. If examination of the water amount reveals the presence of any foreign starch, a small amount of a solution of iodine in potassium iodide is drawn under the cover glass. This reagent stains blue any starchy material that may be present. For further examination a small portion of the powder should be mounted in chloral hydrate solution and gently heated over the flame. This solution serves to dissolve any starch that may be present and clears the tissues generally. Until the microscopist has become thoroughly familiar with the pyrethrum tissues, standard samples ground from the various parts of the flower head should be kept on hand for comparative study."

Whether such an examination would detect the addition of exhausted or extracted flowers is not stated. Fortunately, pyrethrum which has been extracted with kerosene has a strong kerosene odor and pyrethrum residues, from which volatile solvents have been recovered by distillation, are dark in color and have a cooked odor. The addition of exhausted pyrethrum to ground flowers could, of course, be detected by the lowered pyrethrin content, determined chemically.

Labelling Pyrethrum

Under the present rulings of the Food and Drug Administration of the United States Department of Agriculture, pyrethrum containing less than 5 per cent of stems and 2 per cent of acid-insoluble ash can be labelled and sold as "insect powder" without any declaration of active and inert ingredients on the label. If, however, such pyrethrum is sold under a proprietary name, either alone or mixed with other material, the percentage of stems and acid-insoluble ash must be declared as inert ingredients. On the other hand, the Department has ruled that in labelling hellebore and tobacco dust, only the alkaloids may be declared as active ingredients, while the remainder of the plant material, usually about 99 per cent, must be declared as inert.

It is quite evident, that in order to protect the consumer, the Food and Drug Administration should require that pyrethrum be labelled to show the pyrethrin content as active material and the remaining plant material as inert, precisely as it does with hellebore and tobacco dust, permitting, of course, either of the optional methods of stating active and inert ingredients, specified in the Insecticide Act.

Roark, of the Bureau of Chemistry and Soils, another branch of the Department of Agriculture, has said *(749)*: "In buying pyrethrum flowers or extracts thereof, it would be well if the purchaser would insist on a statement from the manufacturer giving the exact percentage of pyrethrins present."

Simulation of High Pyrethrin Content

With the development of the chemical assay methods it has been suggested that some chemical adulterant might be mixed with or sprayed on the flowers, so as to indicate a higher pyrethrin content than is actually present. Aside from the fact that a similar line of reasoning can be applied to many of the methods used for detecting adulteration of foods and drugs, let us consider the properties such a material must have in order to pass all the available tests without detection:

1. It must be soluble in low-boiling petroleum ether at 20°.
2. It must be an ester to pass Tattersfield's acid method.
3. It must not be an acid, for it will be neutralized and

probably removed by the sodium carbonate treatment in the Gnadinger-Corl method.

4. It must reduce alkaline copper tartrate solution to pass the Gnadinger-Corl method and, moreover, the appearance of the reduced copper solution must be like that produced by the pyrethrins, which is characteristic; any abnormal appearance of the reduced solution will be immediately noted by an experienced analyst.
5. It must not be destroyed, in petroleum ether solution, by washing with potassium permanganate solution.
6. It must not be an aldehyde or ketone which can be removed from petroleum ether by sodium bisulfite solution.
7. It must not be precipitated from 80 per cent alcohol by basic lead acetate.
8. It must form an insoluble semicarbazone to pass Tattersfield's semicarbazone method.
9. It must be odorless and colorless, or yellow.

While a compound having these properties might be added to ground pyrethrum to increase the apparent pyrethrin content, the adulteration of whole flowers in this way would be more difficult. Gnadinger and Corl have shown that the ray florets, disk florets, involucral scales and receptacles contain only traces of pyrethrins; that is, the outer parts of the flower head contain very small amounts of pyrethrins, while the greater part is contained in the achenes, which are invisible from the outside. The low pyrethrin content of the receptacles and scales is shown in Table LXVIII.

TABLE LXVIII. PYRETHRIN CONTENT OF RECEPTACLES AND SCALES OF PYRETHRUM FLOWERS (GNADINGER AND CORL)

Kind of flower	Pyrethrin content Of orig. flowers %	Of recept. and scales %	Receptacles and scales in flowers %	Percentage of total pyrethrins in receptacles and scales
Japanese	0.96	0.27	25.0	7.0
Japanese	0.84	0.20	22.8	5.4
Japanese	0.76	0.26	30.8	10.5
Dalmatian	0.53	0.17	31.1	10.0
Dalmatian	0.52	0.17	30.0	9.8

Hence any chemical which, dusted or sprayed on whole flowers, would increase the apparent pyrethrin content of the outer parts could readily be detected. It would be difficult to apply such a chemical to the achenes without increasing the apparent pyrethrin content of the outer parts.

CHAPTER X

POWDERED AND GROUND PYRETHRUM

Production of Fine Powder

The earliest method of producing finely powdered pyrethrum was by means of the mortar and pestle, worked by hand. Then the mortar was enlarged and the pestle was raised and lowered by a wheel and crank. The next improvement was the employment of burr mills. These were later replaced by chaser mills, which were still in occasional use when the demand for fine powder declined about 1927. In a chaser mill 2 granite millstones are mounted like wheels on an axle. The center of this axle is connected with a vertical shaft, which is driven by a motor and revolves the granite "wheels" on a block of granite. The whole arrangement is enclosed to prevent the escape of the fine powder.

The modern mill, now generally used for fine powdered pyrethrum, is the high speed steel mill. This type of mill consists of an outer shell (usually corrugated or roughened on the inner surface) in which a number of beaters revolve on a central shaft at 3000 to 4000 revolutions per minute. These beaters may be steel disks with lugs attached to the edge, or may consist of a central hub from which are suspended spokes or "hammers" of various designs. In either type, the flowers are mechanically fed into the mill, where they are battered to powder by impact with the beaters and the inner surface of the mill. These mills have the same effect as a fan, producing considerable air pressure, which is released through "dust collectors" that hold back the fine dust while permitting the air to escape. The dust collector may be of the canvas tube or metal "cyclone" type. The main part of the powder does not go through the dust collector, but is conveyed direct from the mill to a mechanical sifter, where it is bolted over a fine silk cloth; 120 mesh cloth is used for very fine powder. The tailings from the sifter are automatically fed back to the mill in a constant stream, insuring that the composition of the finished powder will represent the whole flowers.

The fineness of the powder can be controlled to some extent by the rate at which the mill is fed. The feed must, however, be so regulated that the mill does not become hot, otherwise the powder will have a cooked odor and dark color and the pyrethrins will be partly oxidized. The flowers should not be allowed

MODERN MILL FOR POWDERING PYRETHRUM.

to become warmer than 40° to 50°. When properly powdered, no loss of pyrethrins occurs in the process. A sample of Japanese pyrethrum assayed 0.92 per cent pyrethrins when coarsely ground for analysis; the same lot when finely powdered assayed 0.94 per cent. The mechanical loss in milling fine powdered pyrethrum varies from 7 to 12 per cent, depending principally on the moisture content of the flowers; most of this loss is moisture.

Fine powder can be separated by means of an air current

instead of by sifting. The fineness of the "air-floated" powder so produced depends on the velocity and volume of the air current used.

Flowers which are shipped in highly compressed bales must be broken into small pieces before grinding. This is accomplished by disintegrators, which are mills operated at slower speeds; they break the compressed flowers into a very coarse, granular powder, usually about 10 mesh, which is suitable for feeding into the high speed mills.

Pyrethrum flowers sometimes contain pebbles, nails, scraps of metal, coins and occasionally tools, sandals, etc., lost by workmen in the producing countries. Some of these materials can be removed by electro-magnets; others are removed by mill-tenders. Flowers which contain excessive sand will rapidly destroy the beaters in a high speed mill.

Formerly all the pyrethrum imported was converted into fine powder, in which form it was used principally as a household insecticide and to some extent as a horticultural insecticide. Small amounts were used in the manufacture of "mosquito sticks." At the present time powdered pyrethrum is rarely used in the household and its use on plants is restricted to those sections where aqueous sprays cannot be used, or to certain special crops where the use of sprays is not advisable.

Grinding for Percolation

Most of the pyrethrum now imported is used in manufacturing liquid insecticides of various kinds. For this purpose the coarsely ground flowers are employed because the powder is not suitable for percolation or extraction. The difference between ground and powdered flowers is shown by the following typical screen analyses:

Ground for percolation (GFP)		%
Retained	on 30-mesh sieve	2.6
	on 40-mesh sieve	13.7
	on 60-mesh sieve	38.5
	on 80-mesh sieve	11.2
	on 100-mesh sieve	11.2
	on 120-mesh sieve	6.1
Through	120-mesh sieve	16.7

Fine powder		%
Retained	on 100-mesh sieve	0
	on 200-mesh sieve	39
Through	200-mesh sieve	61

G.F.P. pyrethrum should be ground on a mill especially adapted for the purpose. Uncompressed flowers can be fed directly into the mill, but compressed flowers must first be broken

WHOLE ACHENES, PROPERLY GROUND FLOWERS AND IMPROPERLY GROUND FLOWERS

into small pieces. This is best accomplished by a breaker which does not grind the flowers but merely breaks the compressed slab into pieces suitable for feeding.

The ideal grind for percolation would be a uniform No. 30 powder, but it is impossible to obtain this in practice, because the ray florets, disk florets and achenes are easily ground, while the receptacles and involucral scales are more refractory. The ground flowers always contain some fine material which is very rich in pyrethrins; the amount of this fine material must be kept as low as possible to prevent clogging of the extractors. At the same time, the achenes or "seeds" which contain most of the pyrethrins, must be cracked to permit complete extraction.

A sample of G. F. P. pyrethrum was sifted on a 200-mesh sieve and the separated portions were assayed with the following results:

Through 200-mesh sieve 4%; pyrethrin content 1.91%
Left on 200-mesh sieve 96%; pyrethrin content 1.16%

In grinding pyrethrum for percolation it is essential, therefore, that no part be removed by sifting, so that the ground flowers will have the same pyrethrin content as the original whole flowers.

The necessity for cracking the achenes was shown by this experiment: A sample of achenes from mature flowers was thoroughly mixed; one portion was finely ground and extracted with petroleum ether for 24 hours in a Soxhlet extractor. A second, equal portion of the unground achenes was extracted in the same way; analysis showed:

Ground achenes, 1.00 per cent pyrethrins
Whole achenes, 0.05 per cent pyrethrins

This proves that the pyrethrins cannot be extracted from mature achenes unless they are ground and emphasizes the importance of correct grinding of G. F. P. pyrethrum. On page 185 are shown whole achenes, properly ground pyrethrum and improperly ground flowers, all considerably magnified. As the name indicates, G. F. P. pyrethrum is used entirely for percolation or extraction in the manufacture of various kinds of pyrethrum extracts.

Both ground and powdered flowers are ordinarily packed in burlap bags or slack barrels holding from 100 to 200 pounds; plain or waterproof liners may be used. A grain trier is convenient for sampling such packages, but it should be remembered that the fine material, rich in pyrethrins, has a tendency to sift to the bottom of the package. Ground or powdered pyrethrum should be stored in a cold, dry place.

CHAPTER XI

MANUFACTURE OF PYRETHRUM EXTRACTS

Most of the pyrethrum imported into this country is used in the manufacture of various kinds of extracts or liquid sprays. These products are best classified, according to the purpose for which they are used, as household insecticides, horticultural sprays and cattle or live stock sprays.

The household insecticides are extracts of pyrethrum in mineral oil. The better commercial extracts contain the active principles from about 1 pound of high-test flowers to the gallon. The oil is usually the kerosene fraction, sometimes highly refined to improve the odor. Generally the odor is further masked by a suitable perfume. Recently, hydrogenated mineral oil suitable for insecticides has been patented.

The horticultural sprays are usually made with alcohol or acetone; occasionally oleoresin of pyrethrum emulsified with soap is used. Oil extracts of pyrethrum are sometimes used on plants. As a rule, the horticultural sprays contain a higher proportion of pyrethrins than household sprays and are diluted with water for use on plants.

Live stock sprays are usually made with the heavier mineral oils and contain the pyrethrins from about 1 pound of flowers to the gallon, when properly made. Various chemicals, reputed to be more or less repellent to flies, are added to these sprays.

The household, horticultural and livestock sprays are described in greater detail in succeeding chapters.

SELECTING THE FLOWERS—STANDARDIZATION

In manufacturing a pyrethrum extract of any kind the first consideration must be the pyrethrin content of the flowers used. When the first pyrethrum household sprays were put on the market, about 1919, nothing was generally known regarding the nature and percentage of the active principles in the flowers. Manufacturers accordingly standardized their extracts on the basis of the weight of flowers used per gallon of spray, the only method of standardization then available. The work describing the isolation and identification of the pyrethrins was published in 1924. In 1929 the development of methods for determining the pyrethrins showed that the percentage of pyrethrins in commercial flowers varies from 0.4 to 1.2 per cent.

Until 1929, Dalmatian closed flowers were considered the best obtainable. In that year it was found that Japanese flowers are superior to Dalmatian and that the pyrethrin content increases as the flowers mature. Recent work has shown that high grade flowers can be grown in Dalmatia and Kenya. Hence the selection of the flowers is no longer based on trade beliefs and customs, but on the pyrethrin content, accurately determined by chemical methods.

At the present time foreign dealers will not sell on a guaranteed pyrethrin content unless the buyer accepts or rejects the lot before it is loaded on the steamer. This makes it necessary for the American importer to assay all of the pyrethrum purchased, before it leaves the producing country, unless he is content to buy fair average quality of the season, without regard to pyrethrin content. When samples representing definite lots are sent to this country, a considerable time elapses before analysis can be made (about 30 days in the case of Japan) ; in the meantime the price may have greatly increased or declined. It is known that flowers lose a considerable part of their pyrethrin content during storage, consequently, as the end of the crop year is reached, it becomes increasingly difficult to find lots rich in pyrethrins. In 1932 about 30 per cent of the lots offered in Japan, during 1 year, contained less than 0.90 per cent of pyrethrins.

It is evident, therefore, that the manufacturer can produce a uniform product only when he assays each lot of flowers immediately before extraction and in addition he must be sure that the extraction process is efficient. If these precautions are taken, it is now possible to manufacture an insecticide having a definite and constant pyrethrin content and toxicity. On the other hand, the manufacturer who merely uses a fixed weight of flowers to the gallon of spray is certain to produce an insecticide of varying pyrethrin content and toxicity. This is perfectly obvious from the analyses of the several crops of Japanese flowers reported on page 161. Some manufacturers have attempted to standardize their products by biological tests. Such biological standardization is not as satisfactory as chemical standardization because the biological tests are not as accurate as the chemical assays and, furthermore, it is sometimes difficult to tell whether variations in the biological tests are due to lack of uniformity in the different batches of insecticide or to lack of uniformity in the different cultures of test insects.

Direct Extraction Process

Pyrethrum extracts may be conveniently classified as direct extracts or concentrated extracts, according to the process of manufacture.

The direct extraction process is limited almost entirely to the preparation of mineral oil extracts for use as household insecticides. Direct extractions can be made by any of the processes ordinarily used for the extraction of botanical drugs: maceration, percolation, or combinations of the two. In one process, for example, G. F. P. pyrethrum is macerated with refined kerosene in a tank equipped with an agitator, using about 1 pound of flowers to each gallon of oil. The maceration is prolonged from a few days to several weeks, the longer, the better; the mixture is then pumped through a filter press. The solution held by the flowers is rich in pyrethrins and can be recovered by washing with additional kerosene, by centrifuging, by pressing in an hydraulic press or in an Anderson oil expeller or similar equipment.

Direct extractions can also be made by simple percolation. The ground flowers are moistened with the oil and are then packed in a cylindrical or slightly tapering percolator, which can be made of iron. The flowers are now covered with oil and allowed to stand for 24 hours. The percolation is then conducted slowly, in the usual way. A detailed description of the percolation process is given in U. S. Pharmacopoeia X. The percolators must be properly packed and the rate of flow must be slow, otherwise extraction will be incomplete, as Richardson has shown *(716)*.

In a modification of this process the ground flowers are placed in an extractor and the oil is circulated through the flowers by a rotary pump. An extractor of the kind described on page 192 is very satisfactory. Fifteen hundred pounds of flowers can be used for each batch and these are extracted with 5 portions of oil, each of the 5 extractions requiring 1 day. The following practical example illustrates the process (Table LXIX).

The nearly exhausted flowers are extracted with 400 gallons of oil, which is used as the first percolate for the next batch. The loss of oil is about 0.75 pound for each pound of flowers extracted; this oil contains only traces of pyrethrins and is usually not recovered. Such an extract contains 99.4 per cent of the pyrethrins from 1.5 pounds of flowers in each gallon of spray and can easily be standardized. Assume that the standard adopt-

TABLE LXIX. DIRECT EXTRACTION OF GROUND PYRETHRUM, 1500 POUNDS OF FLOWERS IN 1000 GALLONS OF OIL

Percolate	Amount of oil taken gal.	Total volume of oil solution in extractor gal.	Amount of oil extract drawn off gal.	Amount of oil extract left in extractor gal.	Proportion of soluble material extracted %
1	400	407*	227	180	55.8
2	200	380	204	176	23.7
3	200	376	207	169	11.3
4	200	369	204	165	5.1
5	150	315	158	157	2.0
Total	1150	...	1000	...	97.9
6**	400	557	400	157	1.5

*Including 7 gal. of soluble material from the flowers.

**Used as first percolate for next batch.

ed is 1 pound of flowers, of 0.90 per cent pyrethrin content, to each gallon of spray and that the flowers in the above example contained only 0.60 per cent pyrethrins; then the 1000 gallons of extract would be used without further dilution. If, however, the pyrethrin content of the flowers were 1.10 per cent, then the 1000 gallons of extract would be diluted to 1833 gallons. The weight of flowers per gallon of spray would vary, therefore, from 1.5 to 0.82 pounds, according to their pyrethrin content.

In any of these processes gentle heat may be applied.

MANUFACTURE OF CONCENTRATED PYRETHRUM EXTRACTS

In general it is not commercially practicable to produce an extract stronger than about 4 pounds of flowers to the gallon by direct extraction; somewhat stronger extracts can be made by re-percolation, but only at considerably increased expense, or with appreciable loss of pyrethrins because of incomplete extraction. Pyrethrum extracts said to be made by direct extraction and to contain the active principles from 20 to 40 pounds of flowers to the gallon have been offered to the trade. The difficulty of making such a product by direct extraction lies in the fact that 20 pounds of flowers will absorb about 2.5 gallons of kerosene or alcohol without leaving any free liquid; one gallon of solvent to 20 pounds of flowers will scarcely wet the latter thoroughly. Direct extraction is also not very satisfactory for the preparation of livestock sprays which are made with the more viscous oils.

It became evident about 1929 that a highly concentrated, standardized product, of uniform pyrethrin content and toxicity, which could be used for any of the three classes of sprays, would find a ready market. In that year the writer developed a process for the manufacture of a standardized, concentrated pyrethrum extract; in 1931 about one-third of all the pyrethrum imported into the United States was extracted by this process, a description of which follows *(341, 348)*.

The pyrethrins are viscous liquids, soluble in petroleum ether, ether, alcohol, acetone, ethylene dichloride and most other organic solvents. When pyrethrum flowers are extracted, the pyrethrins are dissolved, together with inert materials, the amount of which depends on the solvent used. Petroleum ether extracts less inert matter than any other solvent and methanol extracts more; the former extracts about 4 per cent of the weight of the flowers and the latter about 19 per cent (page 201). The petroleum ether extract will dissolve almost completely in kerosene to form a clear yellow solution exactly like that produced when pyrethrum is extracted directly with kerosene. The alcohol extract, however, cannot be dissolved completely in kerosene, and the solution is dark green, making it unsuitable for a household spray. Acetone yields an extract which is only partly soluble in kerosene, and the kerosene solution is also dark green. Other solvents such as benzene are too toxic to the workmen in the plant. Some, such as carbon tetrachloride and most other chlorinated compounds, decompose with the formation of hydrochloric acid when distilled in the presence of moisture, thus corroding the equipment. Still others, such as ether, petroleum ether, and naphtha, are highly explosive. Petroleum ether and naphtha have the additional disadvantage of wide boiling ranges, although certain petroleum fractions can now be obtained with a narrower distillation range than was formerly available. The low-boiling fractions of the petroleum solvents are expensive to recover during the process, resulting in high solvent losses and increased costs. The solvent which comes nearest meeting all of the requirements is probably ethylene dichloride. It readily dissolves the pyrethrins, extracting only 7 per cent of the weight of the flowers; it yields an extract which dissolves in kerosene with a yellow color and is also soluble in alcohol, acetone, and nearly all other organic solvents. Its toxicity to the workmen is about the same as that of gasoline. It is the most stable of all the chlorinated hydrocarbons when distilled in the presence of moisture. Its flash point is comparatively high (14°), and it does not support combustion readily. Its boiling point is constant (83.5°) and sufficiently low, under vacuum distillation, to prevent decomposition of the pyrethrins by overheating; at the same time, it

VIEW OF PLANT FOR MANUFACTURING CONCENTRATED PYRETHRUM EXTRACT.

is high enough to permit easy and almost complete recovery of the solvent.

Ethylene dichloride also has the advantage of being immiscible with water, permitting easy separation from the small amount of water which distils over in one step of the process. Finally, its cost is low.

The extractors consist of covered vertical tanks mounted on trunnions. They are fitted with false bottoms so constructed that the solvent passes through readily but the ground flowers are held back. Each extractor is charged with 1500 pounds of ground flowers, and sufficient ethylene dichloride is pumped in from a storage tank to fill the extractors. The solvent is then circulated through the flowers with a 2-inch rotary pump for 8 hours and is allowed to stand overnight. The next morning it is circulated 2 hours and is then drawn off into a collecting tank. A second portion of fresh ethylene dichloride is then pumped into the extractor and is circulated in the same way as the first extraction. This process is repeated until five extractions have been made. The last three extracts, which are weak, are kept separate from the first two strong extractions, and are used for the first two extracts of the succeeding batch. That is, the first two extractions (for every charge after the first) are made with the weak extracts from the preceding batch, while the last three extractions are made with fresh ethylene dichloride. In this way the amount of solvent which must be distilled is reduced by half.

The exhausted flowers in the extractors are now dumped by rotating the extractor on its trunnions and are conveyed to a rotary drier. The ethylene dichloride in the marc is distilled off and conducted back to the storage tank, after separating the small amount of water which comes from the flowers. The exhausted flowers, free from ethylene dichloride, can be burned as fuel. The exhausted material has approximately the following composition: moisture 7 per cent, pentosans 18 per cent, cellulose 43 per cent, lignin 21 per cent, ash 6 per cent.

The ethylene dichloride extract containing the pyrethrins is run into a vacuum still and distilled at a temperature below 60° until all ethylene dichloride is driven off and conveyed back to the ethylene dichloride storage tanks. The loss of solvent in the entire process is less than 3 per cent. The residue in the still is a dark, viscous oily liquid which becomes semi-solid on cooling. About 100 pounds of this oleoresin are obtained from the 1500-pound charge of flowers, the amount varying slightly with different lots of flowers. This oleoresin contains practically all the pyrethrins that were present in the flowers. The pyrethrins are easily dissolved from the oleoresin by almost any organic solvent. The oleoresin is transferred to a mixing kettle where it is treated with the desired solvent. This may be any cut of mineral oil, from petroleum ether to heavy lubricating oil, or any of the common solvents such as alcohol, acetone, carbon tetrachloride, etc. From the mixing kettle the solution is pumped to a narrow, calibrated tank, where it is made up at 20° to the desired volume, which is calculated from the pyrethrin content of the flowers.

In Table LXX are shown the various lots of flowers used in a complete year's run. These flowers were bought in Japan in 1929, with no guarantee as to pyrethrin content.

TABLE LXX. YIELD OF CONCENTRATED EXTRACT FOR PERIOD OF ONE YEAR

Lot	Pyrethrin content of flowers %	Gallons of concentrated extract produced from 3000 pounds of flowers
2909	0.97	159
2910	1.08	177
2911	1.10	180
2912	1.00	164
3001	0.91	149
3002	0.95	156
3003	0.99	162
3004	0.84	138
3007	0.82	134
3008	0.73	120
3009	0.68	112
Average	0.91	150

It is readily seen that the number of gallons of concentrated extract produced from a 3000 batch of flowers varies greatly with the pyrethrin content of the flower, being highest with new crop flowers and gradually decreasing as the end of the crop year is reached.

The completed solution is now pumped to a cooling tank, where it is held three days at 0°, to throw out traces of resinous material. It is then treated with Filter-Cel and pumped through a filter press to a storage tank, from which it is drawn off into drums as required.

This process has been thoroughly investigated, both on laboratory and plant scale, to determine whether there is a loss of pyrethrins at any point. Chemical analysis has shown that the extracted flowers contain less than 0.02 per cent pyrethrins, corresponding to an average extraction of 97 to 98 per cent of the pyrethrins originally present in the flowers. In a laboratory test a sample of flowers assaying 0.744 per cent pyrethrins yielded an oleoresin containing pyrethrins equivalent to a content of 0.731 per cent in the flowers, a recovery of 98.2 per cent. Similarly, 2970 pounds of flowers containing 0.72 per cent pyrethrins yielded 232 pounds of oleoresin contain 8.99 per cent pyrethrins, recovery of 97.5 per cent.

It has also been proved that the degree of heat used in the process does not decompose the pyrethrins. Although these compounds are decomposed by heat when isolated in the pure condition, they are apparently not injured by heat when dissolved in ethylene dichloride or mineral oil.

Three thousand pounds of flowers were extracted with 810 gallons of ethylene dichloride. A sample of the ethylene dichloride extract was reserved, and the remainder was distilled in vacuum in the regular way, yielding 191 pounds of oleoresin. A portion of the oleoresin was dissolved in sufficient ethylene dichloride to make the concentration of the solution the same as before distillation; this was established by determining the solids in the two solutions and adjusting the volume of the oleoresin solution accordingly. The two ethylene dichloride solutions were then diluted with seven volumes of kerosene and tested on flies by the Peet-Grady method. The results are shown in Table LXXI.

TABLE LXXI. EFFECT OF DISTILLATION ON TOXICITY OF ETHYLENE DICHLORIDE EXTRACT (GNADINGER AND CORL)

Ethylene dichloride-kerosene extract made from:	Extractive g. per 100 cc.	Pyrethrins mg. per 100 cc.	Tests	Kill Max. %	Kill Min. %	Kill Av. %
Ethylene dichloride soln. before distn.	0.3061	46	10	36	20	28
Ethylene dichloride soln. of oleoresin	0.3061	..	10	35	18	25

The difference in toxicity between the two extracts is well within the error of the method and indicates little or no loss of pyrethrins during distillation of the ethylene dichloride solution. In Table LXXII the results of heating kerosene extracts of pyrethrum are given, as determined on flies by the Peet-Grady method.

In this experiment a kerosene extract of pyrethrum was prepared in the laboratory from assayed flowers. The extract was divided into three parts, one of which was kept at room temperature while the other two were heated in sealed pressure flasks at the temperatures indicated in Table LXXII. There was little or no loss of toxicity caused by subjecting the extracts to heat.

Finally, the process has been checked by preparing kerosene extracts directly from flowers which were carefully sampled, assayed, and put through the process. Kerosene extracts were then prepared from the oleoresin, containing the same proportion of pyrethrins (based on the assay of the flowers) as the direct extracts. The following two examples of these experiments are typical:

TABLE LXXII. EFFECT OF HEAT ON TOXICITY OF KEROSENE EXTRACT OF PYRETHRUM (GNADINGER AND CORL)

Pyrethrins mg. per 100 cc.	Temp. °C	Time heated hours	Tests	Kill Max. %	Kill Min. %	Kill Av. %
75	26	.	9	62	34	45
75	70	5	10	54	37	46
75	98	5	10	60	30	42

(1) A 3000-pound batch of flowers containing 1.03 per cent pyrethrins was carefully sampled and then extracted, yielding 227 pounds of oleoresin. This oleoresin was dissolved in sufficient kerosene to make 180 gallons of concentrated extract, which would contain 2.05 grams of pyrethrins per 100 cc. if no loss had occurred; 87.7 cc. of this solution were diluted to 2000 cc. with kerosene, yielding a solution containing 90 mg. of pyrethrins per 100 cc., again assuming no loss to have occurred in the process. The toxicity of this solution was compared with that of the extract made from the sample of flowers reserved for assay, as mentioned above, prepared in the following manner: 50 grams of the ground flowers were macerated with 498 cc. of the kerosene overnight. The mixture was then shaken in a mechanical agitator for 8 hours, macerated for 2 days, and filtered. This solution contained 103 mg. of pyrethrins per 100 cc.; consequently, 87.4 cc. were diluted to 100 cc. with kerosene to yield a direct extract containing 90 mg. of pyrethrins per 100 cc.

(2) A 2984-pound batch of flowers also containing 1.03 per cent pyrethrins yielded 232 pounds of oleoresin which were made up to 180 gallons with kerosene. The resulting solution was diluted with kerosene (88.1 cc. diluted to 2000 cc.) to a concentration of 90 mg. of pyrethrins per 100 cc., provided no pyrethrins were lost in the process. This extract was compared for toxicity with a kerosene extract, made by direct extraction of the flowers, also containing 90 mg. of pyrethrins per 100 cc.

The results of these two experiments, which were made several weeks apart, are given in Table LXXIII.

The extracts made by diluting the concentrated extract have the same toxicity as extracts made by extracting equivalent weights of flowers directly with kerosene, proving that no measurable loss of pyrethrins occurs in the manufacturing process.

The toxicity of the exhausted flowers from the 3000 pounds

used in the first experiment was also determined. A part of the extracted flowers was air-dried for 7 hours at room temperature;

TABLE LXXIII. TOXICITY OF EXTRACTS MADE FROM CONCENTRATES COMPARED WITH DIRECT EXTRACTS OF SAME FLOWERS, PEET-GRADY METHOD (GNADINGER AND CORL)

Sample	Description of spray	Pyrethrin content of Flowers %	Spray mg. per 100 cc.	Tests	Kill Max. %	Kill Min. %	Kill Av. %
5135	Direct extract	1.03	90	12	73	40	55
5135	Extract made from concentrate	1.03	..	12	71	39	53
5145	Direct extra	1.03	90	9	63	37	48
5145	Extract made from concentrate	1.03	..	10	59	41	49

very little odor of ethylene dichloride remained. Ninety-six grams of the air-dried material was extracted with 400 cc. of kerosene in the manner described above. This solution was compared with kerosene alone for toxicity, with the result shown in Table LXXIV.

TABLE LXXIV. TOXICITY OF EXTRACTS MADE FROM MARC (GNADINGER AND CORL)

Description of spray	Tests	Kill Max. %	Kill Min. %	Kill Av. %
Kerosene extract of exhausted flowers (2 lb. per gallon)	3	5	4	4
Kerosene only	5	7	3	5

The toxicity of the extract made from the marc is no greater than that of kerosene alone.

EFFECT OF STORAGE ON CONCENTRATED EXTRACTS

Concentrated pyrethrum extract made by this process retains its strength when properly stored. Such an extract, containing 1800 mg. of pyrethrins per 100 cc. was stored indoors, in tin cans and iron drums, for 13 months at 26° to 35°. The toxicity of this extract to flies was determined at the beginning and end of the storage period by the Peet-Grady method. The concentrated extract was diluted for testing with 19 parts of kerosene yielding a solution containing 90 mg. of pyrethrins per 100 cc. The results are given in Table LXXV.

A second extract, containing 2.49 per cent pyrethrins, determined by the Seil method, was stored at 35° for 9 months. At the end of the storage period the pyrethrin content was found to be 2.43 per cent. For a further discussion of the effect of storage on pyrethrum products see Chapter VIII.

By the above process concentrated extracts containing about 4 g. of pyrethrins per 100 cc. can easily be produced. The extract which has proved most popular is one containing 2.15 g. of pyr-

ANOTHER VIEW OF A PYRETHRUM EXTRACTION PLANT.

ethrins per 100 cc. This is the same as 2.57 per cent pyrethrins and is also equivalent to 20 pounds of flowers, containing 0.90 per cent pyrethrins in each gallon. When diluted with 19 parts of oil this concentrate yields an insecticide containing 108 mg. of pyrethrins per 100 cc. With most oils the diluted solution will be brilliant and need not even be filtered. With certain highly refined oils, however, a very slight precipitate of inert resin may form; this is easily removed by filtration after the addition of about 0.15 per cent of Filter-Cel.

TABLE LXXV. EFFECT OF STORAGE ON CONCENTRATED PYRETHRUM EXTRACT (GNADINGER AND CORL)

Kind of container	Pyrethrin content of concentrate mg./100 cc.	Pyrethrin content when diluted 1+19 for testing mg./100 cc.	Av. kill First test, 12/26/30	Av. kill Second test, 1/26/32	Loss in toxicity during storage for 13 months
Tin can	1800	90	53	53	None
Iron drum	1800	90	53	58	None

Hopkins *(438)* prepares a concentrated extract of pyrethrum by extracting the ground or whole flowers with petroleum ether, boiling below 70°, under pressures as high as 200 to 500 pounds per square inch. The solution is then passed into a distillation zone where it is mixed with a higher boiling solvent, such as a kerosene, having an initial boiling point above 200°. The petroleum ether is thereby distilled off at a temperature below 100°, leaving the material extracted from the flowers dissolved in the kerosene. The amount of kerosene used is varied according to the degree of concentration desired in the finished product.

Sańkowsky *(768)* has patented a process for extracting pyrethrum which comprises placing pyrethrum flowers in cylindrical extraction tanks of special design and distributing a suitable solvent (e.g. kerosene) over the flowers at a rate less than sufficient to flood the interstices between flowers, whereby the solvent is caused to flow substantially only over the exterior surfaces and through the intercellular surfaces of the flowers. The extraction tanks are superimposed; as many as twelve are used in a series. As soon as the flowers in the upper tank are exhausted, it is taken off, recharged with fresh flowers and placed at the bottom of the series. Instead of tanks, a battery of centrifugal machines may be used. In one example cited, 70 pounds of flowers yielded 0.7 pound of extract after 33 days extraction with kerosene; this is said to be the maximum concentration obtainable. The application was filed in 1928 before the pyrethrin content of flowers was accurately known. Seventy pounds of flowers should yield from 2 to 3 pounds of non-volatile extractive, even with petroleum ether

Sankowsky and Grant *(770)* prepare a colorless pyrethrum extract by shaking a concentrated mineral oil extract with methanol, cooling and filtering the latter, distilling off the methanol and dissolving the remaining material, which contains pyrethrins, in water white kerosene.

The many advantages of a properly manufactured, standardized, concentrated extract of pyrethrum have resulted in a constantly increasing demand for the product.

The following are analyses of a number of concentrated pyrethrum extracts offered in the United States in 1935. It is interesting to note that only three of the samples assayed contained the amount of pyrethrins guaranteed by the manufacturer.

Origin	Guaranteed pyrethrin content	Actual pyrethrin content (Seil method) I	II	Total
	%	%	%	%
U. S. A.	2.4	0.40	0.53	0.93
Spain	2.5	0.20	0.90	1.10
U. S. A.	2.4	0.72	0.91	1.63
U. S. A.	2.4	0.76	0.97	1.73
U. S. A.	2.4	0.71	1.12	1.83
U. S. A.	2.4	1.00	1.28	2.28
U. S. A.	2.4	0.98	1.30	2.28
U. S. A.	2.4	0.93	1.38	2.31
U. S. A.	2.4	0.93	1.41	2.34
U. S. A.	2.4	1.01	1.43	2.44
U. S. A.	2.4	1.14	1.30	2.44
U. S. A.	2.4	1.14	1.48	2.62
Japan	9.0	3.37	4.76	8.13
Germany	15.0	5.20	8.42	13.62

Oleoresin of Pyrethrum

Different solvents remove different amounts of extractive materials from pyrethrum flowers. Gersdorff *(310)* found that 13 solvents completely extract the active principles. The percentages of material extracted by these solvents at the boiling point of the solvent and at room temperature, as determined by Gersdorff, are tabulated on page 201. The percentages extracted by acetone and petroleum ether, as determined by the writer, are also given. For convenience the number of pounds of flowers equivalent to one pound of oleoresin and the pyrethrin content of the oleoresin are calculated on the basis of room temperature extraction (Table LXXVI).

Neu *(637)* has determined the amount of extractive obtained

TABLE LXXVI. EXTRACTION OF PYRETHRUM WITH VARIOUS SOLVENTS (GERSDORFF)

Solvent	Material extracted from pyrethrum flowers At boiling point of solvent %	At room temperature %	Pounds of flowers equivalent to one pound of oleoresin	Pyrethrin content of oleoresin if flowers contain 0.90% pyrethrins %
Methanol	24.6	19.2	5.2	4.7
Ethyl alcohol, 95%	22.1	18.4	5.4	4.9
Acetone (Gnadinger)	. . .	14.0	7.1	6.4
Isopropyl alcohol	14.7	10.2	9.8	8.8
Tertiary butyl alcohol	12.6	9.5	10.5	9.5
Normal butyl alcohol	17.4	9.0	11.1	10.0
Secondary butyl alcohol	13.7	8.8	11.4	10.2
Chloroform	8.1	7.1	14.1	12.7
Xylene	6.9	6.9	14.5	13.0
Ethylene dichloride	7.2	6.6	15.1	13.6
Benzene	6.9	6.2	16.1	14.5
Carbon tetrachloride	6.3	5.3	18.9	17.0
Tetrachloroethylene	8.5	5.2	19.2	17.3
Di-ethyl carbonate	9.7	. . .	. . .	. . .
Petroleum ether (Gnadinger)	. . .	3.5	28.6	25.7

from Dalmatian flowers with different solvents, using a Soxhlet extractor. His results are reported in Table LXXVII.

TABLE LXXVII. EXTRACTIVE OBTAINED FROM DALMATIAN PYRETHRUM WITH VARIOUS SOLVENTS (NEU)

Solvent	Extractive %	Solvent	Extractive %
Methanol	22.7	Ethyl acetate	8.3
Denatured alcohol	20.5	Chloroform	7.8
Trichlorethylene	11.9	Carbon tetrachloride	5.0
Acetone	10.0	Petroleum ether	4.9
Benzene	8.4	Ether	4.2

It is obvious from Tables LXXVI and LXXVII that the quality of the oleoresin depends on the solvent used and that oleoresin should be bought only on the basis of pyrethrin content. The amount of material extracted by a given solvent varies slightly with different lots of flowers.

The approximate solubility of oleoresin of pyrethrum in a number of solvents is given in Table LXXVIII. The oleoresin used in these tests was prepared by the ethylene dichloride process; 15 g. of oleoresin were shaken with 100 cc. of solvent at room temperature (25°).

TABLE LXXVIII. SOLUBILITY OF OLEORESIN OF PYRETHRUM IN VARIOUS SOLVENTS (GNADINGER AND CORL)

Degree of solubility and solvent:	Color of solution
Almost completely soluble	
Butyl acetate	dark
Coal tar creosote oil	dark
Diacetone alcohol	dark
Dibutyl pnthalate	dark
Dichlor ethyl ether	dark
Diethyl sulfate	dark
Ethyl acetate	dark
Hexalin	dark
Isopropyl acetate	dark
Methyl amyl acetate	dark
Methyl cyclohexanone	dark
Methylene chloride	dark
Octyl alcohol	dark
Oil of camphor sassafrassy	dark
Ortho-dichlorbenzene	dark
Pine oil, steam distilled	dark
Propylene dichloride	dark
Safrol	dark
About two-thirds soluble:	
Butyl lactate	medium
Octyl acetate	dark
Tetralin	dark
Toluene	dark
Xylene	dark
About one-half soluble:	
Butyl acetyl ricinoleate	medium
Decalin	light
Diacetin	light
Dibutyl ether	medium
Diethylene glycol monobutyl ether	medium
Diethylene glycol monoethyl ether	medium
Ethylene glycol monobutyl ether	medium
Ethylene glycol monoethyl ether	medium
Ethylene glycol monoethyl ether acetate	medium
Kerosene	medium
Pseudocumene	medium
Less than one-half dissolved	
Diethylene glycol	light
Ethylene glycol	light

The ethylene dichloride oleoresin used consists of about one-half oily material and about one-half solid resins. The first group of solvents dissolved both the oils and resins. The second group dissolved the oils and part of the resins. The third group dissolved the oils but very little of the resins. The last group did not dissolve all of the oils. The pyrethrins are in the oily part of the oleoresin, hence they were dissolved by all of the solvents, excepting possibly the last group.

CHAPTER XII

PYRETHRUM HOUSEHOLD INSECTICIDES

Pyrethrum is particularly adapted for controlling household insects because it is non-poisonous to man, although to insects its active principles are among the most toxic substances known. For more than a hundred years finely powdered pyrethrum has been used to destroy bed-bugs, roaches, ants and similar insects. The use of such a dust in the household is, however, incompatible with neatness and its presence serves notice to all comers that insect pests are present.

Kerosene extracts of pyrethrum were recommended for controlling insects on plants more than 40 years ago. It was not until about 1919, however, that mineral oil-pyrethrum sprays, which are much less objectionable than powdered pyrethrum, were commercially offered for household use. In 1923 Fred D. Terry of Pasadena, California, applied for a patent on a liquid household insecticide "consisting of kerosene and the volatile active principles of pyrethrum." The patent was granted in 1926 and some time afterward the Terry Fly Spray Company brought suit for infringement against a California corporation engaged in marketing a similar product. This suit was defended by means of a fund to which a number of insecticide manufacturers contributed. The Terry patent was held invalid in 1930. In the course of the litigation it was brought out that a similar insecticide was used by the United States Navy during the World War and that several similar commercial products were on the market more than two years prior to Terry's application. With the threat of the Terry patent removed, the number of firms marketing pyrethrum insecticides increased. Educational campaigns showing the danger of household insects as carriers of various diseases increased the demand for pyrethrum sprays, so that today these insecticides occupy the position of household necessities.

Correct Pyrethrum Content for Household Insecticides

Household sprays can be made by direct extraction of the flowers or from standardized concentrated extract as described in the preceding chapter. In either case a product of uniform pyrethrin content can be produced. The pyrethrin content of commercial, pyrethrum household insecticides varies greatly.

Extracts containing less than 50 mg. of pyrethrins per 100 cc., equivalent to about ½ pound of high test flowers to the gallon, are still fairly common. Reference has been made to the attempt of the National Insecticide and Disinfectant Manufacturers Association to establish a minimum standard of 60 per cent kill by the Peet-Grady method. Unfortunately, as Gothard has pointed out, the method is not sufficiently accurate to be used in this way. It has been shown (Fig. XI, page 126) that the percentage of kill does not increase in direct proportion to the pyrethrin content, hence, after a certain point is reached, a large increase in pyrethrin content results in a comparatively small increase in kill. Tests of a large number of commercial sprays in comparison with sprays of known pyrethrin content have proved that a pyrethrin content of 100 to 120 mg. per 100 cc. is satisfactory. One pound of flowers, containing 0.90 per cent pyrethrins, to a gallon of spray is equivalent to 108 mg. of pyrethrin per 100 cc. and is a very satisfactory standard. Experience has verified that an extract containing the pyrethrins from one pound of high test flowers in a gallon of oil can be marketed with entirely satisfactory results to the consumer.

Proper Oil Base

Next in importance to the pyrethrin content is the kind of oil used. There are four types of oils to be considered:

1. Straight run distillate from paraffin base crude.
2. Cracked distillate from paraffin base crude.
3. Straight run distillate from naphthenic base crude.
4. Cracked distillate from naphthenic base crude.

Gothard *(366)* compared types 1 and 3 and found: "As far as oils from naphthenic base crudes are concerned, our experience has been with Gulf Coast crude, that a straight run distillate from it shows but very little more killing power than a similar distillate from a paraffin crude and the killing power is so small as to be negligible as an effective killing agent in insecticides."

The cracked distillates, however, have a definitely higher toxicity to insects than straight run distillates. This toxicity is due to the presence of unsaturated compounds which, unfortunately, have very disagreeable odors and are so irritating as to render the cracked distillates unsuitable for household insecticides. According to Gothard, when the cracked distillate is sufficiently refined with sulphuric acid to make it usable, its toxicity is lost. Gothard also suggests that the unsaturated compounds in the cracked distillates may polymerize on standing with the formation of varnish-like gums. The writer tested such a distillate by the Peet-Grady method and obtained 96 per cent knock-down and 70 per cent kill; three months later the knock-down was 24 per cent and the kill was no higher than that of a refined kerosene, that is, about 5 per cent.

Gothard concludes that the most satisfactory oil is a straight run distillate from a paraffin base crude.

The volatility of the oil must also be considered. The flash point of the oil should be not less than 49° (120° F.). A lower flash point is objectionable chiefly because the product might be dangerous to the user; a secondary consideration is the increased freight and insurance rate on more inflammable products. On the other hand, the oil must be sufficiently volatile to evaporate without leaving an oily residue on furniture, draperies and other objects with which the spray may come in contact.

Gothard *(366)* has recommended an oil having an initial boiling point of 177° (350° F.), a 50 per cent point of 210° (410° F.) and an end point of 265° (510° F.). Richardson *(714)* found that a distillate having a boiling range from 192° (378° F.) to 269° (516° F.) was more toxic than three others, having lower boiling ranges. He compared the speed of paralytic action and killing power of this distillate with that of a pyrethrum extract made from a low grade pyrethrum powder and an oil distilling at 154° (310° F.) to 202° (395° F.), and concluded that "The kerosene base used in preparing pyrethrum extract is at least as important, if not more so, than the amount of pyrethrum powder used in preparing the extract." Few who are experienced in the manufacture of pyrethrum extract will agree with Richardson's conclusion; it can easily be demonstrated by a practical household test or by the Peet-Grady method that the oil alone is responsible for very little of the toxic action of pyrethrum-oil sprays, under the conditions of actual use.

In co-operation with Dr. Erich Meyer of L. Sonneborn Sons, Inc., New York, the writer attempted to determine the relative insecticidal value of eight fractions obtained from the distillation of a paraffin base crude. The crude oil had the following properties:

API gravity	44.0
Saybolt viscosity at 37.8° (100° F.)	36.
Initial boiling point	53° (128° F.)
Iodine number	15.8

The different fractions were prepared and their physical properties were determined under the supervision of Dr. Meyer and were as follows (Table LXXIX):

TABLE LXXIX. ANALYSES OF OIL FRACTIONS (MEYER)

Fraction	Sp. gr. API	Viscosity Say., 100°F	Viscosity Thermo. 60°F	Flash, open cup °C	Flash, open cup °F	Fire, open cup °C	Fire, open cup °F	Boiling range Initial °C	Boiling range Initial °F	Boiling range End °C	Boiling range End °F	Iodine number
1	73.1	...	97	...	...	...	...	34	94	172	342	3.9
2	63.2	...	120	...	...	...	...	54	130	214	417	6.4
3	56.3	...	155	...	...	...	...	93	200	249	480	9.1
4	51.8	...	215	24	75	32	90	119	246	272	522	10.8
5	48.8	...	285	41	105	52	125	138	280	296	564	11.9
6	45.8	30.5	405	68	155	79	175	175	347	318	605	13.5
7	42.7	34.0	...	93	200	107	225	204	399	338	640	15.4
8	37.8	46.0	...	132	270	168	335	262	503	377	710	18.3

Solutions having the same pyrethrin content were prepared by dissolving 2 per cent of a very strong pyrethrum extract in each of the eight fractions. The relative toxicity of these solutions was then determined by the Peet-Grady method. In the first series of tests a No. 164 De Vilbiss sprayer having a slow rate of discharge was used; in the second series a faster sprayer

TABLE LXXX. TOXICITIES OF VARIOUS FRACTIONS OF OIL HAVING SAME PYRETHRIN CONTENT (PER CENT KILL, PEET-GRADY METHOD)

No. 164 De Vilbiss sprayer (slow).

Culture No.	Fraction	1	2	3	4	5	6	7	8
	Spray time (sec.)	19.6	19.6	20.4	21.2	22.4	25.2	30.3	51.0
1/30		64	76	79	76	77	63	83	69
2/1		63	59	65	65	62	66	65	50
2/2		82	69	79	81	72	77	77	70
2/3		78	62	66	64	56	64	71	66
2/14		75	77	70	61	67	70	84	84
Average		72	69	72	69	67	68	76	68

No. 164 De Vilbiss sprayer (fast).

Culture No.	Fraction	1	2	3	4	5	6	7	8
	Spray time (sec.)	10.0	10.0	10.2	10.0	10.2	11.2	12.2	18.0
2/14		82	72	56	67	..	..	..	..
2/15		..	..	..	..	62	58	53	62
2/15		82	80	69	44	58	57	59	30
2/16		72	66	62	40	52	51	53	44
2/17		76	62	69	48	54	62	51	43
Average		78	71	64	50	56	57	54	47

was used. The percentage kill and the time required to discharge 12 cc. of the different fractions are given in Table LXXX. With the slow sprayer there was no significant difference in the toxicities of the 8 different fractions. With the fast sprayer, however, the heavier fractions were much less toxic than the lighter ones. The toxicities of the various fractions of oil alone, that is, containing no pyrethrins, are shown in Table LXXXI.

TABLE LXXXI. TOXICITIES OF VARIOUS OIL FRACTIONS CONTAINING NO PYRETHRINS (PER CENT KILL, PEET-GRADY METHOD)

	No. 164 De Vilbiss sprayer (slow).							
Culture No.	Fraction 1 Spray time (sec.) 19.2	2 19.4	3 20.2	4 21.0	5 22.4	6 24.8	7 31.6	8 48.8
2/2	0	0	0	2	..	..	..	..
2/4	0	0	0	..	..	..	..	..
2/8	..	..	..	0	0	9	9	1
2/8	..	..	..	..	5	0	9	7
2/10	0	0	0	0	6	12	6	1
2/13	0	0	0	0	2	5	8	2
Average	0	0	0	1	3	6	8	3

New solutions of pyrethrins were now prepared with fractions 2 and 6 in the following manner: A high test pyrethrum powder was extracted with petroleum ether which was then filtered and evaporated in vacuum to constant weight. The resulting oleoresin was weighed out on the analytical balance into two equal parts which were dissolved in equal volumes of fractions 2 and 6 respectively. These solutions were tested by the Peet-Grady method using the two No. 164 De Vilbiss sprayers

TABLE LXXXII. EFFECT OF USING DIFFERENT SPRAYERS ON RELATIVE TOXICITY OF FRACTIONS HAVING SAME PYRETHRIN CONTENT (PER CENT KILL, PEET-GRADY METHOD)

		No. 164 sprayer (slow)		No. 164 sprayer (fast)		No. 153 sprayer	
Culture No.	Fraction Spray time (sec.)	2 19.9	6 25.9	2 9.9	6 11.0	2 18.9	6 20.5
2/20		55	79	52	31	51	53
2/20		..	77	56	40	57	56
2/21		54	53	40	50	60	51
2/21		..	60	42	46	61	50
2/24		45	65	45	26	47	42
2/24		65	71	35	28	59	53
2/25		50	..	..	..	..	..
2/25		50	..	..	..	..	..
Average		53	67	45	37	56	51
Oil only		0	6	0	6	0	6
Corrected kill		53	61	45	31	56	45

previously employed and also a No. 152 De Vilbiss sprayer. The results of these tests are given in Table LXXXII.

It is fairly obvious from these experiments that the relative toxicity of sprays made from different fractions of petroleum, with the same pyrethrin content, will vary according to the type of sprayer used and more or less independently of the physical constants of the oil used. The experiments also emphasize the fact that a sprayer which is suitable for one oil-pyrethrum spray may be unsuitable for a second spray, even though the pyrethrin content of the two sprays be the same.

Murphy and Peet *(628)* concluded that the toxicity of pyrethrum-oil insecticides varies inversely with the viscosity of the oil. The following is a summary of their results, obtained by testing several oils, containing the same percentage of pyrethrins, by the Peet-Grady method.

Solvent	Relative viscosity	Kill, %
Water	1.00	..
Naphtha	0.98	81
Kerosene	1.07	69
Kerosene plus 10% lubricating oil	1.08	55
Kerosene plus 20% lubricating oil	1.10	56
Kerosene plus 40% lubricating oil	1.17	57
Kerosene plus 75% lubricating oil	1.48	48
Lubricating oil	2.51	41
Stock spray base	2.62	37

The Steam Sprayer

When pyrethrum-oil household insecticides were first introduced, a small inefficient "mouth-sprayer" was furnished with each bottle or can. The importance of an efficient sprayer was soon realized and various kinds of hand sprayers were developed. The electric sprayer was the next type introduced. Recently the steam sprayer has attracted considerable attention. The following excellent description of the various steam sprayers on the market in 1933 is taken from "Soap" *(58)*.

"For those who are not familiar with the steam application of insecticides, a brief explanation of the principle behind the machines will be given, as those which have come to our attention thus far, operate on the same general basic principle (Fig. XII). In place of a stream of air being passed over the top end of the capillary tube immersed in the liquid insecticide, as in the case of an ordinary hand or electric sprayer, the steam vaporizer shoots a fine stream of steam across the tip of the tube. This throws out the usual mist of liquid insecticide, at

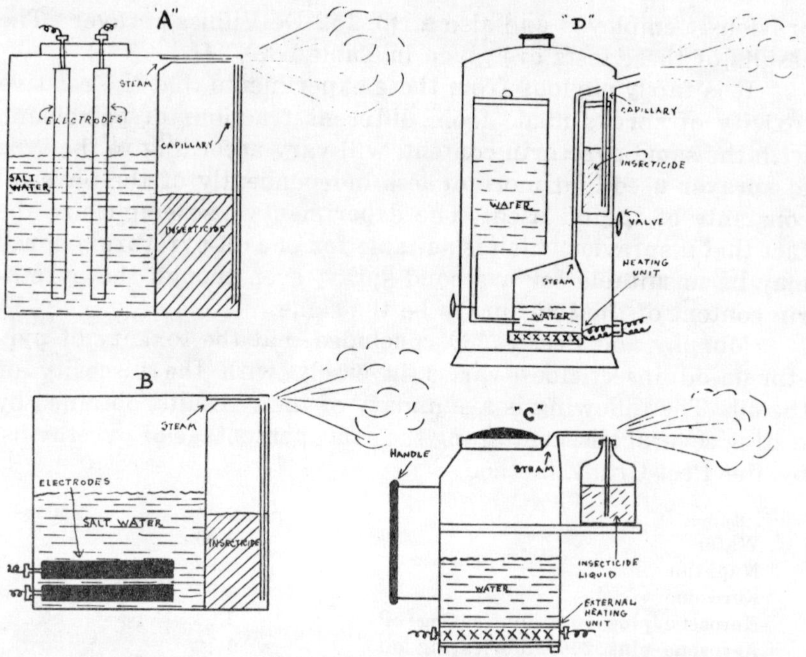

FIG. XII. TYPES OF STEAM INSECTICIDE VAPORIZERS. A, VERTICAL ELECTRODE TYPE. B, HORIZONTAL ELECTRODE TYPE. C, EXTERNAL HEATING UNIT TYPE. D, HEAVY-DUTY INDUSTRIAL TYPE. (FROM "SOAP").

the same time mixing the mist with steam and raising its temperature considerably. The steam pressure in the average steam vaporizer is stated to run from five to nine pounds, varying with the machine and conditions. In the small machines, the spray is thrown some five to six feet from the spray nozzle, and in the larger types from ten to fifteen feet.

"There are several different types of vaporizers being made today, some with external heating units and others with internal heating units in the form of carbon, bronze, or similar electrodes. The external units consist merely of a small electric stove outside and underneath the miniature boiler for generating the steam. The electrode type has two electrodes either vertically or horizontally immersed in an electrolyte, such as ordinary salt solution. The resistance to the passage of the electric current heats the water to boiling, generating steam which is forced from the nozzle and across the tip of the capillary by its own pressure. The use of alternating current is apparently best. The use of direct current is reported accompanied by a disintegration of the positive electrode, and electrolysis of the salt solution with the liberation of chlorine in small quantities mixed with the steam.

"The electrode type of heating unit has the advantage of automatically shutting itself off as soon as the water has boiled away to a point below the bottom of the electrodes or below the top electrodes where they are in a horizontal position. The external heating type requires a time clock or other control device to throw a switch. The electrode type, however, is slow in steam generation. Also, if too concentrated an electrolyte is used, a fuse may blow. In the smaller units, 24 to 32 ounces of water with a small pinch of salt are used as a charge in the steam generator, with three to eight ounces of insecticide liquid. In the larger units, two to three gallons of water are used with a gallon of insecticide.

STEAM SPRAYER FOR PYRETHRUM INSECTICIDES. (PHOTO (COURTESY OF THE TANGLEFOOT CO.)

"The insecticides themselves have been found to vary considerably. Pyrethrum extract forms the base of the liquids. Petroleum distillate (special kerosene) is the chief solvent, although ethylene dichloride, carbon tetrachloride, and others are also present in some of the liquids. In one case, chloroform was present in appreciable quantity. For ordinary use to which the atomizers are put, that is against roaches, moths and bed bugs, a concentrated pyrethrum extract representing five to eight pounds of pyrethrum flowers per gallon is used."

Burgess and Golley *(145)* state that one ounce of oil extract of pyrethrum containing one per cent of pyrethrins will effectively treat a room having a capacity of 10,000 cubic feet. The addition of ethylene dichloride, propylene dichloride or carbon tetrachloride to such an extract is said to increase the efficiency somewhat. The chlorinated hydrocarbons should be used cautiously, however, especially in occupied houses.

"Against flies and mosquitoes, a regular one pound to the gallon pyrethrum product does the work, but inasmuch as these atomizers are not ordinarily used against flies, the results do not have any material significance.

"The advantages of the steam vaporizer, as worked out in

actual use, are the apparent greater effectiveness and penetration of the insecticide at the higher temperature, and the ability to set the machines up and permit them to operate for some time without an attendant, a material saving in labor. In a Peet-Grady chamber, we are informed, they give a high kill and show evidence of considerable effectiveness. Among the disadvantages appear to be the difficulty with the small machines in securing and maintaining a sufficient, uniform concentration of insecticide mist, due to the fact that the average machine does not run long enough or produce sufficient volume. Others are the concentrating of the spray in a single spot or direction, causing accumulation of condensing steam and oil.

"For industrial uses, larger units are reported being developed which can fill a volume of several thousand cubic feet with mist within a few minutes. One type it is stated, will generate enough steam within one minute after hooking up with the current to enable it to throw a spray fifteen feet. This type uses a large unit with very small boiler and auxiliary water tank.

"Several sidelights of the steam vaporizer for use in the application of insecticides have appeared. They include its use for the application of disinfectants and deodorants where the higher temperature of a steam jet may have advantages. It is understood that these latter are receiving some experimental attention at this time. A parallel case of interest is the use of a special nozzle attached to a live steam line in plants, mills, and warehouses for larger scale extermination work against roaches, beetles, moths, weevils, etc. Reports state that this latter is now being tried out with some success under certain conditions."

In September, 1933, and July, 1935, patents were granted to Burgess and Golley *(145, 146)* for a method of destroying insects by steam atomization of pyrethrum extract. In February, 1935, a patent was granted to Breuer *(132)* for a steam sprayer for insecticides.

Perfume Materials and Other Chemicals

The first pyrethrum-oil household sprays were perfumed with methyl salicylate, citronella, safrol, cedar oil and similar materials. Strong perfumes were required to mask the objectionable odors of the oil bases then available. Oils which are nearly odorless can now be obtained and fly spray perfumes of the floral type are now available in almost endless variety and at reasonable cost. Obviously each manufacturer must determine the type of spray most suitable for his class of trade. Sprays made with the highly refined, odorless oil bases are excellent when faintly perfumed. The special oil bases are, however, considerably more expensive than ordinary refined kerosene and some classes of trade prefer a strong perfume, erroneously associating the strength of the odor with high toxicity. Recently compounds have been offered for deodorizing or neutralizing the odor of kerosene and other spray oils. The action of these compounds is purely physical, their odors masking those of the odoriferous materials in the oils. It should be remembered that any chemical compound which would destroy the odor of kerosene by chemical action would probably also destroy the pyrethrins.

The ideal perfume must have an agreeable odor and must cover the odor of the oil base while the spray is being used and after the spray is used; it should evaporate as soon as possible after the oil has evaporated so that the perfume will not linger in the room indefinitely.

A large number of odorous materials has been added to fly sprays, partly to serve as perfumes and partly to increase the toxicity of the sprays. In Table LXXXIII are given the toxicity tests, by the Peet-Grady method, on some of these materials. The various substances were dissolved in a mineral oil known to have a very low kill by the Peet-Grady method. In some cases it was necessary to add about 10 per cent of acetone to the oil in order to obtain a clear solution. Most of these substances have little or no toxicity of their own. Whether they have any effect toward increasing, preserving or destroying the toxicity of the pyrethrins has not been fully investigated and the evidence thus far obtained is not conclusive. The writer has seen a few cases where either the perfume or the oil base itself had apparently caused a loss of toxicity.

Color of Household Sprays

The color of household sprays has also been the subject of considerable discussion. The coloring matter of pyrethrum flowers is decomposed as the flowers age. It is much more readily oxidized than pyrethrins; for example, if a petroleum ether extract of pyrethrum be washed with 2 per cent potassium permanganate solution, most of the color is destroyed but the pyrethrins are scarcely attacked. A given lot of flowers will yield a yellower oil extract and a higher pyrethrin content when

TABLE LXXXIII. TOXICITY OF OIL SOLUTIONS OF ODOROUS MATERIALS TO FLIES (PEET-GRADY METHOD)

Substance	Concentration* %	Kill %	Substance	Concentration* %	Kill %
Acetophenone	5	5	Ethyl cinnamate	5	6
Aldehyde C 14	5	26	Ethyl oenanthate	5	14
Aldehyde C 16	4	15	Ethyl phenyl acetate	5	5
Aldehyde C 18	4	8	Eucalyptol	5	16
Aldehyde C 20	4	7	Eugenol	4	8
Amyl acetate	5	2	Geraniol	4	7
Amyl benzoate	5	16	Geranyl acetate	5	7
Amyl cinnamic aldehyde	5	25	Geranyl butyrate	5	5
Amyl formate	5	11	Hydroxy citronellal	5	11
Amyl furoate	5	11	Ionone	5	12
Amyl salicylate	5	5	Linalool	3	11
Amyl valerianate	5	5	Linalyl acetate	5	8
Anethole	4	17	Menthol	1	15
Anise aldehyde	4	25	Methyl anthranilate	5	6
Anisole	5	4	Methyl cinnamate	5	9
Anisyl alcohol	3	14	Methyl eugenol	5	10
Apiole	5	6	Methyl ionone	5	4
Benzyl acetate	4	4	Methyl iso-eugenol	5	6
Benzyl alcohol	4	7	Methyl salicylate	5	6
Benzyl benzoate	4	28	Naphthalene	5	3
Benzyl butyrate	5	5	Nitrobenzene	5	8
Benzyl cinnamate	5	13	Nonyl alcohol	5	2
Benzyl formate	5	9	Nonyl aldehyde	5	5
Benzyl phthalate	4	6	Oil of Almond	5	4
Benzyl propionate	5	5	Oil of Bay	4	35
Benzyl valerianate	5	7	Oil of Bergamot	5	24
Borneol	5	7	Oil of Birch tar	4	6
Bornyl acetate	5	4	Oil of Cade	4	9
Camphor	5	18	Oil of Cajeput	5	4
Carbon disulfide	2	8	Oil of Calamus	5	18
Carvone	5	30	Oil of Camphor, Jap., white	5	7
Cinnamic aldehyde	4	27	Oil of Camphor,		
Cinnamic alcohol	3	4	sassafrassy	5	11
Cinnamyl acetate	5	10	Oil of Cananga	5	5
Citral	4	20	Oil of Caraway	4	16
Citronellal	5	7	Oil of Cassia	4	30
Citronellyl butyrate	5	3	Oil of Cedar Leaf	4	16
Citronellyl formate	5	4	Oil of Chamomile, Ger.	5	20
Cresyl acetate, para	5	8	Oil of Chenopodium	5	45
Decyl alcohol	5	4	Oil of Citronella	4	23
Decyl aldehyde	5	4	Oil of Cloves	4	19
Dichlorbenzene, ortho	5	14	Oil of Copaiba	5	7
Dichlorbenzene, para	5	5	Oil of Coriander	4	7
Dimethyl anthranilate	5	4	Oil of Cubeb	5	4
Duodecyl alcohol	5	6	Oil of Cypress	5	5
Duodecyl aldehyde	5	7	Oil of Dill	5	12
Ethyl benzoate	5	20	Oil of Erigeron	4	4

TABLE LXXXIII (CONTINUED). TOXICITY OF OIL SOLUTIONS OF ODOROUS MATERIALS TO FLIES (PEET-GRADY METHOD)

Substance	Concentration* %	Kill %	Substance	Concentration* %	Kill %
Oil of Eucalyptus	4	8	Oil of Pinus pumilionis	5	5
Oil of Ginger, dist.	3	3	Oil of Pinus sylvestris	5	6
Oil of Hemlock	4	14	Oil of Rosemary	5	9
Oil of Hyssop	5	10	Oil of Santal	5	8
Oil of Juniper berries	4	6	Oil of Sassafras	4	11
Oil of Lavender	4	9	Oil of Savin	5	4
Oil of Lemon	5	8	Oil of Spruce	5	5
Oil of Lime	5	4	Oil of Tansy	5	6
Oil of Mustard, volatile	5	10	Oil of Thuja	5	7
Oil of Nutmeg	5	3	Oil of Wormwood, Am.	4	4
Oil of Olibanum	4	3	Phenyl ethyl alcohol	2	22
Oil of Opoponax	5	3	Phenyl isocyanate	2	6
Oil of Orange	5	8	Phenyl salicylate	1	19
Oil of Origanum	4	8	Piperonal	1	6
Oil of Parsley	5	27	Piperonyl alcohol	1	9
Oil of Patchouly	5	9	Pyridine	2	6
Oil of Pennyroyal	5	7	Rhodinol	5	4
Oil of Pepper	5	7	Rhodinyl butyrate	5	4
Oil of Peppermint	5	7	Safrol	5	14
Oil of Pettitgrain	5	3	Terpineol	5	6
Oil of Pine, D. D.	4	11	Terpinyl acetate	5	16
Oil of Pine, S. D.	4	19	Thymol	1	3
Oil of Pine cones	4	10	Turpentine	10	3
Oil of Pine needles, Siberian	5	4	Undecyl aldehyde	5	8
			Vanillin	1	6

*By weight for solids, by volume for liquids.

freshly harvested and dried than will be obtained from the same flowers one year later. On the other hand, closed, half-open and open flowers, harvested at the same time, will yield extracts whose color decreases in the order given, but the pyrethrin content will be greatest in the light colored extract from the open flowers. Hence it is impossible to judge the toxicity or pyrethrin content of an extract from its color.

The normal color of an oil extract of pyrethrum flowers is yellow. If the flowers contain added stems the extract will have a greenish tint. Concentrated pyrethrum extracts made as described in Chapter XI also yield a yellow solution when diluted with oil. Concentrated extracts made with acetone, alcohol and certain other solvents have a green tinge when diluted with oil. The color of pyrethrum extracts allowed to stand in contact with zinc or copper becomes greenish. The metal is attacked by the acids in the extract forming a precipitate. The acidity of the average household spray is equivalent to about 2.4 cc. of 0.1*N* acid per 100 cc. of spray.

Bake *(99)* cautions against the use of lead liners in caps used on cans containing pyrethrum insecticides. He found that lead, solder, zinc and copper are attacked by pyrethrum-oil sprays of the household type; nickel, tin, aluminum and iron were not attacked. He does not present data to prove that the pyrethrins are affected by the different metals.

Hoyer *(446)* reported that tin-plate and flux used by can manufacturers did not cause appreciable loss in pyrethrum-oil sprays during ten months' storage. Confectioners' glaze caused a loss of 41% in pyrethrin content under the same conditions. No decomposition of pyrethrins was found in sprays stored for two weeks in contact with iron, tin, zinc and aluminum, but losses of 13% and 8% were found with copper and lead; these losses were accompanied by the formation of a precipitate. No relation was found between the weight lost by the metals and loss of pyrethrins. Hoyer concluded that tin cans whose interiors are free from an excess of confectioners' glaze, flux, copper, lead or lead solder can be safely used for pyrethrum sprays.

For a description of experiments with antioxidants in concentrated extracts, see Chapter VIII. The use of antioxidants with pyrethrum powder is also discussed in Chapter VIII.

Addition of Rotenone to Household Sprays

Recently various manufacturers of derris extracts have recommended the addition of small amounts of rotenone to pyrethrum-oil sprays, implying that the toxicity of the spray will thereby be greatly increased. Gnadinger and Corl *(349)* conducted a series of experiments to determine the value of rotenone as a constituent of fly sprays of the pyrethrum-mineral oil type. Since pyrethrum flowers and pyrethrum extracts may deteriorate appreciably under certain conditions of storage, it was considered advisable to work with the isolated pyrethrins and pure rotenone rather than with extracts of pyrethrum and derris or cube.

The pyrethrins were isolated from Japanese flowers by a modification of the method of Staudinger and Ruzicka and were purified as previously described (page 124). The mixed pyrethrins consisted of 66.8 per cent pyrethrin I and 33.2 per cent pyrethrin II, calculated from the total pyrethrins determined by the copper reduction method and pyrethrin I determined by the acid method. The isolated pyrethrins were dissolved in petroleum ether, in which they were completely soluble, so that the concentration was 0.888 g. per 100 cc. This stock solution was kept in the dark in a completely filled, stoppered flask. Pure rotenone was obtained from the Bureau of Chemistry and Soils, U. S. Department of Agriculture, through the courtesy of Dr. R. C. Roark, Chief of the Insecticide Division; it melted at 163° C.

Solutions of the pyrethrins and rotenone in mineral oil were prepared and their toxicity to flies was determined by the Peet-Grady method. The mineral oil used was a highly refined Pennsylvania oil giving a kill of less than 6 per cent by the Peet-Grady method. Because of the low solubility of rotenone in the oil, 5 per cent of acetone was added to the solutions. Allowance was also made for the petroleum ether introduced when the stock solution of pyrethrins was diluted with the mineral oil. All of the solutions tested, therefore, contained 5 per cent acetone and 7.2 per cent petroleum ether, by volume. Neither acetone nor petroleum ether is toxic to flies in these concentrations. The pyrethrin and rotenone content of the solutions was so adjusted as to yield approximately 50 per cent kill by the Peet-Grady method, since comparisons of toxicity are best made at the 50 per cent mortality point. The solutions were all prepared at the same time, about three days before the first tests, and were stored in corked, amber glass bottles in the dark, when not in use. Each solution was tested on six cultures of flies. A description of the solutions tested and the kills obtained is given in Table LXXXIV.

The pyrethrin solution, No. 1, gave a much higher knock-down and a considerably greater kill than the rotenone solution of the same strength, No. 2. Replacement of half the pyrethrin content by rotenone lowered both knock-down and kill (No. 3). Solutions No. 4 and No. 5 had the same pyrethrin content as No. 1 and contained in addition ⅓ and ⅙ as much rotenone as pyrethrins, respectively. The kill obtained with No. 4 was slightly greater than with No. 5, which in turn was slightly more toxic than No. 1, that is, the addition of rotenone to the pyrethrin solution increased the toxicity somewhat. Reference to the work of Grady in collaboration with the writer (page 126) shows that the increase in kill, due to the additional rotenone, is not quite as great as would be caused by the addition of the same amount of pyrethrins. The sudden drop in toxicity obtained

TABLE LXXXIV. TOXICITY OF OIL SOLUTIONS OF PYRETHRINS AND ROTENONE TO FLIES (GNADINGER AND CORL, USING PEET-GRADY METHOD)

No.	Composition mg. per 100 cc.		Kill: Culture 9/26 %	Kill: Culture 9/27 %	Kill: Culture 9/28 %	Kill: Culture 9/29 %	Kill: Culture 9/30 %	Kill: Culture 10/3 %	Flies used	Av. knock-down %	Av. kill %
1	Pyrethrins	63.4	40	51	40	39	44	45	1000	94	46
				54	48	44		52			
2	Rotenone	63.4	39	28	22	33	25	7*	1006	76	30
				35	25	33		9*			
3	Pyrethrins	31.7	43	30	66	49	29	38	977	85	42
	Rotenone	31.7		43		49	37	40			
4	Pyrethrins	63.4	54	36	61	48	52	48	979	93	52
	Rotenone	21.1			66	54	55	48			
5	Pyrethrins	63.4	51	36	58	42	44	36	986	94	50
	Rotenone	10.6			67	63	48	52			

*Not included in averages.

with rotenone solution No. 2 on culture 10/3 could not be explained. The solution showed no signs of decomposition and the culture gave a normal kill with the other solutions. Two additional tests with solution No. 2 on a seventh culture 10/4 gave 7 per cent and 6 per cent kill. The experiments extended from September 26th to October 4th and some decomposition may have occurred in solution No. 2, although it was treated, in every respect, exactly like the other four.

It is apparent, therefore, that an oil solution of rotenone is considerably less toxic to flies than a pyrethrin solution of the same concentration. The addition of small amounts of rotenone to oil solutions of pyrethrins does not increase the toxicity as much as the addition of the same amount of pyrethrins.

Birdsall *(112)* has criticized the experiments of Gnadinger and Corl. Referring to the solutions described on page 217, he expresses the opinion that "There was little or no rotenone present in the solution at the time of spraying, or the rotenone present was in the process of crystallization dispersed in such large particles that it had no efficiency as a contact insecticide in the liquid spray."

Birdsall's opinion, which is not supported by experimental evidence, is incorrect. The solutions used by Gnadinger and Corl contained from 10.6 mg. to 63.4 mg. of rotenone per 100 cc., remained perfectly clear and showed no evidence of crystallization. Moreover, these solutions were not made by dissolving the rotenone in acetone and adding it to the mineral oil, as Birdsall appears to assume, but, as stated in the original paper *(349)*, the acetone was added last; no heat was used to dissolve the rotenone. Roark has stated that "rotenone dissolved in kerosene at the rate of about one gram per liter" makes a good fly spray. Other investigators of the Department of Agriculture have been unable to determine accurately the solubility of rotenone in kerosene but estimate it at "less than one-tenth of one per cent at 20°."

Birdsall *(112)* states that combinations of pyrethrins and rotenone dissolved in hydrocarbon oils, with the aid of certain solvents (not named), make efficient fly sprays. The concentrations of pyrethrins and rotenone used are not given.

Badertscher *(94)* compared the toxicity of three series of samples containing pyrethrins and rotenone; the average kills obtained, by the Peet-Grady method, are reproduced on page 220.

Unfortunately, the pyrethrin content of the sprays was unknown and Badertscher was unable, therefore, to draw definite

Badertscher's Tests	Sample A Kill %	Sample B Kill %	Sample C Kill %	Sample D Kill %
I. Commercial sprays, pyrethrin content unknown	81	83	54	71
II. Same sprays as I, except diluted with 50 per cent base-solvent	44	56	26	39
III. Same sprays as I, except diluted with 50 per cent base-solvent containing 0.1 per cent rotenone	85	87	74	80

conclusions regarding the relative toxicity of the pyrethrins and rotenone as fly spray ingredients.

Richardson *(717)* prepared a kerosene extract of derris using 23.8 g. of the powdered root to 100 cc. of oil; the rotenone content of the derris was 2.9 per cent, hence the extract was a saturated solution of rotenone. He also made a kerosene extract from a low grade pyrethrum powder containing 0.26 per cent pyrethrin I, using 11.9 g. of the pyrethrum to 100 cc. of oil. One part of the derris extract was mixed with 5 parts of the pyrethrum extract and the toxicity of this mixture was compared with that of a solution composed of 1 part of kerosene and 5 parts of the pyrethrum extract, using Richardson's method (page 105). The pyrethrum-derris mixture gave 59 per cent kill and the pyrethrum-kerosene mixture gave 46 per cent kill. Since both the total pyrethrin content and the rotenone content of the extracts were unknown, Richardson's experiment was of no value in determining the relative toxicity of the pyrethrins and rotenone in fly sprays.

A patent for an insecticidal solution comprising kerosene, pyrethrins and rotenone has been granted to Fulton *(288)*.

Campbell, Sullivan and Jones *(152)* found a rotenone solution in cyclohexanone-kerosene (1:9) "decidedly more effective" than a similar solution in acetone-kerosene (1:9). Whether or not the difference in toxicity could be accounted for by the profound effect which cyclohexanone has on flies is not stated.

Campbell, Sullivan and Jones did not report any direct comparison of toxicity tests of rotenone and pyrethrins, but made a number of comparisons between extracts of derris and pyrethrum *(152, 153)*. They summarize their results as follows:

"Kerosene extracts of derris seem to be superior to those of pyrethrum in several respects:

1. A given weight of derris root is likely to yield a larger volume of effective kerosene extract than the same weight

of pyrethrum flowers. Since the present cost per pound of the two ground plant materials is about the same, derris extracts should be more economical. Further experiments are needed to determine the volume of effective extract that can be obtained from a unit weight of ground derris root in relation to its rotenone content and percentage of total extractives.

2. Where it is possible to sweep up and destroy flies knocked down by a spray, kerosene extracts of pyrethrum would probably give better control than those of derris, but where flies are not, or cannot be, collected and destroyed, kerosene extracts of derris, because of their persistent effect, should give better control.

3. Extracts of derris prepared with a colorless and almost odorless kerosene are practically colorless and odorless, whereas those of pyrethrum are deep yellow and when sprayed have the characteristic but not unpleasant pyrethrum odor. Sprays of kerosene extracts of derris, like those of pyrethrum, have no unpleasant or harmful effects on the operator.

"Kerosene extracts of pyrethrum have certain advantages over those of derris:

1. Kerosene extracts of pyrethrum paralyze flies more rapidly and completely than do those of derris. The former thus give quicker relief and have a good psychological effect on users who expect immediate results.

2. In the kerosene extracts of derris used in this investigation precipitation of solids was noted when the extracts were kept in a refrigerator. The quantity of solids separating from chilled kerosene extracts of pyrethrum was apparently not so great."

Their conclusions were based largely on mortality counts made 48 to 72 hours after spraying. On the basis of counts made 24 hours after spraying, pyrethrum was much more effective than derris. Extracts of derris containing no rotenone were as effective as extracts of derris rich in rotenone. (See also page 266.)

Chemical Methods for Assaying Pyrethrum-oil Sprays

Vollmar *(935)* has described a method for determining the pyrethrin I in kerosene extracts. Unfortunately Vollmar assumes that the ratio of pyrethrin I to pyrethrin II in pyrethrum flowers is fairly constant, which is not the case (page 118). Hence the method is not very accurate as an index of the toxicity of pyrethrum-oil sprays and, of course, is not applicable to certain perfumed sprays. The following is a description of Vollmar's method:

"Solutions Required. — Sodium hydroxide in methanol, approximately 1*N*. Fiftieth normal sodium hydroxide, 0.02*N*. The strength of this solution should be checked occasionally. Sulfuric acid 1*N*. Petroleum ether, maximum boiling point 60°.

"Apparatus.—For refluxing the sample and subsequent distillation, use a long-necked flask of 100 cc. capacity.

"Procedure.—Measure 100 cc. of extract into a long-necked flask of the type described above. Add 5 cc. alcoholic soda solution and reflux for 1½ to 2 hours, the bulb of the flask being immersed in a beaker of hot water. Shake occasionally. Remove from the beaker and pour off as much kerosene as can be decanted without disturbing the alcohol layer into another similar flask which contains a second 5 cc. portion of alcoholic soda. Reflux as before, and when completed, pour off the kerosene layer, rejecting it. A few cubic centimeters of kerosene will remain in each flask. Add to the second flask 11 cc. 1*N* sulfuric acid, shake, and transfer to the first flask. Rinse the second flask with two 25-cc. portions of petroleum ether, and after transferring these washings to the first flask, confirm its acidity with phenolphthalein. Distil in a current of steam, using a long condenser cooled with ice water. Do not apply a flame to the distillation flask until all the petroleum ether has been distilled. Collect the petroleum ether and 50 cc. of water in a separatory funnel, then continue the distillation until an additional 50 cc. of distillate has been collected. Keep the volume low in the distillation flask, but do not allow it to go dry.

"The titration is carried out in an Erlenmeyer flask to which have been added about 20 cc. carbon dioxide-free water, a few cubic centimeters of ethyl alcohol, and 1 cc. of phenolphthalein indicator. Add to this enough 0.02*N* sodium hydroxide to give a definite pink color, usually one to two drops will be sufficient. Shake the contents of the separatory funnel vigorously and add the water layer to the second 50 cc. of distillate. Wash the petroleum ether layer with about 20 cc. of water, then add it to the titration flask. Titrate to a definite pink, shaking well and allowing the layers to separate after each addition. Extract the combined distillates with a second 50-cc. portion of petroleum ether, and add this to the titration flask. Continue the addition of alkali until the pink color is the same as it was originally. A blank determination should be run, using 100 cc. of kerosene, and the proper deduction should be made from each determination. This will usually be about 0.3 cc.

"Calculation.—From the molecular weight of 330, we get the relation, 1 cc. 0.02*N* caustic is equal to 0.0066 grams pyrethrin I."

Seil *(794)* has suggested a modification of his acid method (page 67) for assaying pyrethrum-oil sprays. The method is applicable to ordinary household sprays or to concentrated extracts, either perfumed or unperfumed. The accuracy of the method is greater for unperfumed sprays than for perfumed sprays. The method is also more satisfactory for concentrated extracts than for household sprays. All pyrethrum-oil sprays should be filtered before analyzing, to remove any altered pyrethrins present.

According to Seil, if the spray contains a perfume, which may consist of esters, that will interfere with the estimation of the pyrethrum acids, the perfume can be removed as follows: Place 100 cc. of the filtered household extract in a 500 cc. flask with 50 cc. of water. Heat the flask and distil with steam until the distillate has no perfume odor. Cool the contents of the flask and transfer to a separatory funnel; allow the liquids to separate, draw off the water into a second separatory funnel and wash with 25 cc. of kerosene. Transfer the oil in the first funnel to a 250 cc. Erlenmeyer flask. Wash the first funnel with the kerosene in the second, which is then added to the flask. Add 20 cc. of 0.5*N* ethyl alcoholic sodium hydroxide solution and reflux the mixture on a hot plate for 1 to 2 hours. Transfer to a 600 cc. beaker, washing the flask with water, finally adding enough water to make the aqueous layer measure 200 cc. Add a few beads and boil the solution until the aqueous layer is 150 cc. Cool the solution and transfer to a 500 cc. separatory funnel and separate the kerosene. Transfer the alkaline aqueous solution to a 250 cc. volumetric flask. Wash the kerosene once with 10 cc. of water, which is also added to the volumetric flask. Add one gram of filter-cel and an excess of 10 per cent barium chloride solution, usually 10 to 15 cc. are sufficient.

Make to the mark with water and mix again. After the precipitate has settled filter through a fluted paper, transfer 200 cc. of the clear filtrate to a 500 cc. Erlenmeyer flask, add 1 cc. of concentrated sulfuric acid and distil with steam, using an efficient trap and condenser. Continue the distillation until the liquid in the distilling flask is between 15 and 20 cc., collecting the distillate (usually about 250 cc.) in a 500 cc. Squibb separatory funnel. Allow the distilling flask, containing the dicarboxylic acid, to cool.

To the separatory funnel add 50 cc. of neutral petroleum ether, and shake thoroughly for one minute. After the liquids have separated, draw off the aqueous layer into a second 500 cc. separatory funnel to which a second 50 cc. of neutral petroleum ether has been added. Shake for one minute and after the liquids have separated, discard the aqueous layer. Wash the petroleum ether in the first separatory funnel with 10 cc. of water using the same wash water for the petroleum ether in the second funnel. Repeat with a second wash water of 10 cc. as before. Combine the petroleum ether extracts. Neutralize 15 cc. of water containing 1 drop of phenolphthalein indicator solution with 0.02*N* sodium hydroxide solution and add it to the combined petroleum ether solutions and titrate with 0.02*N* sodium hydroxide solution, shaking after each addition until the aqueous layer is just pink. Each cc. of 0.02*N* sodium hydroxide solution consumed is equal to .0066 g. of pyrethrin I.

Filter the solution containing the dicarboxylic acid through a Gooch crucible, washing the flask with a little water. Make the clear filtrate alkaline with bicarbonate of soda and transfer to a separatory funnel; wash it twice with chloroform. Wash the first chloroform extract with water, using the same wash water for the second chloroform extract. Combine the aqueous solutions, acidify strongly with hydrochloric acid, saturate with sodium chloride and extract with 50 cc. of ether, shaking for about 1 minute. Repeat the extractions with three more portions of ether, using 50 cc. for the second and 25 cc. each for the third and fourth extracts. Wash the ether of the first extraction with 10 cc. of water using the same wash water for the other ether extracts successively. Repeat with a second wash with 10 cc. of water as before. Combine the ether solutions, draw off any water separating and filter the ether into a flask. Distil the ether on a water bath and dry the residue at 100° for 10 minutes. Add 2 cc. of neutral alcohol, warm gently, then add 20 cc. of distilled water and heat to dissolve the acid. If a residue remains undissolved, cool and filter through a Gooch crucible. Add a drop of phenolphthalein indicator solution and titrate with 0.02*N* sodium hydroxide solution, of which 1 cc. is equivalent to .00374 g. pyrethrin II.

Seil's method gives low results when applied to perfumed

sprays, probably due to decomposition of pyrethrins during the steam distillation.

Ripert *(745)* has adapted his acid method (page 69) to the analysis of pyrethrum-oil sprays. He states that the method can be applied also to alcoholic pyrethrum extracts and pyrethrum-soap mixtures; he does not, however, provide any means for eliminating the oxidized or altered pyrethrins, which may be present in these alcoholic and soapy products, hence his method is inaccurate when applied to them. Ripert's procedure for the analysis of pyrethrum-oil sprays follows:

For household sprays: if the spray is perfumed, place 500 cc. in a 1 liter flask, add 200 cc. of water and distil off the water. Repeat the addition and distillation of the water, in 200 cc. portions as long as the product has an appreciable odor. Cool the contents of the flask. Treat the oil solution according to the following procedure, which is also used for non-perfumed household sprays. As in the Seil method, the steam distillation causes a considerable loss of pyrethrins.

Neutralize the pyrethrum-oil solution exactly, with normal aqueous potassium hydroxide solution; draw off the aqueous solution and discard. If an emulsion forms add a few cc. of barium chloride solution and saturated sodium chloride solution to break the emulsion. Discard the aqueous layer and filter the oily layer. Transfer the oil solution to a flask fitted with a reflux condenser, add 50 cc. of normal methyl alcoholic potassium hydroxide solution and boil for 1½ hours. Cool the contents of the flask and transfer to a separatory funnel; add 50 cc. of water, shake, allow to settle and draw off the aqueous layer. Wash the oily layer twice with 50 cc. portions of water. Combine the aqueous solutions and proceed with the determination of the pyrethrins as directed on page 70, line 1.

For concentrated pyrethrum-oil extracts: dissolve a volume of extract, equivalent to about 50 g. of flowers, in 100 cc. of ethyl ether and neutralize exactly with normal aqueous potassium hydroxide solution. Draw off the aqueous layer, wash it with 50 cc. of ethyl ether and combine the latter with the ether solution of the concentrate. Wash the combined ether solutions once with distilled water, transfer the ether solution to a flask and remove the ether by distillation. Saponify the oily solution remaining in the flask with 50 cc. of normal methyl alcoholic potassium hydroxide and proceed as above.

The comparative value of pyrethrum-oil household insecticides can also be determined by biological tests. The Peet-Grady method more nearly duplicates the actual conditions under which household insecticides are used than any other of the biological methods described in Chapter V. It can be used to distinguish considerable differences in the pyrethrum content of sprays, but it is by no means as accurate as generally supposed for comparing commercial sprays.

Notes on the Analysis of Pyrethrum Extracts

In applying Seil's method to concentrated extracts it is best to weigh the extract, rather than measure it. For kerosene extracts containing about 2.5 per cent pyrethrins, weigh 5 g. of the filtered extract on the analytical balance into a tared 150-cc. Erlenmeyer flask; add 15 cc. of 0.5*N* alcoholic sodium hydroxide and reflux the mixture for 1½ to 2 hours as above. Extracts which are very concentrated or which are made with other solvents than kerosene may contain altered pyrethrins. Unless these are removed the method gives high results.

Mineral oil extracts containing 10 to 15 per cent pyrethrins also contain a considerable proportion of fatty oils and acids from the flowers, in which altered pyrethrins may be dissolved. In such cases weigh accurately 1 to 2 g. of the extract into a tared 150-cc. Erlenmeyer flask, add 100 cc. of low boiling petroleum ether, cork the flask and allow the mixture to stand in a refrigerator over night. Then add 0.4 g. of Celite analytical filter aid (Johns-Manville), mix, cork the flask loosely and allow to stand for about 15 minutes with frequent shaking. Filter the solution through an 11 cm. No. 42 Whatman filter paper (or equivalent) into a 500 cc. Erlenmeyer flask and wash the cork, flask and filter thoroughly with petroleum ether. Add a few grains of sand to the petroleum ether solution and evaporate on a steam bath; the last traces of petroleum ether may be removed under vacuum. Saponify the residue in the flask with 15 cc. of 0.5*N* alcoholic sodium hydroxide and determine the chrysanthemum acids as above, by Seil's method.

The following analyses of special concentrated hydrocarbon extracts of pyrethrum show the pyrethrins found with and without the petroleum ether separation; the pyrethrin contents by the copper reduction method are also given (page 227). The copper reduction method is not applicable to ordinary kerosene extracts, but can be applied when the pyrethrin content is 8 to 15 per cent.

It has been shown that the Seil method gives high results when applied to pyrethrum flowers (pages 77, 78). When applied

Sample	Pyrethrins Seil method I	II	Total	Copper reduction method Total
	%	%	%	%
1. Not treated with petroleum ether	6.63	8.36	14.99	....
1a. Treated with petroleum ether	6.10	8.03	14.13	14.06
2. Not treated with petroleum ether	6.29	7.78	14.07	....
2a. Treated with petroleum ether	5.73	7.71	13.44	13.53
3. Treated with petroleum ether	5.49	7.17	12.66	12.35
4. Treated with petroleum ether	4.98	6.28	11.26	10.91

to certain types of concentrated pyrethrum extracts, however, the Seil method, modified as above, gives concordant results with the copper reduction method. This fact was discovered in the course of experiments on the production of highly concentrated extracts, made by the processes described in Chapter XI. Extracts containing from 10 to 15 per cent of pyrethrins were prepared, using decalin, kerosene and a mixture of kerosene-octyl alcohol as solvents (page 169). In some cases the flowers, from which the extracts were made, were assayed by both the Seil and copper reduction methods. The analyses are given in Table LXXXV.

TABLE LXXXV. ANALYSES OF FLOWERS AND EXTRACTS BY SEIL AND GNADINGER-CORL METHODS

No.	Pyrethrin content of flowers: Gnadinger method Total	Seil method I	II	Total	Pyrethrin content of concentrate made from same flowers: Gnadinger method Total	Seil method I	II	Total
	%	%	%	%	%	%	%	%
1.	0.92	0.42	0.66	1.08	14.4	6.2	8.2	14.4
2.	0.92	0.43	0.64	1.07	10.9	5.0	6.3	11.3
3.	...	...	...	...	11.5	5.5	6.7	12.2
4.	...	...	...	...	11.0	4.9	6.1	11.0
5.	...	...	...	...	9.8	4.3	5.6	9.9

In the first two experiments the pyrethrin content of the flowers was about 17 per cent higher by the Seil method, but the concentrated extracts made from these flowers showed the same percentage of pyrethrins by the two methods. This indicates that the interfering material, which causes the Seil method to yield high results, is removed in the manufacturing process; whether it is removed by precipitation or by vacuum distillation is not known.

The ratio of pyrethrin II to pyrethrin I is higher in the flowers than in the concentrated extracts. This indicates that the high results obtained on flowers by the Seil method are probably caused by substances that interfere with the pyrethrin II determination.

Gnadinger and Corl have also found that the copper reduction method, Seil acid method and Gnadinger-Corl acid method (page 73) give concordant results on highly concentrated extracts of this type, as the following analyses indicate:

No.	Copper reduction method	Pyrethrin content					
		Seil acid method			Gnadinger acid method		
	Total	I	II	Total	I	II	Total
	%	%	%	%	%	%	%
XII-1	11.0	4.9	6.1	11.0	4.8	5.8	10.6
XII-2	9.8	4.3	5.6	9.9	4.1	5.4	9.5

The procedure for applying the copper reduction method to highly concentrated extracts, containing from 8 to 15 per cent pyrethrins, is as follows:

Tare accurately, on the analytical balance, a 125 cc. Erlenmeyer flask. Weigh into the flask approximately 1 g. of the concentrated extract; this is conveniently done by means of a quick-draining pipette. The weight taken should be between 0.900 and 1.100 g. Add 100 cc. of petroleum ether (b.p. 20° to 60°), mix thoroughly, cork the flask and place in a refrigerator over night. Then add to the cold solution, 0.4 g. Celite analytical filter aid, mix, let stand for 15 minutes, loosely corked, and filter through a No. 42 Whatman filter-paper (or equivalent) into a 500 cc. Erlenmeyer flask, washing the cork, flask and filter thoroughly with petroleum ether. The filtrate should be perfectly clear. Add a few grains of sand and evaporate the petroleum ether on a steam bath, removing the last traces with a vacuum pump if desired, taking care not to heat the flask above 75°. From this point, proceed with the copper reduction method as directed on page 60, line 25.

Rogers and Calamari *(757)* have described methods for detecting and estimating rotenone in liquid insecticides, in the presence of pyrethrins.

Substitutes for Pyrethrum

The work of Staudinger and Ruzicka was undertaken with the object of synthesizing the active principles of pyrethrum. They concluded that the extremely complicated structure of the pyrethrins makes it unlikely that they can be synthesized on a commercial scale. This has led to many attempts to find synthetic substitutes for the pyrethrins. Among the compounds for which patents have been granted are dibutyl phthalate, terpinyl acetate, acetylated pine oil, oxalic acid esters, aliphatic thiocyanates, aromatic thiocyanates, benzophonone, and many others. None of these products has so far displaced pyrethrum to any great extent. One synthetic which was very toxic to flies was also found to be extremely toxic to chickens. Hence in using such a product the manufacturer should first assure himself that the consumer will suffer no injury.

Labelling Household Insecticides

The following opinions on the labelling of pyrethrum-oil household sprays are taken from a letter issued March 30, 1936, by the Food and Drug Administration, U. S. Department of Agriculture:

"Insecticidal uses. While these products are commonly referred to as "fly sprays" they are often recommended for use against a number of other household insects, as for example, mosquitoes, roaches (water bugs), bed bugs, ants and clothes moths. They are contact sprays, that is, in order to be effective they must be applied in such a manner as to hit the insects to be killed. Since the habits and life cycles of different insects vary the directions must in each case be adapted to the particular varieties of insects to be controlled, as for example:

"For flies and mosquitoes. The directions should provide for closing all doors and windows and thoroughly spraying the material in all parts of the room, particularly toward the ceiling, filling the room with a fine mist. The room should be left closed for 10 to 15 minutes, and the fallen insects then swept up and destroyed. This latter precaution is necessary because some of them will be only paralyzed and will later recover. Preparations of this type applied to the face and hands have some effect for a short period in repelling mosquitoes.

"For ants, roaches and bed bugs. The product should be sprayed thoroughly with force into all parts of the room, paying special attention to cracks and crevices and hitting as many of the insects as possible. For the control of bed bugs the bed, all tufts and seams in the mattress and all places in the room where the bugs may hide should be thoroughly sprayed. Directions should be given for repeating the treatment as often as may be necessary.

"For clothes moths. The directions should provide for first cleaning all articles to be protected followed by a thorough spraying, paying particular attention to the seams and folds. The interior of all containers should also be thoroughly sprayed. Unless the articles are to be stored immediately after treatment in moth-tight containers, directions should be given for repeating the sprayings at least once a month. Preparations of this type should not be recommended for use on upholstered furniture except where explicit directions are given for opening up the upholstery and heavily spraying or saturating the interior fabric, as well as the outside surfaces, and repeating the treatment when necessary.

"Unwarranted claims. These preparations cannot be relied upon to repel mosquitoes when used in the open such as on the porch or about camps. They are not effective against all household insects and claims of 'extermination,' the abbreviation 'etc.,' and statements such as 'all other insects,' 'all crawling insects' and 'all flying insects' are unwarranted and should not be made. These sprays cannot be relied upon to control any insect that cannot be reached by the spray. This applies also to the eggs, which are often placed where they are inaccessible.

"Products of this type are injurious under certain conditions to both man and animals. Therefore their labels must not bear such unqualified claims as 'non-poisonous,' 'non-injurious,' or 'harmless to man and animals.'

"They are of no value in disinfecting and will not prevent diseases.

"Inflammability. Kerosene is, of course, inflammable and fly sprays containing it should not be sprayed in the presence of open flames. A warning statement on the label is desirable.

"Deterioration. Mineral oil-pyrethrum sprays, if exposed too long to the light of the sun in ordinary glass bottles, may lose much of their efficiency due to the decomposition of the active ingredients. It has also been reported that deterioration may occur due to decomposition of the pyrethrins through contact with the solder or lining of the can when packed or stored for considerable periods of time in metal cans.

"Ingredient statement. The Insecticide Act requires that any preparation containing an inert ingredient must bear on its label a statement of the name and percentage amount of each and every such inert ingredient and the fact that they are inert, or, in lieu of this, a statement of the name and percentage amount of each and every active ingredient and the total percentage of the inert ingredients. Preparations of the type described above ordinarily consist entirely of active ingredients and no statement regarding the ingredients is required to appear on the label. However, such claims as 'Active Ingredients 100%' and 'Contains no inert ingredients' are unobjectionable provided, of course, they are true. The statement '100% active' is objectionable as it may be taken to imply that the product is 100 per cent effective." (See also page 240.)

Pyrethrum-oil sprays can be used where food is handled or stored, without danger to the consumer. In flour mills and food warehouses it is best to use a spray made with one of the odorless oil bases, adding little or no perfume. Pyrethrum sprays made with carbon tetrachloride, ethylene dichloride and similar solvents have also been employed in warehouses where food is stored but they should be used with caution.

In many states it is necessary to declare on the label the net contents of package insecticides. The name and address of the manufacturer should appear on the label, or if the goods are marketed by someone other than the manufacturer the words "packed for", "packed by", or "distributed by" should precede the name and address of the packer or distributor. Several states require the registration of insecticides sold within their boundaries.

CHAPTER XIII

PYRETHRUM LIVESTOCK SPRAYS

The flies which cause annoyance and injury to livestock can be conveniently considered in two groups; those which suck the blood and those whose larvae feed on the flesh of the animal. The most important of the blood sucking flies is the "stable fly", *Stomoxys calcitrans* L. It is frequently confused with the common house fly, *Musca domestica* L., which it resembles. The house fly does not bite, but the stable fly has mouth parts which are adapted for piercing the skin and sucking the blood and its bite produces acute pain. It attacks nearly all warm blooded animals. The injury inflicted on live stock by the stable fly is due to the annoyance caused by its bite and to loss of blood. During severe outbreaks of the stable fly, animals are greatly reduced in flesh and are even killed. A severe outbreak in 1912 reduced the milk yield, in dairying sections, from 40 to 60 per cent. The horn fly, *Haematobia irritans* L., another blood sucking species, is somewhat smaller than the stable fly and usually attacks only cattle. Stock is also tormented by various species of mosquitoes, especially on the sea coast or in swampy areas. There are many other blood sucking species of flies and gnats less important than those mentioned. Most of these blood suckers are carriers of disease germs, which are dangerous to man or livestock. The transmission of malarial fever and yellow fever by mosquitoes and the spread of sleeping-sickness by the tsetse fly of Africa are well known examples.

The blow flies deposit their eggs in scratches or wounds on the skin of the animal. The screw-worm fly, *Cochliomyia macellaria* Fab., is said to be the most destructive blow fly in this country. Less important are the black blow fly, *Phormia regina* Meig., the green bottle flies, *Lucilia sericata* Meig., and *L. cuprina* Meig., and numerous other species. The heel fly or warble fly, *Hypoderma lineatum* and *Hypoderma bovis,* attacks cattle, depositing its eggs on the hair of the legs. When the eggs hatch the larvae burrow into the skin and spend the next seven or eight months in various parts of the cow's body. The grubs then work their way to the skin of the back, which they puncture, forming inflamed, swollen lumps or warbles. This fly is said to cause damage to the extent of $50,000,000 annually. The house fly does not attack men or animals, but is dangerous as a carrier of typhoid and other diseases. It is especially undesirable in dairy barns.

Nelson *(634)* has pointed out that in New Jersey alone there are at least 155 families of insects partially or completely parasitic on domestic animals; these insects are the following:

Order	Families
Mallophaga (bird lice)	13
Parasitica (lice)	7
Hemiptera (bedbug)	1
Siphonaptera (fleas)	3
Diptera:	
Chironomidae (midges)	4
Culicidae (mosquitoes)	38
Simuliidae (gnats)	5
Tabanidae (horse flies and gad flies)	75
Oestridae (bots and warbles)	5
Muscidae (house fly type)	3
Pupipara (sheep tick)	1

SPECIFICATIONS

Fly sprays for protecting livestock have been in use for more than thirty years. The earlier sprays were designed to act as repellents and the search for a material that would keep flies off of animals for long periods has occupied the time of many investigators. Until about 1926 the best of the sprays on the market were mixtures of petroleum oils with varying amounts of coal-tar creosote oil, naphthalene, paradichlorbenzene, pine oil, pine tar oil and other similar materials. Such sprays contaminated milk and stained the animals. Moreover, their repellent action was rather limited, though in many cases extravagant claims were made, and their killing action was very slight. It is only within the last few years that the use of pyrethrum in livestock sprays has led to the development of really efficient killing sprays.

The ideal livestock spray should meet the following specifications:

1. It should have little or no physiological effect on the animal.
2. It should kill flies as efficiently as a good household spray.
3. It should repel flies for 5 to 8 hours.
4. It should not stain the animals.

5. Its odor should be as inoffensive as possible to prevent milk contamination.
6. It must be inexpensive.

Physiological Effects of Sprays on Stock—Type of Oil

The physiological effect of the spray is due almost entirely to the kind of oil base used, since pyrethrum has no action on the animal. When the pyrethrum stock sprays were first introduced they were made by mixing a household kerosene-pyrethrum spray with a heavy petroleum oil. This was necessary because pyrethrum flowers cannot be readily extracted by percolating with the heavier oils. The resulting mixture had the disadvantage of kerosene, which blisters the hide when applied alone, as well as those of the heavy oil. Standardized concentrated pyrethrum extracts are now available which make it possible to prepare solutions of uniform pyrethrum content in the heaviest oils, without introducing more than five per cent of kerosene.

Freeborn, Regan and Folger *(278, 279)* have investigated the physiological action of different sprays on dairy cows. Their earlier experiments were of questionable value because of the excessive amounts of spray applied, 250 cc. per cow. Later experiments were conducted with 20 to 80 cc. of spray per animal. They found that "The characteristics of the oil are extremely important from the standpoint of the welfare of the cows. An oil having a viscosity of 38 seconds (S. U. 100° F.) was toxic even though the unsulfonated residue was nearly 100 per cent. On the other hand, an oil with a viscosity of 70 seconds was toxic in connection with a 90 per cent unsulfonated residue and safe with a 100 per cent unsulfonated residue. With a 100 per cent unsulfonated residue it is safe to go as high as 105 seconds viscosity, providing excessive amounts are not used. In other words, an oil with a viscosity below 40 seconds is apt to burn the cow and one above 65 seconds may also burn if the unsulfonated residue is much below 90 per cent. The natural inference is that the oil of the higher viscosity holds the unsaturates on the body of the animal long enough to produce toxic effects, while one of a lower viscosity, such as 50 seconds, allows for the volatilization of the unsaturates before serious damage is produced.

"In this connection it must be borne in mind that a wide blend made of a very light and a very heavy oil may give a viscosity reading within the safe range (40-65) but with the rapid volatilization of the lighter fraction, the residual oil left on the cow may have a much higher viscosity than the original blend, in which case any percentage of unsaturates might prove dangerous. After all is said, however, the matter of application itself becomes of extreme importance. With a careful, uniform application where the spray does not wet the animal's skin but hangs in the hair, a mixture that violates all the precepts mentioned above can be applied without injury."

Freeborn and Regan *(276)* also investigated the effect of sprays on the body temperature of cows. The cows were held in a psychrometric chamber at a temperature of 26.7° (80° F.) and 68 per cent humidity; under these conditions their normal temperature was from 38.3° (101° F.) to 38.9° (102° F.). Raising the temperature of the chamber to 29.4° (85° F.) increased the body temperature of the cows to 39.6° (103.2° F.) and also increased respiration and decreased milk yield. Lowering the chamber temperature again to 26.7° (80° F.) lowered the body temperature and respiration to normal. The cows were then sprayed, every day for 10 days, with a blended oil base containing pyrethrum and pine oil. Their temperature rose to 39.3° (102.8° F.), respiration increased and milk yield decreased. The cows were allowed to rest at 26.7° (80° F.) until body temperature, respiration and milk yield were normal. They were then sprayed with a straight cut oil, (boiling range 288°-349° (550°-660° F.), viscosity 54 seconds, unsulfonated residue 93 per cent) containing 12 ounces of pyrethrum to the gallon. Their body temperature rose slowly to a maximum of 39.1° (102.3° F.) or 0.17° above normal. From these experiments it is evident that the effect of these sprays on body temperature, respiration and milk yield was not greater than an increase from 26.7° to 29.4° (80° to 85° F.) in the temperature of the air in the test chamber. Freeborn and Regan advanced the tentative assumption that oil sprays exercise an effect on the skin of the cows that interferes with the cooling function of the skin.

Freeborn, Regan and Berry *(277)*, in later experiments, found that spraying cows with oil caused a loss of 46 per cent in the cooling power produced by moisture evaporation from the skin. They suggest that the deleterious action of oils may be due to chemical combination with the body tissues. Believing that the oil was the principal injurious factor in the spray, they developed two formulas for effective fly sprays which do not cause a rise in the body temperature of the animals but do cause a loss in transpired moisture. These formulas were:

No. 1:

Petroleum oil (unsulfonatable residue 90; viscosity 97)	84 cc.
Pyrethrum extract (kerosene extract; 2% pyrethrins)	48 cc.
Pine oil, steam distilled	48 cc.
Triethanolamine oleate	16 g.
Water	100 cc.

Dilute one part of this stock emulsion with four and one third parts of water, for spraying.

No. 2:

Petroleum oil (as above)	50 cc.
Pyrethrum extract (as above)	50 cc.
Pine oil, steam distilled	50 cc.
Diglycol oleate	29 cc.

Dilute slowly with eight parts of water, agitating vigorously.

Regan and Freeborn *(708)* confirmed the repellent action of pine oil. They also found that the loss in milk production caused by house flies and horn flies was negligible, but stable flies caused a milk loss of about 10 per cent. Certain petroleum oil sprays caused a milk loss of 22 per cent. Oils having a viscosity lower than 40 seconds were dangerous regardless of the amount of unsulfonated residue. Oils containing less than 90 per cent of unsulfonated residue were dangerous when viscosity was higher than 65 seconds. Water emulsions of pyrethrum and pine oil with a little petroleum oil were as efficient as repellents as petroleum oil sprays and were less injurious to the cows.

Freeborn does not describe the type of crude petroleum from which his oils were obtained. It should be remembered that the viscosity and gravity of an oil of given distillation range are determined largely by the type of crude from which the oil is distilled. Oils of all types should be compared for their physiological action on cows before any general conclusions are drawn. The writer has known of one cattle spray on the market more than 20 years, which has caused little or no complaint because of burning. The base oil used is a "gas oil" of 35 gravity. The base oil should be as light as can be used without burning the animals.

Wilson, Pearson and Cannon *(972)* determined the effect on cows of twelve petroleum oils, of different viscosities and degrees of refinement, containing added chemical repellents (not named). The sprays were applied once a day, at the rate of 60 cc. per cow. Five of the twelve oils caused visible skin injury about 35 days after spraying started. Injury was shown, in some cases, by unusually large amounts of loose hair, with some bare spots (about 1 inch in diameter) where the hair had completely fallen out. In other cases, the skin was reddened and roughened. Injury was most evident on the portions of the body exposed to the sun. The remaining seven oils caused no visible skin injury at the end of two months. There was a slight increase in the body temperatures of the sprayed cows.

The effect of flies and fly sprays on cattle has also been investigated by Melvin *(611)* whose conclusions were: "All of the petroleum oil sprays studied caused a measurable rise in body temperature and respiratory rate of both heifers and producing cows under certain weather conditions. Both air temperature and intensity of the sun influence the rise in body temperature of oil sprayed animals. The elevation of body temperature due to oil spray is not marked until the air temperature exceeds 26.7° to 29.4° (80° to 85° F.). Oil sprays cause a greater rise in body temperature of producing cows than of heifers. In oil sprayed animals exposed to direct rays of the sun there was a greater rise in the body temperature of the dark colored animals than in the light colored ones. House flies did not affect the body temperature of heifers or producing cows. Stable flies caused a measurable rise in body temperature and respiratory rate of both heifers and producing cows under certain weather conditions."

Nelson *(634)* found that severe infestations of stable flies not only reduced the total milk production but also reduced the butter fat content of the milk. He concluded that "with the proper type of spray, properly applied during the height of the fly season, it was possible to maintain a normal milk flow considerably higher than that obtained when the cows were left to the mercies of the biting flies and also possible to maintain a higher butter fat content as well."

Repellent Action

The repellent action of a fly spray varies with different species of flies. Freeborn found horn flies much easier to repel than stable flies and the investigators of the Department of Agriculture observed differences in the responses of the various blow flies and the house fly to repellents. Repellent action also varies with temperature and humidity, factors which influence the activity and vitality of flies, as well as the volatility of the repellent. Repellent action is more prolonged indoors than when animals are exposed to wind and sun. The best repellents are pyrethrum, pine tar oil, pine oil, either steam-distilled or destructively distilled, pine tar, crude turpentine, wood naphtha, clove oil, terpinyl acetate and diethyl phthalate, the last two being patented *(535, 665, 666)*. Pyrethrum, in addition to its repellent action, is by far the best of these materials for killing flies, the pine products having very little toxicity except when the concentration is high.

Moore *(622)* determined the repellent action of a number of substances to house flies. He confirmed the superior repellent properties of the pyrethrins and dialkyl phthalates and ranked the other materials in the following order of decreasing effectiveness: santalyl acetate, santalol, terpinyl acetate, eugenol acetate, eugenol, terpineol, menthol, pinene hydrochloride, oil of cedar wood.

Mail *(574)* and Cleveland *(181)* found that compounds and esters of salicylic acid are valuable as repellents.

Freeborn and Regan *(276)* concluded that pine oil adds to the effectiveness, as a repellent, of a pyrethrum-oil spray. In their tests they used 2 to 4 per cent of pine oil. Repellent action is very difficult to measure and this has led to claims for effectiveness over long periods of time. According to the Department of Agriculture, "A claim for a period of effectiveness for over eight hours is open to question" *(46)*. When conditions are favorable for the flies the period of protection will be much shorter.

The following method of measuring repellent action is recommended by Pearson, Wilson and Richardson *(673)*: "Thirty-five selected cows were used in determining the relative efficiency of 6 fly sprays. The cows were scrubbed with soap and water, then staked individually in a pasture, being removed only for watering and milking. The normal fly susceptibility of each cow was obtained from the average of two counts of the number of flies present, made hourly from 7 A. M. till 3 P. M. for a period of three days. The cows were then placed in 7 groups of 5 each, the maximum difference in the number of flies per cow per count between the groups being about two. The 6 spray materials were then assigned by chance to each of 6 groups of cows; the remaining group served as the control. Each cow was sprayed at 6 A. M. daily for 4 consecutive days with 2 fluid ounces of spray material applied with an electric sprayer. The fly susceptibility of the sprayed and control cows was determined as before. The results show that close individual observations of a relatively few cows of known fly susceptibility give more consistent and dependable results than less accurate observations on a large number of cows. The population of stable flies on unsprayed dairy cattle increases from 7 A. M. till about mid-day after which it tends to become stationary."

Pearson *(671)* has described the following improved method of determining repellence:

"The most satisfactory procedure discovered consists of hourly observations for eight consecutive days on cows staked individually. On the day preceding a series of tests, all of the test cows are thoroughly washed. The first four days constitute a preliminary period, during which all of the cows are sprayed with a base oil alone at 6:00 A. M. daily, and hourly fly counts are made from 7:00 A. M. to 4:00 P. M. From the results of these fly counts the cows are then placed in groups of five each, on a basis of their individual fly susceptibility. The average number of flies per cow per count for each group should not vary more than about two. During the following four days, the cows are sprayed daily at 6:00 A. M. with the same base oil, in which has been incorporated the repellent ingredient or ingredients under test. Fly counts are made as before. The preliminary period of four days gave more consistent results than one of shorter duration, but a longer period was not necessary in order to obtain reproducible results. By this method, various materials may be tested for relative repellence without the necessity of allowing for the repellence of the base oil, since it is used throughout an entire series."

Doty *(234)* and an anonymous writer *(79)* have described other methods of measuring the repellent action of fly sprays.

Killing Power

The killing power of a pyrethrum oil spray is due almost entirely to the pyrethrum. Many think that the petroleum oil used has considerable killing power but this is not the case. Kerosene has a very slight toxicity to flies, under the conditions of actual use, and the heavier oils used in cattle sprays are even less toxic than kerosene. The better cattle sprays contain from 12 to 16 ounces of high test pyrethrum per gallon of spray. Since extraction of pyrethrum flowers by the heavy oils used in cattle sprays is not easily accomplished, most manufacturers use a standardized, concentrated pyrethrum extract which mixes easily with the heavier oils and permits the production of a spray of uniform toxicity. Five per cent of pine oil may be added to increase repellent action and to mask the odor of the base oil. Safrol, citronella, oil of camphor sassafrassy, and other aromatics may be used instead of pine oil. Such a spray will kill and repel flies and can be used on the animals and also to kill flies in the barn, especially before milking time.

If the base oil is properly selected a spray of this kind will have an amber color and will not stain the animals. If the base oil is too heavy the killing power of the pyrethrum will be considerably reduced. A spray containing one pound of high test flowers to the gallon may give a kill below 20 per cent by the Peet-Grady method if a heavy paraffin oil is used. Pyrethrum-oil sprays will kill blowflies and blood sucking flies as well as fleas and lice. Extracts of pyrethrum in kerosene, alcohol or carbon tetrachloride, containing 2 pounds of flowers per gallon, have been found effective against cattle grubs or ox warbles when injected into the holes with an oil can. An ointment composed of one part of pyrethrum powder and two parts of petrolatum has also been used for this purpose. Experiments are being conducted with different pyrethrum washes.

Dibble *(226)* recommends the use of pyrethrum for winter control of lice and ticks on cattle and sheep. He employs a mixture of one part powdered pyrethrum and three parts of flour.

Pearson *(672)* reports that pine oil in pyrethrum cattle sprays increases ("activates") the toxicity of pyrethrum extracts as well as the knockdown. High grade pine oils are more effective than those of low grade. He also states that pine oil retards the decomposition of oil-pyrethrum extracts by sunlight and increases the repellence to flies. Sprays containing as much as 25 per cent of pine oil caused no skin injury or other ill effects to cattle.

Doty *(234)* was unable to confirm all of Pearson's conclusions.

Labelling Livestock Sprays

The following comments on labelling livestock sprays are the official opinions of the U. S. Department of Agriculture *(46)*:

"Fly sprays for animals are subject to the provisions of the Federal Insecticide Act if shipped in interstate commerce, exported from or imported into the United States, or sold in the District of Columbia or any territory of the United States. This act is explicit in the requirement that the labelling of any articles within its purview shall not be false or misleading in any particular. The responsibility under the law for all labelling statements rests with the manufacturer or shipper.

"Sprays for killing flies. In order to destroy flies with a spray the material must be atomized in a fine mist so that the flies will be enveloped in it. There is no known substance that, when applied to animals, will kill flies lighting on the treated animals. Furthermore, it cannot be truthfully claimed that the use of even a killing fly spray will effectively control flies in stables unless the breeding places of the flies are also treated. Therefore, any claims for the control of flies in stables or similar places should include directions for use of the spray so as to envelop the insects in the mist and for the treatment or removal of their breeding places.

"Sprays as fly repellents. The value of fly sprays as repellents is limited. Experiments conducted by the Bureau of Entomology and Plant Quarantine of this Department have shown that certain materials applied as sprays to animals have temporary repellent properties against stable and horn flies, but none have been found that are effective against other species of flies, such as horse flies, bot flies, grub flies and warble flies which so commonly attack and annoy livestock. Repellency claims should therefore be limited to those for the particular varieties of flies against which the product is known to be effective. Claims for repellent action against 'flies' without qualification constitute misbranding under the act.

"Preparations for the screw worm fly. Fly sprays that consist essentially of mineral oil will not repel screw worm flies. Pine-tar oil (of a specific gravity 1.065 and 1.085) applied over wounds, cuts, scratches or abrasions of animals will prevent screw worm flies from depositing eggs therein so long as the covering is maintained. Benzol applied in wounds which have already become infested with screw worms is effective in killing the worms. While coal tar phenols and other caustic or irritating substances may kill screw worms in wounds, they should not be used for this purpose since they damage healthy tissue and thereby create a favorable environment for further screw worm infestation or may even cause general poisoning of the treated animal.

"Poisonous properties of fly sprays. Oils in general are injurious to animals and, particularly if heavily applied, may do more damage than good. Therefore, the time, frequency and manner of application, as well as amount of product used, should be specified.

"Unqualified statements to the effect that oils are 'safe,' 'non-poisonous,' 'harmless,' and 'will not injure the hair or hide of animals' are not generally true and should not be made.

"Tainting the milk and effect on milk production. An unqualified assertion that a given product will not taint the milk is always questionable because this can be truthfully said of few, if any, fly sprays which are on the market, unless they are very carefully used. The directions should include a warning to spray long enough before milking to allow the spray to settle, to keep the spray away from milk or milk utensils; and to wash the udder and teats with warm water and soap before milking.

"No statement which would lead a purchaser to believe that the use of a preparation will increase milk production or butterfat above the normal yield should be made. The fact that individual cases may be cited where increases were apparent does not justify the general conclusion that similar results will be obtained regardless of conditions. In fact, it has been shown that under some conditions milk production has decreased following treatments with oil sprays.

"Disinfectant claims. Mineral oils are not disinfectants and oils in general are not recognized as satisfactory disinfectants. Therefore, claims for disinfectant value of fly sprays should not be made unless the type of oil and method of application are such that disinfection will be accomplished under all conditions for which it is so recommended.

"Disease prevention claims. Representations that fly sprays will prevent insect-borne diseases—as, for example, 'Will drive away flies and thus prevent infectious diseases'—are objectionable and unwarranted because they imply a promise of benefit in preventing the spread of disease which fly sprays cannot be relied upon to fulfill.

"Ingredient statement. The Insecticide Act requires that any preparation containing an inert ingredient must bear on its label a statement of the name and percentage amount of each and every such inert ingredient and the fact that they are inert, or, in lieu of this, a statement of the name and percentage amount of each and every active ingredient and the total percentage of inert ingredients. If the preparation consists entirely of active ingredients, no statement regarding the ingredients is required. However, such claims as 'Active Ingredients 100%'

or 'Contains No Inert Ingredients' are unobjectionable, provided, of course, the product is composed entirely of active ingredients. The statement '100% Active' is objectionable as it may be taken to imply that a spray is 100 per cent effective." (See also page 229.)

CHAPTER XIV

PYRETHRUM HORTICULTURAL POWDERS, DUSTS AND SPRAYS

Pyrethrum was first used as a horticultural insecticide in the form of a fine powder, applied either dry or suspended in water. Emulsified kerosene extract of pyrethrum was recommended for use on plants as early as 1890. Alcoholic extract of pyrethrum was employed on plants about 1895 and pyrethrum-soap preparations were used in the vineyards of Europe many years ago. Therefore, the increasing use of pyrethrum for protecting plants is not due to new types of products but rather to improvements in old forms. These improvements became possible with the isolation and identification of the active principles in 1924 and the development of quantitative methods for determining the pyrethrins in 1929. The standardization of the pyrethrin content of horticultural pyrethrum insecticides by means of the chemical assay methods has been the greatest improvement effected in these products. It should be kept in mind that many of the entomological investigations conducted with pyrethrum prior to 1930 are of questionable value because the pyrethrin content of the materials used was unknown and because experiments were made without taking the proper precautions to prevent decomposition of the pyrethrins.

POWDERS AND DUSTS*

Finely powdered pyrethrum is an excellent horticultural insecticide, although today it has been replaced to a large extent by pyrethrum sprays. An outstanding example of the successful use of powdered pyrethrum is in the control of the celery leaf-tier *(859)*. Stone and his associates found that pyrethrum applied with dusting machines gave about 90 per cent control. A fifty per cent mixture of powdered pyrethrum with lime or with sulphur was just as effective as pure pyrethrum. Two applications were made at intervals of 30 minutes at the rate of 20 to 25 pounds of powder per acre. Pyrethrum powder was effective against the larvae but failed to kill the eggs.

*The term *powder* is applied to finely ground pyrethrum without admixture. The term *dust* is applied to mixtures of pyrethrum or its active principles, with diatomaceous earth, talc, gypsum, and similar materials.

Pyrethrum dusts have also been successfully used against different insects. These dusts are prepared by extracting pyrethrum flowers with various solvents and then mixing the solution with diatomaceous earth, talc, bentonite, powdered charcoal and similar materials. Exhausted pyrethrum has been used as a filler or diluent for pyrethrum powder and as an

AEROPLANE APPLYING PYRETHRUM DUST TO BEANS.
(COURTESY B. G. THOMPSON, DEPT. OF ENTOMOLOGY, OREGON AGRICULTURAL COLLEGE).

absorbent in the preparation of pyrethrum dusts. Pyrethrum dusts are conveniently prepared by mixing the concentrated extracts of pyrethrum in alcohol, acetone or kerosene, which are commercially available, with diatomaceous earth or talc. Solutions of pyrethrum in steam distilled pine oil, safrol, other essential oils and pyridine have also been employed for making dusts. The mixing can be done in a ball mill or in a barrel mixer containing pebbles or steel balls. Mixers of the ordinary screw agitator type are not so effective as the ball or pebble mills because the material has a tendency to lump until thoroughly mixed.

Tattersfield has investigated the loss of toxicity of dusts prepared by extracting pyrethrum flowers with petroleum ether which was then mixed with talc and stirred until the solvent evaporated (page 156).

Mote and Thompson*, Oregon Agricultural College, have experimented for several years with dusts prepared from pyreth-

* Unpublished report.

rum extract and diatomaceous earth or bentonite. They found that a dust containing 8 per cent of kerosene extract or alcohol extract of pyrethrum (the extract having a pyrethrin content of 2 g. of pyrethrins per 100 cc.) was effective against the Western spotted cucumber beetle, *Diabrotica soror*, the Syneta beetle, *Syneta albida*, and certain lepidopterous larvae, but it was not effective against the pea aphid.

A pyrethrum dust consisting of one part of powdered pyrethrum with nine parts of gypsum is said to be effective against the false blossom leaf hopper when applied to cranberries at the rate of 100 pounds to the acre. Pyrethrum dusts have been successfully used on a large scale for controlling various species of insects in the forests of Germany. For this purpose the dust, prepared with talc, is applied by means of aeroplanes *(57)*. The dust is said to contain one per cent of pyrethrum oleoresin containing 13 per cent pyrethrins and is applied at the rate of 50 to 100 kg. per hectare (45 to 90 pounds per acre).

De Long *(220)* has found that a dust composed of 10 per cent pyrethrum and 90 per cent sulfur is especially effective because the pyrethrum immediately kills the insects present while the sulfur kills insects hatching or appearing later.

Evans *(250)* obtained control of *Thrips imaginis* Bagnall with a sulfur-pyrethrum dust.

Pyrethrum dusts have been found effective by Hervey and Palm *(427)* for controlling the cauliflower worm.

A mixture of 3 parts of powdered pyrethrum and 1 part naphthalene has been recommended in Germany for dusting. Naphthalene-pyrethrum dust has also been used by MacLeod and Maughan *(568)* for control of onion thrips.

Badertscher *(96)* has patented a method for making horticultural dusts which includes mixing an inert powder with the pyrethrins dissolved in a volatile solvent, while agitating and heating the mixture to volatilize the solvent.

The different types of pyrethrum dusts may be classified as follows:

I. Powdered pyrethrum flowers—give very satisfactory results but the cost is prohibitive because a large part of the pyrethrin content is held within cell walls and does not come in contact with the insects. Pyrethrin content 0.4 to 1.0 per cent.

II. Powdered pyrethrum mixed with diluent. Examples: 50 per cent pyrethrum with 50 per cent lime; 10 per cent pyrethrum with 90 per cent sulfur. Mixtures of powdered pyrethrum flowers and powdered pyrethrum stems are also used. Cheaper than type I, but not so effective and otherwise open to same objections as type I. Pyrethrin content 0.1 to 0.5 per cent.

III. Pyrethrum extract mixed with suitable diluent or filler. Pyrethrins are dissolved in a solvent which is then mixed with filler, thus coating each particle of filler with pyrethrins; the result is to make most of the pyrethrin content available for contact with insects, hence a lower pyrethrin content can be used than with type I dusts with equally satisfactory results and lower cost. Examples:

(a) 8 per cent of kerosene extract of pyrethrum (containing 2½% pyrethrins) mixed with 92 per cent diatomaceous earth.

(b) 8 per cent alcohol extract of pyrethrum (containing 2½% pyrethrins) mixed with 92 per cent diatomaceous earth.

(c) pyrethrins extracted with a volatile solvent which is added to filler while warming and stirring to drive off volatile solvent (Badertscher patent *(96)* and Tattersfield experiments, page 156).

Example (a) is the most effective of this type because kerosene is not very volatile and the pyrethrins remain dissolved in it and are thus protected from decomposition; this is not the case with (b), in which the alcohol soon evaporates, leaving the pyrethrins unprotected; this is also true of (c). The usual pyrethrin content of Type III dusts is 0.1 to 0.5 per cent.

IV. Oleoresin of pyrethrum mixed with suitable filler. Not very satisfactory because of difficulty of mixing and because decomposition of pyrethrins is rapid.

V. An exceptionally highly concentrated pyrethrum extract in a non-volatile solvent is mixed with filler of high absorptive capacity to yield a product, which is not a dust (because it is too moist for dusting) containing 2 to 10 per cent pyrethrins. This product can be shipped to the point where it is to be used, at which point it is diluted with any suitable filler in the proportion of 10 to 3 pounds of concentrate to 90 or 97 pounds of filler. Suitable oil can be added to the mixture to facilitate mixing. A dust of this type,* freshly prepared with a suitable antioxidant, contained 5.43 per

* Patent pending.

cent pyrethrins. After storage at 35° C. for 180 days the pyrethrin content was found to be 5.60 per cent.

A dust of this kind, diluted with talc so as to contain 0.2 per cent pyrethrins, was more effective on cabbage loopers than powdered pyrethrum flowers containing 0.45 per cent pyrethrins.

Pyrethrum dusts must be labelled to show the percentage of active and inert ingredients. The active ingredients are the pyrethrins; the inert ingredients are the diatomaceous earth, or talc and the extractive matter, other than pyrethrins, obtained from the pyrethrum flowers. The solvent used for the pyrethrum may be active in some cases (kerosene) and inert in others (acetone, alcohol). Unfortunately the Food and Drug Administration, of the Department of Agriculture, has ruled that the petroleum ether extract of pyrethrum is to be considered active. This is an erroneous and purely arbitrary ruling, which has resulted in some products being labelled to show a higher active principle content than they contain. For a further discussion of labelling powdered pyrethrum see page 180.

Pyrethrum-Soap Sprays

One of the oldest types of pyrethrum sprays is that made from oleoresin of pyrethrum emulsified with soap. The oleoresin of pyrethrum may be made by extraction with alcohol, acetone, ethylene dichloride, naptha, or other solvent by a process similar to that described in Chapter XI. Before the nature of the pyrethrins was understood, excessive amounts of alkali were used in making the soaps employed for the emulsification, and this resulted in the saponification of the active esters, the pyrethrins. In recent years, the stability of this type of spray has been improved by avoiding excessive free alkali in the soap. Nevertheless there has been a considerable difference of opinion in regard to the keeping qualities of soap-pyrethrum sprays. It is generally agreed that the saponification products of the pyrethrins are insecticidally inert, but the rate at which saponification takes place has not been accurately determined. Roark *(749)* has condemned the use of pyrethrum with soap on purely theoretical grounds. Tutin *(920)* also stated that pyrethrum should not be used with soap or other alkaline emulsifier. Walker *(939)* found that oleoresin of pyrethrum in combination with a coconut oil soap with a pH of 9.5 retained its strength for at least a month, while a similar mixture with a pH of 12.0 lost practically all of its toxicity within 12 hours. Further, the toxic properties of oleoresin stored alone, and stored in combination with coconut oil soap for 11 months were not appreciably different.

Badertscher *(93)* concluded, from tests on honey bees and aphids, that concentrated mixtures of oleoresin and soap did not materially decrease in toxicity during storage for three years. Tattersfield and Hobson *(879)* found that emulsions of the pyrethrins in alkaline spray fluids of varying pH "proved more permanent than usually supposed."

It seems reasonably certain therefore that concentrated mixtures of soap and pyrethrum oleoresin retain their strength fairly well although there is probably a gradual slight decrease in toxicity. In dilute mixtures, however, the loss is fairly rapid. Hence when the concentrate is diluted for spraying it should be used immediately.

One way of avoiding the possible decomposition of the pyrethrins by soap is to dissolve the oleoresin of pyrethrum in acetone or alcohol and omit the soap entirely. Solutions of the pyrethrins in acetone or ethyl alcohol are very stable if protected from the light. Staudinger and Harder *(840)* have pointed out that there is some risk that methanol or ethyl alcohol may replace pyrethrolone in the pyrethrin molecule, thus destroying the toxicity, but Tattersfield and Hobson *(879)* were unable to detect any decrease in the toxicity of ethyl alcohol extracts kept for several years.

The alcohol or acetone extracts of pyrethrum mix readily with water and form excellent, stable emulsions when diluted with water and soap can be added to the dilute solution just before spraying. This permits greater flexibility in adjusting the soap content with relation to the pyrethrin content. For example, in using a certain soap-oleoresin mixture against aphids, a dilution of 1 to 800 will be employed, while if the same spray is used against the Japanese beetle, a 1 to 100 dilution must be used. If the soap content is such as to be correct for the 1 to 800 dilution it may be too strong at the 1 to 100 dilution. On the other hand, if the soap content is correct for the 1 to 100 dilution, it may be too weak at the 1 to 800 dilution. Hence it is impossible to adjust the soap content of the soap-oleoresin mixture so that it will be at the optimum concentration for all dilutions. With the alcohol or acetone extracts this is not the case, since the soap is added independently of the pyrethrum extract and is correctly adjusted for the dilution employed.

OTHER "SPREADERS"

The purpose of using soap in pyrethrum sprays is not merely to emulsify the pyrethrins. The activity of the pyrethrins is greatly increased by the use of soap or some other suitable wetting agent. These "wetting agents" or "spreaders" lower the surface tension of the spray and otherwise change its physical properties, resulting in more efficient coverage of the insect and increased tracheal penetration. The New Hampshire Experiment Station, Durham, N. H., has published a series of studies of "wetting agents" by O'Kane and his associates.

Among the materials other than ordinary soap, which have been used as spreaders, are: ammonium caseinate, blood albumin, calcium caseinate, coconut-oil soap, dextrin, emulsified oil, fish-oil soap, flour, glue, skim milk, sodium oleate, sulfonated castor oil, sulfonated oleic acid, sulfonated oxidized petroleum oil, sulfonated petroleum oil, sulfonated sperm oil, triethanolamine abietate, triethanolamine oleate, and various waxes.

Within the last five years a large number of wetting out agents has been developed for the textile industry. Some of these seem to be promising for use with insecticides. They have the distinct advantage of not being affected by the calcium and magnesium salts in hard waters. Most of these wetting out agents are prepared by treating hydrocarbons, alcohols, fatty acid esters, or fatty acid amides with sulfuric acid and neutralizing the resulting product.

The products obtained by sulfating and neutralizing lauryl, myristyl and oleyl alcohols are known as Gardinols. A brief description of the various types of these sulfated compounds has been given by Kritchevsky *(531)*.

Ginsburg *(329)* has investigated the toxicities of several of the new wetting agents. He found Gardinol (sodium lauryl sulfate) more effective than Igepon (sulfated oleic acid). Arescap (sulfated butyl hydroxy diphenyl) and Aresket (sulfated butyl diphenyl) (Monsanto) were also better wetting agents than Igepon, and had the additional advantage of being soluble in acetone, alcohol, and other organic solvents, as well as in water. Gardinol, Arescap and Aresket had considerable toxicity to aphids and mosquito larvae. Ginsburg did not determine the effect of these spreaders on the toxicity of pyrethrum.

Ginsburg *(326)* has investigated the action of waxes emulsified with triethanolamine oleate as carriers for pyrethrum. The waxes tested included bay, bee, candelilla, carnauba, Chinese, Japan, montan, myrtle, spermaceti and various paraffins. A kerosene extract of pyrethrum was mixed with the previously melted wax, which was then stirred into an aqueous solution of triethanolamine oleate, held at 90° to prevent solidification of the wax. The final emulsion contained 25 per cent of wax, 5 per cent triethanolamine oleate and the active material from 6 per cent of pyrethrum flowers, equivalent to 0.05 per cent of pyrethrins. This concentrated solution was diluted with 25 parts of water for spraying. The best results were obtained with spermaceti and two paraffin waxes melting at 42° and 55°, respectively. These waxes slightly increased the toxicity of pyrethrum; there was no injury to the plants.

A formula for a pyrethrum-soap spray, typical of those formerly used, is given by Van Leeuwen *(929)* as follows:

Oleoresin of pyrethrum flowers	9.5 ounces
Oleic acid U. S. P.	5 pounds
Sodium hydroxide C. P.	11.2 ounces
Water	2.5 gallons

This mixture, when diluted with 47.5 gallons of water, was recommended for controlling Japanese beetle.

A more recent soap-pyrethrum formula is the following given by Badertscher *(50)*:

Oleoresin of pyrethrum	40%
Coconut oil-potash soap	15%
Sulfonated oleic acid	15%
Alcohol	5%
Water	25%

According to Badertscher, 1 pound of the oleoresin used in this formula represents 5 pounds of flowers. Reference to Table LXXVI indicates that the oleoresin was probably made by alcoholic extraction.

Other formulas for horticultural sprays have been described by Badertscher *(95)* and Grant *(371)*.

Newer Type of Pyrethrum-Oil-Soap Spray

The ideal soap-pyrethrum horticultural insecticide should have the following properties:

1. It should contain a high concentration of pyrethrum, on which the toxicity principally depends.

2. It should contain the greatest possible amount of soap so that sufficient wetting and spreading is obtained, when the product is used at high dilutions.
3. It should contain the minimum amount of water to prevent hydrolysis of the soap and saponification of the pyrethrins, which destroys the toxicity to insects.
4. The product should be a homogeneous liquid readily miscible with cold water. Solid soap-pyrethrum insecticides are not commercially practicable because they dissolve with great difficulty in cold water.
5. The pyrethrins should be dissolved in an oil, preferably one having marked ovicidal and insecticidal properties.
6. Heat should not be employed in making the product after the pyrethrum extract has been mixed with the soap. Of course, the use of excess alkali in making the soap should be avoided. Potassium soaps should be used in preference to sodium soaps.
7. It should not injure plants.

The writer has investigated a large number of solvents and emulsifiers in order to produce a product meeting these specifications. Certain oils such as steam distilled pine oil, destructively distilled pine oil, safrol, oil of camphor sassafrassy, and coal tar creosote oil are good solvents for oleoresin of pyrethrum. However, the amount of soap which can be incorporated with these different solvents varies greatly. Moreover, the type of product obtained when these solvents (or extracts of pyrethrum made with them) are mixed with soap or other emulsifier varies markedly. For example,—petroleum oils and turpentine do not form homogeneous mixtures with potassium oleate but the oils are emulsified, forming opaque mixtures which gradually separate on prolonged standing.

There are certain oily liquids which, when mixed with soap, become readily soluble in water. The best example is cresol, which is almost insoluble in water, but which becomes easily soluble when mixed with soap, as in the liquor cresolis compositus of the U. S. Pharmacopoeia. Other examples are coal tar creosote oil and pine oil, which form milky emulsions when the degree of dilution is great. Soap emulsions of these oils are widely used as disinfectants.

Pine oil, coal tar creosote oil, oil of camphor and safrol form homogeneous mixtures with potassium oleate. These mixtures are clear and have the appearance of true solutions and do not separate on standing. The amount of soap which can be incorporated is greatest for pine oil and decreases in the order given.

The amount of soap which can be incorporated also depends on the type of soap. Rosin soap is the least soluble of the soaps investigated; potassium oleate is intermediate in solubility while potassium-castor oil soap is the most soluble.

As a result of this investigation an insecticide of standardized, high pyrethrin content has been developed containing less than 14 per cent water, more than 44 per cent of soap and 30 per cent of toxic oils.*

Effectiveness of Pyrethrum Against Various Insects

The alcohol and acetone pyrethrum sprays contain the active principles from 5 to 60 pounds of flowers to the gallon. Sprays of this type, of uniform, standardized pyrethrin content, are easily produced and are commercially available. The value of any pyrethrum spray depends primarily on the pyrethrin content.

Spraying with pyrethrum-oil mixture for Syneta beetle (Oregon).

Pyrethrum sprays will kill most insects. Some, such as mealy bugs, red spiders, cyclamen mites, scales and a few others are very resistant to pyrethrum. Many insects are readily killed by pyrethrum sprays in laboratory experiments but cannot be controlled under commercial conditions because of the difficulty of getting the spray in contact with the insect,—for example the codling moth. The following chart (Table LXXXVI) shows some of the commoner insects that can be controlled with pyrethrum. This chart is applicable to an alcoholic or acetone extract of pyrethrum containing 0.90 g. of

* Patent applied for.

pyrethrins per 100 cc. The proper dilutions for an extract of different pyrethrin content can be readily calculated.

It will be noted that pyrethrum kills both chewing and sucking insects. Some insecticides, such as lead arsenate, are

TABLE LXXXVI. SHOWING PROPER CONCENTRATIONS OF PYRETHRUM EXTRACT FOR VARIOUS INSECTS (EXTRACT CONTAINING 0.90 G. OF PYRETHRINS PER 100 CC.)

Crop	Insect	Dilution	Amt. of soap* per gal of spray
Apple	Aphis	1:800 or 1:400	1½ oz.
	Canker worm	1:200	1½ oz.
Asparagus	Asparagus flea beetle	1:400	1 oz.
Bean	Aphis	1:800 or 1:400	1½ oz.
	Mexican bean beetle	1:400	1½ oz.
Cabbage, cauliflower and Brussels sprout	White fly	1:200	1½ oz.
	Cabbage aphis	1:800 or 1:400	1½ oz.
	Cabbage worm	1:300	1½ oz.
Celery	Leaf tier	1:400	1 oz.
Cranberry	Cranberry fireworm	1:800 or 1:400	1½ oz.
Cucumber	Aphis	1:800 or 1:400	1½ oz.
	Striped beetle	1:100	1½ oz.
Currant and gooseberry	Currant worm	1:300 or 1:200	1½ oz.
Egg plant	Aphis	1:800 or 1:400	1½ oz.
	Flea beetle	1:400	1½ oz.
General foliage	Japanese beetle	1:100	2 oz.
	Spittle bug	1:250	1½ oz.
	Aphis	1:800 or 1:400	1½ oz.
Grape	Aphis	1:800 or 1:400	1½ oz.
	Leaf hopper	1:400	1½ oz.
	Leaf roller	1:300 or 1:200	1½ oz.
Grass (golf greens)	Sod web worm	1:400	none
Lettuce	Aphis	1:800	1 oz.
Mushroom	Maggot	1:400	none
Pear	Slug	1:200	1½ oz.
Pea	Aphis	1:400 or 1:200	1½ oz.
Pepper	Pepper weevil	1:200	1½ oz.
Potato	Potato aphis	1:400	1½ oz.
	Flea beetle	1:200	1½ oz.
	Colorado beetle	1:100	1½ oz.
Prune	Tent caterpillar	1:200	1½ oz.
Squash	Aphis	1:800 or 1:400	1½ oz.
	Stink bug	1:200	1½ oz.
Strawberry	Leaf beetle	1:200	1½ oz.
Tomato	Cutworm	1:100	none
	Hornworm	1:200	1½ oz.
	Aphis	1:800 or 1:400	1½ oz.
Watermelon	Aphis	1:800 or 1:400	1½ oz.
	Watermelon beetle	1:100	1½ oz.

TABLE LXXXVI (CONTINUED). SHOWING PROPER CONCENTRATIONS OF PYRETHRUM EXTRACT FOR VARIOUS INSECTS (EXTRACT CONTAINING 0.90 G. OF PYRETHRINS PER 100 CC.)

Crop	Insect	Dilution	Amt. of soap* per gal. of spray
Aster	Aphis	1:800 or 1:400	1 oz.
	Aster beetle	1:100	1 oz.
Carnation	Aphis	1:800 or 1:400	1 oz.
	Cutworm	1:100	none
Chrysanthemum	Aphis (green fly)	1:800 or 1:400	1½ oz.
	Leaf tier	1:300	1 oz.
	Cutworm	1:100	none
Cineraria	Leaf tier	1:300	1 oz.
Chinese lantern	Striped cucumber beetle	1:100	1½ oz.
Geranium	White fly	1:300	1 oz.
	Leaf tier	1:300	1 oz.
Lily and calla	Aphis	1:800 or 1:400	1 oz.
Nasturtium	Black aphis	1:400	1½ oz.
Sweat pea	Aphis	1:800 or 1:400	lt. suds
	Cutworm	1:100	none
Rose	Aphis	1:800 or 1:400	1 oz.
	Leaf roller	1:300	1 oz.
	Rose slug	1:200	1 oz.
	Rose chafer	1:100	1½ oz.
Violet	Aphis	1:400	lt. suds

*This chart is based on the use of coconut oil-potash soap containing 38 to 40 per cent soap solids.

toxic only to insects which eat the foliage or fruit and are not toxic to sucking insects. Such insecticides are commonly called stomach poisons to distinguish them from insecticides, such as pyrethrum, which kill on coming in contact with insects, even if not taken into the stomach.

A coconut oil-potash liquid soap containing 40 per cent of soap solids makes a convenient and satisfactory spreader. However, any good grade of laundry soap or soap flakes, not containing excessive free alkali, can be used. Solid soaps should be used at half the concentrations given in the chart for liquid soaps. If too much soap is used the plant will be injured or even killed; if too little is used the kill of insects will not be satisfactory. As in all kinds of spraying, the type of sprayer and pressure at which the spray is applied are important; the higher the pressure the better the kill. For this type of insecticide a coarse driving spray is better than a fine mist or fog. Care should be taken to hit as many insects as possible since pyrethrum is primarily a contact spray. Spraying should be started when insects first appear and the application should be repeated as often as necessary. Sprays containing soap should not be used with lead or calcium arsenate because a soluble arsenate is formed which will burn the plant. Pyrethrum sprays should not be used with lime-sulphur, sodium-sulphur, barium-sulphur, or hydrated lime because the alkali may decompose the pyrethrins. The acetone or alcohol extracts, without soap, may be combined with lead arsenate, neutral Bordeaux and oil emulsion sprays.

Unlike nicotine, the effectiveness of pyrethrum sprays against most insects is not greatly influenced by atmospheric temperature changes. Fleming *(271)*, however, states that Japanese beetles are much more susceptible to pyrethrum sprays in warm, sunny weather than in cool, cloudy weather. Hartzell and Wilcoxon *(406)*, using rose chafer adults, found that both the forces involved in causing death and in bringing about recovery are accelerated by an increase in temperature; if the insects have received a sublethal dose, recovery is more rapid, but if the dose is lethal, death occurs more rapidly at the higher temperature.

In cool or cloudy weather plants may be sprayed at any time. When the weather is hot and sunny they should be sprayed early in the morning or late in the afternoon, not during the heat of mid-day. Injury to plants is most likely to occur when the temperature is high and the humidity is low.

Pyrethrum-oil Horticultural Sprays

Oil extracts of pyrethrum have been successfully used in several ways as horticultural sprays. A kerosene extract of pyrethrum emulsified with soap is recommended by Haegele *(381)* for controlling *Mineola scitulella* Hulst, a prune worm which has become an important pest in Idaho, causing a loss of 50 per cent of the crop in some orchards. The formula recommended by Haegele is as follows:

Concentrated kerosene extract of pyrethrum..........	14	quarts
Kerosene (dormant oil may be substituted for kerosene)	28	gallons
Neutral soap..	7½	pounds
Water...	15	gallons

The pyrethrum extract contains the pyrethrins from 20 pounds of flowers, having a pyrethrin content of 0.90 per cent, in each gallon, that is, 2.15 g. of pyrethrins per 100 cc.

The soap is dissolved in the water. The pyrethrum extract, dissolved in the kerosene, is then slowly added to the soap solution while mixing. This strong emulsion is diluted for spraying, using 5 gal. to 95 gal. of water containing an additional 3½ pounds of soap. The spray is liberally applied with 250 pound pressure at the rate of about 5 gal. per tree. Trees are sprayed in the spring when the overwintering worms have left their cocoons; in Idaho this is between April 1st and 15th. About 90 per cent control is obtained without injury to the trees. The experiments extended through three seasons.

Better kills are obtained with dormant oil than with kerosene. The pyrethrum extract is first dissolved in the dormant oil which is then emulsified with calcium caseinate or, by the tank-mix method, with albumin.

Oil extracts of pyrethrum emulsified with various sulfonated oils have also been used. A formula of this type recommended by the makers of "Penetrol" is:

Concentrated kerosene extract of pyrethrum	1 gallon
Sulfonated oxidized petroleum oil (Penetrol)	6 gallons

The pyrethrum extract is the same strength as that used by Haegele, that is, 2.15 g. pyrethrins per 100 cc. The above formula is diluted with water for spraying; a dilution of 1 to 50

HIGH PRESSURE SPRAYER USED IN CITRUS GROVES (CALIFORNIA).

ANOTHER TYPE OF HIGH PRESSURE SPRAYER.

is said to be effective against Japanese beetles, while a 1 to 200 dilution is used for aphids and less resistant insects.

Pyrethrum has also been used in combination with the heavier mineral oils. For this purpose the concentrated extract in kerosene may be added to the regular dormant or summer spray oils, or concentrated extracts in the heavier oils can be made by the process described in Chapter XI. In either case, the dormant or summer oils containing the pyrethrum can then be emulsified with ammonium caseinate, calcium caseinate, glue, blood albumin, or other emulsifier and used in the regular way. Pyrethrum extract can also be used with proprietary emulsified oils, provided they are not strongly alkaline, by simply adding the pyrethrum extract to the spray tank containing the diluted proprietary oil and mixing well before spraying.

Pyrethrum has been used experimentaliy with summer oil sprays for codling moth control as part of the Northwest Cooperative Oil Spray Project, the results being reported by Spuler *(832)* in 1931.

According to Spuler the pyrethrum sprays alone had larvicidal properties but no ovicidal value. The oil-pyrethrum sprays were both larvicides and ovicides and, moreover, were effective in controlling other pests such as aphids and red spiders. Oil-pyrethrum spray was as effective as lead arsenate or oil-barium fluosilicate spray, when used in the last two cover sprays for codling moth control, and greatly reduced the red spider damage. The cost of the oil-pyrethrum spray was double that of lead arsenate alone but less than that of the oil-fluosilicate spray. The pyrethrum extract used by Spuler contained 3.8 grams of pyrethrins per 100 cc. and was used at a dilution of 1 to 2600. The finished spray contained 1 per cent of 70 viscosity mineral oil which was emulsified with ammonium caseinate.

Webster *(948)* did not obtain favorable results with oil-pyrethrum spray, nor with oil-pyrethrum-rotenone spray, in codling moth control experiments in 1932. Webster's spray contained 0.8 per cent of mineral oil emulsified with ammonium caseinate, in some experiments, and with blood albumin in others. The pyrethrum extract which he used contained 4.3 grams of pyrethrins per 100 cc. and was employed at a dilution of 1:3200.

Pyrethrins in oil have also been used in attempts to increase the toxicity of oil sprays to the various citrus scales but the results so far have not been very favorable. The use of pyrethrum with dormant and summer spray oils has been limited because of the cost. This has led to the use of too low concentrations of pyrethrins, with disappointing results. An extract of pyrethrum such as that used by Haegele, containing 2.15 g. of pyrethrins per 100 cc., should be used at dilutions of 1 to 200 or 1 to 600 depending on the type of insect.

Experiments by Bouquet and Thompson, of Oregon Experiment Station, have indicated that the squash bug, *Anasa tristis*, can be controlled by means of kerosene extract of pyrethrum diluted 1 to 200 with water and applied by means of a pressure sprayer. The kerosene extract of pyrethrum contained 2 per cent pyrethrins.

When oil extracts of pyrethrum are emulsified, as in the

foregoing examples, the pyrethrins probably remain in the oil phase, since they are very soluble in oil and extremely insoluble in water.

An interesting development in the use of pyrethrum-oil sprays is described by Lamiman *(539)* and also by Parker *(662, 663)*. The formula which Lamiman recommends is:

Highly refined kerosene	87½ gallons
Neutral oil (90 per cent unsulfonatable and 60 viscosity)	10 gallons
Kerosene extract of pyrethrum (2 g. pyrethrins per 100 cc.)	2½ gallons

This spray has been used successfully for controlling grape leaf hoppers in California. No emulsifying agent nor water is used. The spray is vaporized with a paint-gun type of nozzle at the rate of 2 to 5 gal. per acre, making it very economical to use. About 90 per cent control is obtained with no injury to the vines.

Parker *(664)* uses a patented sprayer which not only atomizes the spray material, but dilutes it with an air blast, which drives it upon the plants. Pyrethrum-oil spray has been used in this way for the control of the beet leaf hopper. The material was not sprayed on the sugar beets, but was applied to the hibernating and breeding grounds of the leaf hoppers. A similar solution has given satisfactory control of prune thrips.

Le Pelley *(547, 548, 549)* has used a kerosene extract of pyrethrum containing about 100 mg. of pyrethrins per 100 cc. for the control of *Antestia orbitalis* Westw. and *Lygus simonyi* Reut. on coffee trees in Kenya. The tree is covered with a light cotton cloth and is sprayed with 20 cc. of the extract, using a small hand sprayer. After half an hour nearly all of the bugs on the tree drop to the ground where they are again sprayed with about 3 cc. of the extract to insure satisfactory kill. The average kill of *Antestia* was 95 per cent, for *Lygus* nearly 100 per cent. There was no damage to foliage nor decrease in yield where the proper grade of kerosene was used. The spray was more effective when applied at the rate of 4 gallons per acre than an aqueous pyrethrum spray used at the rate of 1,360 gallons per acre. The cost of applying the oil spray was one-eighth the cost of the aqueous spray.

Notley *(647)*, however, has found that aqueous sprays are more satisfactory in wetter districts. He obtained good control of *Antestia faceta* with a spray containing 1 pound of pyrethrum powder, 1 gallon of kerosene and 0.5 pound of neutral soap in 300 gallons of water.

SPRAYING CALIFORNIA VINEYARDS WITH OIL EXTRACT OF PYRETHRUM FOR LEAF-HOPPERS. (COURTESY J. F. LAMIMAN, DEPT. OF ENTOMOLOGY, UNIVERSITY OF CALIFORNIA).

Ginsburg *(332)* investigated the action of a highly refined kerosene oil, containing pyrethrum, applied to a large number of greenhouse plants infested with various insects. Fog-like sprays caused the least injury to the plants. Very high kills were obtained on red spider, thrips, mealy bug, scale, aphids, white fly and tent caterpillars. Control was better when pyrethrum was used with the oil than when the oil was used alone.

Efforts are now being made to develop oils suitable for application in this way to plants which are very susceptible to oil injury. Research along this line offers a possible solution to the difficult problem of finding a substitute for the dangerous arsenic and fluorine compounds, whose use on food crops has been greatly restricted.

Repellent Action of Pyrethrum

It is well known that pyrethrum has a definite repellent action to leaf eating insects. Cory *(195)* noted that tent caterpillars refused to eat foliage which had been sprayed with soap-pyrethrum and air dried.

Gnadinger and Corl investigated the repellent action of pyrethrum sprays to the Colorado potato beetle, *Leptinotarsa decemlineata* Say. Potatoes were planted in 8 inch pots to facilitate handling. When the plants reached a suitable size they were sprayed and placed in a well ventilated box or frame, fitted with hinged sash, so that the plants could be exposed to direct sunlight or protected from the weather, as required. The frame was kept open during the day, except during rainy weather, and was closed at night. Five series of experiments were made. The plants were sprayed with a pyrethrum solution containing 10 mg. of pyrethrins and 0.45 g. of soap per 100 cc. This solution, when sprayed on adult beetles, killed 61 per cent.

Series I: The plant was sprayed and allowed to stand in the sun half an hour, when the foliage was quite dry. About 25 adult beetles were then placed in a wire cage which was placed over the plant. A similar cage of beetles was placed over a control plant which was not sprayed. Both plants were placed in the frame which was then closed. At the end of 48, 72 and 96 hours, the sprayed plant and control were observed and photographed. On page 264 the plant sprayed with pyrethrum is on the left, the control, on the right. The first picture, (Series I) shows the plants just before the beetles were placed on them. The second picture shows the same plants after the beetles had been on them

48 hours; the sprayed plant was 20 per cent eaten, the foliage of the control was completely devoured. After 72 hours (third picture) the sprayed plant was 60 per cent eaten and after 96 hours (fourth picture) it was 95 per cent eaten.

Series II. One plant was sprayed with the same pyrethrum-soap solution described above; a second plant was sprayed with soap alone; the control was not sprayed. The plants were placed in the frame, which was kept open for 39 hours and closed for 81 hours. At the end of the 120 hours, cages of beetles were placed on the plants. In the upper picture (Series II, page 264) the plants are shown just before the beetles were placed on them; in the lower picture the plants are shown 20 hours later. The beetles devoured 90 per cent of the foliage of the plant sprayed with soap (left), 10 per cent of the plant sprayed with pyrethrum-soap (center) and 100 per cent of the control.

CAGE USED FOR POTATO BEETLE EXPERIMENTS.

Series III. A pyrethrum-sprayed plant and unsprayed control were treated in the same manner as the plants in Series II excepting that they were not placed in the frame, but were exposed to the weather for 144 hours. During this time there was a total rainfall of 1.27 inch; this was apparently sufficient to remove the pyrethrum completely since the foliage of both plants was completely devoured 20 hours after beetles were placed on them. (Series III, page 264.)

Series IV. A soap-sprayed plant and a pyrethrum-sprayed plant were treated as in Series I excepting that the time between spraying and releasing the beetles on the plants was increased to 192 hours,—63 hours with the frame open, 129 hours closed. After the beetles were on the plants 24 hours the foliage of the soap-sprayed plant (left, Series IV, page 264) was 95 per cent eaten, that of the pyrethrum-sprayed plant, (right) 10 per cent eaten.

Series V. In this experiment a pyrethrum-sprayed plant and an unsprayed control were kept in the frame 360 hours,—open 105 hours, closed 255 hours. Beetles were then released on the plants, which were observed 41 hours later. The pyrethrum sprayed foliage (left, Series V, page 264) was about 25 per cent eaten, the unsprayed foliage (right) was completely devoured.

SERIES I.

DURATION OF REPELLENT ACTION OF PYRETHRUM TO POTATO BEETLES.

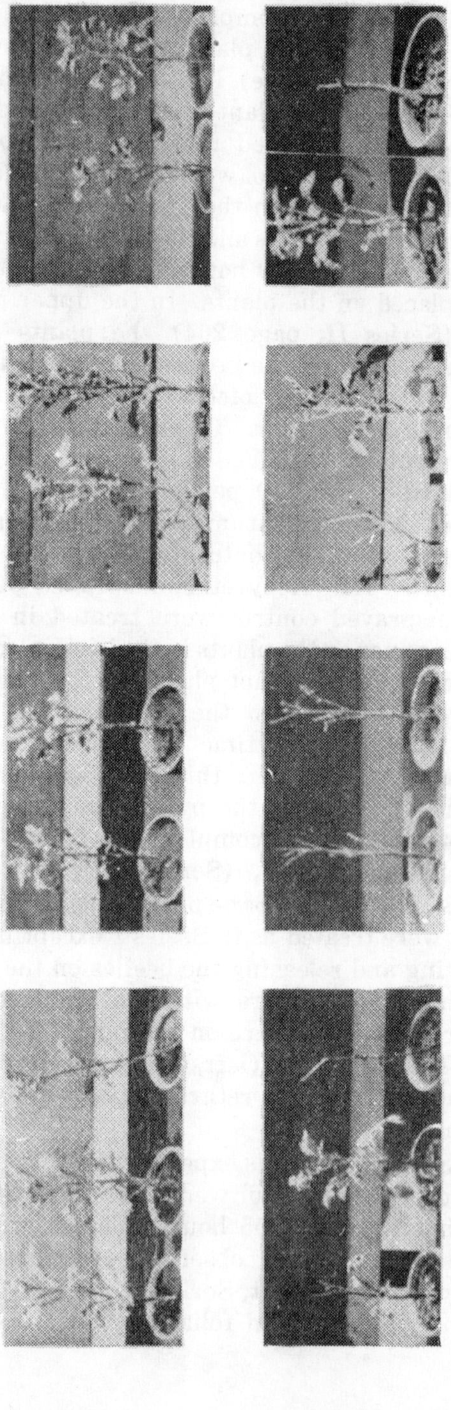

SERIES II. SERIES III. SERIES IV. SERIES V.

EFFECT OF WEATHERING ON REPELLENT ACTION OF PYRETHRUM TO POTATO BEETLES.

These experiments showed that small doses of pyrethrum have a distinct repellent action on potato beetles. In none of these experiments was the mortality of beetles on pyrethrum-sprayed plants greater than 2 per cent. The repellent action persisted for at least 15 days on plants exposed to air and sunlight but exposure to rain destroyed the repellent effect.

In the foregoing experiments the pyrethrum spray was comparatively weak, killing only 61 per cent of the beetles when sprayed directly upon them. The repellent action of a stronger solution was also investigated. Potato plants were sprayed with a solution containing 107 mg. of pyrethrins per 100 cc., or about 10 times the strength previously employed. The sprayed plants were air dried for 1 hour and were then covered with cages containing about 35 potato beetles; unsprayed controls were similarly treated. At the end of 3 days the control plants were completely devoured. After 5 days the sprayed plants were less than 10 per cent eaten. There were no deaths among the control beetles; the mortality among those on the sprayed plants was 11 per cent and the survivors were dark in color and much shrunken. Whether the deaths were due to contact with the pyrethrins or to consumption of them or to starvation could not be determined. All of these experiments were made in July.

Hartzell and Wilcoxon *(406)* sprayed nasturtium foliage with a solution containing 0.2 per cent pyrethrins. The plants were allowed to dry for 24 hours in the greenhouse and were then infested with about 200 *Aphis rumicis*. After an additional 24 hours the kill was 22.7 per cent from which they concluded that satisfactory protection from aphids could not be obtained in this way.

Attempts to develop antioxidants which would prevent decomposition of the pyrethrins when sprayed on foliage have so far not been very successful. This is a much more difficult problem than the stabilization of pyrethrum extracts. See pages 156, 168.

The pronounced repellent action of pyrethrum to chewing insects may explain why some investigators have failed to get results with pyrethrum as a stomach poison. It can readily be appreciated that a stomach poison like lead arsenate might not be a contact insecticide to sucking insects, but it is difficult to understand how a contact insecticide which, like the pyrethrins, can penetrate the integument of insects, either chewing or sucking, can fail to be toxic when taken internally.

Swingle *(864)* employed gelatine films impregnated with

derris and pyrethrum powders to demonstrate the difference in the action of these two insecticides on the imported cabbage worm. Under the conditions of his tests, pyrethrum proved to be a strong contact poison with no action as a stomach poison. Derris, however, was very poisonous when eaten but had no effect as a contact poison.

Labelling Horticultural Sprays

In labelling pyrethrum horticultural sprays of the types just described the pyrethrins, soap and oil are considered as active ingredients. Alcohol, acetone and the extractive from pyrethrum flowers other than pyrethrins are considered inert. Under present rulings of the Food and Drug Administration the amount of petroleum ether soluble extractive (obtained from pyrethrum flowers) can be declared as an active ingredient instead of declaring the percentage of pyrethrins.

Use of Rotenone with Pyrethrum

Publicity has recently been given to mixtures of pyrethrum and rotenone which are said to have great advantages over pyrethrum alone. It is obviously true that a solution containing 1 per cent of pyrethrins and 1 per cent of rotenone would be more effective than a solution containing 1 per cent of either pyrethrins or rotenone. It has not yet been proved that a solution containing 0.5 per cent pyrethrins and 0.5 per cent rotenone is a better general insecticide than a 1 per cent solution of pyrethrins.

The paralyzing action of the pyrethrins is much more rapid than that of rotenone and it is well known that certain insects are more affected by pyrethrins than by rotenone, while with other insects the reverse is true. Darley *(207)* found that rotenone is more toxic to aphids than the pyrethrins, but the latter are much more toxic than rotenone to the cucumber beetle and Mexican bean beetle. Richardson *(713)* concluded that rotenone is slightly more toxic to red spiders than pyrethrins, but satisfactory control of red spiders was not obtained with either. Ginsburg and Schmitt *(331)* showed that the pyrethrins are more toxic to bees but rotenone is more toxic to aphids. Ginsburg* has also stated that rotenone is more toxic to aphids, thrips and red spiders but the pyrethrins are more toxic than rotenone to flies, mosquitoes, squash bugs, Japanese beetles and similar insects. Tattersfield, Hobson and Gimingham *(880)* state: "Pyrethrin I possesses a higher insecticidal value by contact than any compound tested by us; the only substance which in our experience approaches it in toxicity is rotenone." Turner *(915)* considers rotenone less effective than pyrethrum as a mosquito larvicide. Gnadinger and Corl *(349)* found that replacing part of the pyrethrins in fly sprays with rotenone reduced the toxicity to flies.

Walker and Anderson *(941)* obtained better control of the cabbage looper with pyrethrum dusts than with derris dusts, but the latter were superior for control of larvae of the diamond-back moth and adult harlequin bugs.

Hamilton and Gemmell *(392)* found pyrethrum dusts more effective against certain aphids than derris dusts but with water suspensions of the powders the reverse was true. Both pyrethrum dust and derris dust were effective against the grape leaf hopper, but the latter was more effective than pyrethrum dust against the imported cabbage worm.

It would seem from these investigations, therefore, that the only practical advantage in adding rotenone to pyrethrum sprays is to increase the toxicity to aphids. There are, however, less expensive ways of accomplishing this, taking into consideration the present price of rotenone. Both the pyrethrins and rotenone are much more toxic than nicotine. (See also page 216.)

Many of the comparisons between derris and pyrethrum have been made by investigators who considered rotenone the only important toxic constituent of derris. Haller and La Forge *(386)* have pointed out that an active form of deguelin may also be one of the principal toxic constituents of derris.

A patent for "an insecticide comprising pyrethrins and rotenone" has been granted to Fulton *(288)*.

Other Experiments with Pyrethrum Sprays

Many experiments have been made with pyrethrum on various horticultural pests, in addition to those previously cited. The following are some of the more interesting investigations of recent years:

Austin and Martin *(88)* described experiments with pyrethrum and Bordeaux mixture for controlling potato blight.

Beckwith *(106)* recommended pyrethrum for the control of leaf hoppers which cause false blossom of cranberries. He employed a spray containing 22.5 per cent soap and 0.2 per cent pyrethrins; this was diluted 1 to 160 and used at the rate of 400 gallons per acre. A dust consisting of 10 per cent pyrethrum powder and 90 per cent gypsum was effective when applied at the rate of 100 pounds per acre.

Breakey and Miller *(131)* found that a spray containing 0.02 per cent of pyrethrins killed more than 90 per cent of the eggs of the Angoumois grain moth and the blow fly.

Craig* of the West Virginia Department of Agriculture found that "The Mexican bean beetle can be killed with a contact spray. The results with pyrethrum sprays were very gratifying."

Crowley* of the Cranberry Experiment Station, Ilwaco, Washington, established the value of pyrethrum for the control of cranberry fireworm and species of leaf hoppers infesting cranberries.

De Long *(221)* found pyrethrum the only contact insecticide showing promise against the potato leaf hopper.

Doehlert *(232)* concluded that alcohol or kerosene extracts of pyrethrum may be used in combination with Bordeaux mixtures if penetrol is used as spreader, but when fish oil soap is used instead of penetrol the pyrethrins are decomposed in a few hours.

Driggers *(238)* used an oil-pyrethrum spray for the control of the oriental peach moth.

Elliott *(249)* recommends the following formula as the most satisfactory spray for control of the squash bug (*Anasa tristis* DeGeer):

Kerosene extract of pyrethrum containing 2% pyrethrins,	½ pint
Potash soap (40%),	1 pint
Water	50 gal.

The kerosene extract is emulsified with the soap, to which a little water has been added, stirring the extract in, a little at a time, until a thick emulsion is obtained; there should be no separation of oil. The rest of the water is then added. Four applications, at ten day intervals, are necessary.

Emmert* (Kentucky Experiment Station) reported the use of pyrethrum for control of white fly.

Garman and Townsend *(294)* found pyrethrum effective for control of the white apple leaf hopper.

Hamilton *(391)* found pyrethrum dusts and pyrethrum-soap sprays effective for controlling gladiolus thrips on corms. The liquid sprays were not very effective for thrips on flowers and foliage; the dusts were more effective for this purpose.

Headlee and his associates investigated the action of pyreth-

* Unpublished report.

rum on codling moths *(416)* and wireworms *(413)*. A pyrethrum-soap spray controlled the latter without injury to plants.

Hutson *(460)* demonstrated the value of pyrethrum sprays for the control of the celery leaf tier, using 2 applications of spray 30 minutes apart. Pyrethrum was also added to the wash water used in preparing the celery for market; this was necessary because much of the damage is done by the leaf tiers after the celery is shipped. Experiments by other investigators in California indicate that damage in storage or in transit may be prevented by immersing the crated celery in a vat of pyrethrum spray solution. Hutson *(463)* also used pyrethrum for the control of cabbage worms.

Hutson *(462, 467)* recommended the use of pyrethrum dusts for controlling the strawberry leaf roller.

Maercks *(572)* has investigated the action of pyrethrins on the eggs of the codling moth (*Carpocapsa pomonella* L.) and the grape moth (*Polychrosis botrana* Schiff.). He treated the eggs, in different stages of development, with an aqueous suspension containing 0.01 per cent pyrethrins. The eggs were allowed to remain in contact with the solution for 15 seconds; in one series of tests they were then rinsed with water; in a second series the eggs were not rinsed. Mortality of the unrinsed eggs (second series) was much higher than when they were rinsed. Mature eggs were more easily killed than newly-laid eggs. In the experiments with codling moth eggs the mortality of the rinsed eggs varied from 0 to 55 per cent, depending on the degree of maturity; the mortality of the unrinsed eggs was 80 to 98 per cent. Many of the mature eggs hatched but the larvae died shortly afterward. The eggs of the grape moth gave similar results.

McDaniel *(602)* and Thomas *(887, 889, 890)* described the use of pyrethrum for controlling mushroom insects.

Morgan *(624)* obtained better control of the pumpkin beetle, *Aulocophora hilaris*, with pyrethrum than with some other contact insecticides.

Mote*, Oregon Experiment Station, used pyrethrum successfully for control of cucumber beetles.

Noble *(643)* obtained practically complete control of sod web worms using a commercial pyrethrum extract diluted with 500 parts of water and applied at the rate of one gallon per square yard.

* Unpublished report.

Pierstorff and Parks *(697)* employed pyrethrum for the control of the sod web worm in lawns and golf greens.

Price *(702)* used pyrethrum to control larvae of the green June beetle in lawns.

Richardson *(712)* noted that pyrethrum can be used to control species of aphids, thrips and sowbugs.

Richardson and Nelson *(720)* reported pyrethrum-soap spray only partly effective for field control of gladiolus thrips. Dustan *(246)* also found pyrethrum ineffective against this pest.

Roney and Thomas *(758)* report that pyrethrum-sulfur dust, containing 0.125 per cent pyrethrins, was effective against cabbage worms but ineffective against the tomato fruit worm (*Heliothis obsoleta* Fab.). Control of the belted cucumber beetle (*Diabrotica balteata*) and the bean leaf hopper (*Empoasca fabae*) was little better than with sulfur alone.

Schoene *(784)* found pyrethrum extracts effective against the white apple leaf hopper.

Smith and Richardson *(818)* found that immersing gladiolus corms for 1 hour in a solution containing 0.02 per cent pyrethrins and 0.5 per cent of potassium-coconut oil soap gave 83 per cent control of gladiolus thrips.

Stanley and Marcovitch *(839)* found pyrethrum and derris dusts unsatisfactory for control of tobacco flea beetles and hornworms.

Thomas *(888)* observed that pyrethrum, when used for treating seed corn, has some repellent value against wire worms.

Turner *(910)* obtained control of a moderate infestation of Mexican bean beetle with three applications of a dust containing 25 per cent pyrethrum. Injury caused by the bean leaf hopper was reduced.

Walker and Anderson *(942)* recommend pyrethrum sprays for the control of cabbage worms and aphids on cabbage seedlings. The best results were obtained by dipping the plants in a pyrethrum solution before they were set in the field.

Walker and Anderson *(943)* found pyrethrum dust, containing 0.3 per cent pyrethrins, effective against the cabbage looper (*Autographa brassicae* Riley) and the larvae of the diamond backed moth (*Plutella maculipennis* Curtis). The dust was ineffective against the corn ear worm (*Heliothis obsoleta* Fab.), harlequin bug (*Murgantia histrionica* Hahn) and species of aphids.

Walton *(944)* found pyrethrum effective in reducing infestations of raspberry beetles.

Wheeler *(960)* has investigated the action and decomposition of pyrethrum when used for controlling apple thrips (*T. imaginis* Bagnall).

White *(965)* found pyrethrum dust, containing 0.15 per cent pyrethrins, effective on cabbage pests. Talc, clay, diatomaceous earth and sulfur were satisfactory diluents. Best results were obtained with dusts applied late in the afternoon or early in the evening. Fairly satisfactory results were obtained with commercial pyrethrum sprays.

CHAPTER XV

MISCELLANEOUS USES OF PYRETHRUM

Mosquito Control

One of the most difficult problems in connection with mosquito control has been the development of an efficient larvicide which would not be injurious to vegetation, fish and water fowl. Where these forms of life must be protected, fuel oil, the customary larvicide, cannot be applied.

Ginsburg *(323, 324, 325)*, after experimenting with a large number of materials, developed a formula in 1930 for a pyrethrum larvicide. This formula has been slightly modified and a second formula has been recommended for applications on salt water. Ginsburg's formulas are as follows:

Pyrethrum mosquito larvicide to be applied on fresh water:

Kerosene extract of pyrethrum (equivalent to 1 pound of flowers per gallon, the flowers containing 0.90 per cent pyrethrins)	10	gallons
Liquid coconut-oil-potash soap (40 per cent soap solids)	5	pints
Water	4⅜	gallons
	15	gallons

The soap is agitated with the water until the mixture starts to foam. The oil is then gradually added with constant stirring or pumping. When all of the oil has been added the agitation is continued for 5 to 10 minutes, or until a homogeneous emulsion results. This emulsion is the concentrated stock larvicide and should be mixed with 10 parts of water just before spraying. The dilute solution is applied at the rate of about 50 gallons per acre.

Pyrethrum mosquito larvicide to be applied on salt water:

Kerosene extract of pyrethrum (equivalent to 1 pounds of flowers per gallon, the flowers containing 0.90 per cent pyrethrins)	10	gallons
Powdered skim milk (spray process)	4	pounds
Cresylic acid (used only as a preservative)	5	ounces
Water	5	gallons
	15	gallons

The skim milk is first made into a paste with a small amount of water and is gradually thinned with the rest of the water.

The oil is mixed with the cresylic acid and is gradually added to the skim milk with constant stirring or pumping. The skim milk used should be readily and completely soluble in cold water. If the stock larvicide is to be used within 48 hours the cresylic acid may be omitted, since it is used only as a preservative for the skim milk. This stock emulsion, as in the first formula, should be mixed with 10 parts of water before spraying.

The kerosene extract of pyrethrum can be conveniently prepared by diluting a commercial concentrated pyrethrum extract, containing 2.1 grams of pyrethrins per 100 cc., with 19 parts of kerosene.

The concentrated stock emulsion can be stored for months but should not be allowed to become chilled. The dilute larvicide should be used promptly.

According to Ginsburg the dilute larvicide is applied in the same way as the usual oil application. The pyrethrins, being insoluble in water, remain in the oil phase, which is distributed in a very thin film on the surface of the water and becomes toxic to larvae and pupae. Kerosene without pyrethrum is not appreciably toxic when used in this way. There is no injury whatever to plants, fish or water fowl.

Smith *(822)* has used Ginsburg's formulas very successfully and has also employed a modification, using light fuel oil instead of kerosene, as follows:

Light fuel oil	95 gallons
Concentrated pyrethrum extract (containing 2.1 g. pyrethrins per 100 cc.)	5 gallons
Coconut oil-potash liquid soap (40 per cent soap solids)	5 gallons
Water	45 gallons
	150 gallons

This emulsion was prepared in the manner described by Ginsburg, and one quart was mixed with 4 gallons of water for spraying.

The specifications of the light fuel oil were:

Sp. gr.	30°—34°		
Flash	54°—60°	(130°—140° F.)	
Viscosity (Saybolt 100° F.)	50		
Distilling range:			
	149°—204°	(300°—400° F.)	10%
	204°—260°	(400°—500° F.)	40%
	260°—393°	(500°—740° F.)	50%

According to Smith this larvicide was very efficient when

sprayed, killing in 20 to 90 minutes and retaining its toxicity for about two days. The larvicide cost about one-fourth as much as fuel oil and there was a further saving in labor. Other advantages were: the possibility of abolishing the large oil storage tanks, which are not necessary with the pyrethrum larvicide; no oil stains and no oil burns to workmen; no injury to plant or animal life.

An improved formula developed by Ginsburg in 1934 can be used with either fresh or salt water. This formula is:

Light fuel oil or kerosene	94 gallons
Concentrated pyrethrum extract (containing 2.1 g. pyrethrins per 100 cc.)	5 gallons
Defoamer	1 gallon
Gardinol W. Conc.	6 pounds
Water	50 gallons
	150 gallons

The defoamer is made by mixing equal parts of degras wool-grease and fuel oil or kerosene. Gardinol W. Conc. is prepared by sulfating technical lauryl alcohol (page 250); it is commercially available (Du Pont). The defoamer is mixed with the fuel oil. The Gardinol is dissolved in the water, which is then transferred to a pressure sprayer tank. The pump and agitator are started and the oil is then added slowly. Pumping is continued for thirty minutes after all the oil is added and the mixture is then pumped off into drums. This concentrated emulsion is used as directed on the preceding page.

Emulsions of the type recommended by Ginsburg have been successfully used to prevent the breeding of flies and mosquitoes near sewage disposal plants *(330)*.

Kemper *(517)* has studied the use of pyrethrum powder for controlling pests in water works, drainage systems, and sewage disposal plants. He found that 2 mg. of powder per liter controlled the Isopod *Asellus aquaticus*. Mosquito, Cheronomid and Eristalis larvae were killed by 3 to 5 mg. per liter, without injury to fish. Protozoa and microorganisms found in sewage disposal plants were not affected by pyrethrum.

Among others who have investigated the use of pyrethrum sprays for mosquito control are Giemsa *(314)*, Holt and Kintner *(436)*, Baber *(90)*, Covell *(201)*, Swellengrebel and Nykamp *(862, 863)*, Griffiths *(376)*, Sinton and Wats *(815)* and Williams and Dreesen *(970)*. The latter recommend a mixture of one part of kerosene-pyrethrum extract (2% pyrethrins) to four parts of carbon tetrachloride for spraying airplanes to prevent transmission of disease by mosquitoes.

Ant Control

Where it is not necessary to protect vegetation, the kerosene extract of pyrethrum, of the strength ordinarily used as a household insecticide, is very effective for destroying ants. It may be used where the ants infest houses but the relief is only temporary unless the nest is destroyed. This is frequently located in the lawn or near vegetation where kerosene cannot be applied. In such cases the alcoholic extract diluted with water can be used with excellent control and without damage to grass or other vegetation. An alcoholic extract containing 0.90 grams of pyrethrins per 100 cc. should be used at a dilution of 1 to 200 with just sufficient soap to make a light suds. This solution is poured on the ant hills and is liberally sprayed over the infested area.

Pyrethrum Smudges

One of the earliest methods of using pyrethrum was to distribute the powder on the top of a hot stove, in order to volatilize the active material in the room. Somewhat more convenient was the use of cones, consisting of pyrethrum powder, potassium nitrate, and filler such as sawdust or powdered charcoal. Leonardi, Zen and Sardi *(546)* employed candles, consisting essentially of oleoresin of pyrethrum mixed with paraffin, or other wax, and a perfume.

A formula said to be satisfactory for anti-mosquito incense sticks is:

Saw dust	120	grams
Powdered elm bark	90	"
Powdered pyrethrum	100	"
Malachite green	1	"
Benzoic acid	2	"

A consideration of the chemical and physical properties of the pyrethrins would lead one to believe that the use of pyrethrum in any of these ways would be extremely wasteful because the pyrethrins are easily and quickly decomposed by heat, are easily oxidized and are not readily volatilized. A number of experiments with pyrethrum products have confirmed this belief. For example:

One cc. of concentrated pyrethrum extract, equivalent

to 22 milligrams of pyrethrins, was volatilized to carbonization, on an electric hot plate, in a Peet-Grady test chamber. The toxicity to flies was about one-half as great as that of a kerosene extract, containing 22 milligrams of pyrethrins in 12 cc., sprayed in the usual way for the Peet-Grady test.

In a second experiment a concentrated extract of pyrethrum was mixed with exhausted pyrethrum flowers and potassium nitrate in these proportions:

Concentrated kerosene extract of pyrethrum, equivalent to 220 milligrams of pyrethrins	10 cc.
Exhausted pyrethrum powder	72 g.
Potassium nitrate	20 g.

Five grams of this mixture equivalent to 11 milligrams of pyrethrins was burned in a Peet-Grady test chamber. None of the flies was even disabled and the kill was zero. A kerosene spray containing 11 milligrams of pyrethrins in 12 cc. gave 95 per cent knock-down and 50 per cent kill when tested in the usual way. It would appear therefore that pyrethrum smudges are not as effective as other methods of applying pyrethrum.

Pyrethrum for Internal Parasites

The anthelmintic properties of pyrethrum were first reported by Schipulinsky *(780)* in 1854 and Frontali *(284)* in 1858. Noodt *(644)* concluded, in 1858, that pyrethrum powder taken internally was ineffective against tapeworm. More recently the internal use of pyrethrum has been revived in France, where it has been studied by Chevalier; Gaudin and Carron; Anglade, Gaudin and Arcony and Perrot. Chevalier *(167)* claims to have used pyrethrum successfully against *Ascaria lumbricoides* in hogs and *Taenia* in dogs. He also recommends pyrethrum for the treatment of *Trichocephalus, Taenia* and *Ascaris* in human beings, using a dose of 10 milligrams of pyrethrins for children or 20 milligrams for adults, to be repeated every three hours. The eggs are not destroyed.

Gaudin and Carron *(304)* have reported that the pyrethrins have a rapid paralytic action on the muscles of helminths. Anglade, Gaudin and Arcony *(16)* have employed special pills which liberate the pyrethrins gradually in the intestine. These pills (for human use) are said to be effective against *Oxyuris, Ascaris, Trichocephalus* and *Taenia*. Anglade and Gaudin *(15)* employed keratinized granules of pyrethrins for controlling internal parasites.

Rebrassier *(706)* reported that powdered pyrethrum containing 0.8 per cent pyrethrins was effective in doses of 200 mg. in removing *Ascaridia lineata* from chickens.

Urbain and Guillot *(927)* employed pyrethrins in castor oil for destroying internal parasites of horses.

Perrot *(691, 694)* has also reported on the use of pyrethrum as a vermicide.

The writer has conducted a few experiments with pyrethrum against various worms in poultry and hogs but the results were not as promising as those reported by the French investigators. Various pyrethrum extracts have been tried against *Trichomonas vaginalis,* against the liver fluke and in cases of amoebic dysentery without favorable results.

Many of the parasites mentioned above are readily killed by pyrethrum when isolated from the host. Under the normal conditions of their growth, however, they are not so easily controlled.

Minor Uses

Juillet and Diacono *(497)* have recommended the use of pyrethrum for killing body lice. For this purpose the alcohol or acetone extract of pyrethrum can be added to the bath water without danger of injury. An extract containing 0.90 grams of pyrethrins per 100 cc. should be used at a dilution of 1 to 200. A similar solution can be used for head lice. A solution of this type does not kill the eggs, however, so that repeated applications may be necessary. Kerosene extract of pyrethrum emulsified with soap has the advantage of killing the eggs as well as the lice.

Lemaire and Gaudin *(544)* report that a gel containing 0.5 per cent pyrethrins gave good results in the treatment of scabies, but an alcoholic extract was not effective. Pyrethrum ointment and oil extracts have also been suggested for use against the harvest mite.

Smith *(73)* has developed a pyrethrum ointment for the treatment of scabies. This ointment is composed of wool fat, petrolatum, paraffin, and hydrocarbon oil extract of pyrethrum; it contains 0.75 per cent pyrethrins. Sweitzer and Tedder *(861)* claim that this material penetrates the burrows and kills both mites and eggs. Dermatitis is not produced, except in rare instances. This product has been approved by the Council of

Pharmacy and Chemistry of the American Medical Association, as a new and non-official remedy.

The alcoholic extract of pyrethrum, or soap-pyrethrum, can be used for killing fleas on dogs. The material is merely dissolved in the bath water. A special flea soap is available commercially. It consists of a high grade soap, free from excessive alkali, containing about 5 per cent of pyrethrum oleoresin, equivalent to about 0.6 per cent pyrethrins. The treatment must be repeated at intervals. The alcoholic extract properly diluted, can be sprayed without soap, for lice on canaries; it can also be added to the bath water.

Pyrethrum extracts have also been used for controlling the brown dog tick. McIntosh *(610)* recommends a pyrethrum household spray for infested runs and kennels. The following formula is said to give control *(60)*:

Soft soap	1.5 lbs.
Water	2.0 qts.
Kerosene-pyrethrum extract (2.4% pyrethrins)	2.0 pts.

The soap is mixed with the water and the pyrethrum extract is then added. One part of this stock solution is diluted with five parts of water, for use as a sponge bath or dip.

McDaniel *(604)* recommended a kerosene extract of pyrethrum for controlling termites.

Oleoresin of pyrethrum has been recommended as an ingredient of paints to be used on the under water structure of vessels in order to keep the surface free from marine life, but its value for this purpose is not established.

CHAPTER XVI

POSSIBLE SOURCES OF PYRETHRUM FLOWERS IN THE UNITED STATES

The increasing use of pyrethrum has led to many attempts to establish new sources of supply. The plant has been grown experimentally in Australia, Nigeria, India, Ceylon, Malaya *(53)*, Cyprus, Argentine and Greece. In Kenya Colony, British East Africa, production has recently been established on a commercial scale. In North America it has been grown in small quantities in Canada, Mexico and the United States.

The cultivation of species of pyrethrum in the United States was first suggested about 1859. Since then attempts have been made to grow it in almost every State. Recent plantings have been made in Arizona, California, Colorado, Georgia, Idaho, Indiana, Maryland, Mississippi, Montana, Nebraska, North Carolina, Oregon, Pennsylvania, South Carolina, Tennessee, Texas, Virginia and West Virginia. Only in California, however, has commercial production been attained and even there the amount produced was not large. The earlier experiments were confined merely to determining whether pyrethrum would grow under the soil and climatic conditions prevailing in different localities.

Semi-commercial plantings have failed in Florida, Michigan and Kentucky. In the latter state the difficulty is said to have been due to mild winters which failed to throw the plants into dormancy, resulting in injury during sudden cold periods. Other failures have been due to disease and possibly improper cultural methods. The possibility of growing pyrethrum in the Virgin Islands has also been considered *(50)*.

Martin and Tattersfield *(591, 592)* studied the effect of light, temperature and dormancy upon pyrethrum plants. They found that when pyrethrum is grown at a summer temperature of either 15° or 25°, reduction in the amount of light available to the plant results in smaller flowers of poorer quality. The effects of temperature and dormancy upon the production of flowers are closely interrelated. Plants which, following a winter dormant period, were subjected to a high summer temperature, gave large numbers of flowers, at an early date. Plants which were subjected to high temperature during the winter produced only a few flowers the following summer. A dormant period, dependent upon sufficiently low winter temperatures, is necessary for the normal production of large numbers of flowers.

Tattersfield states that the plant fails to produce flowers in the lowlands of tropical countries, as Uganda and Trinidad.

The isolation and identification of the active principles of pyrethrum, by Staudinger and Ruzicka, and the development of quantitative methods for determining the pyrethrins, by Tattersfield and by Gnadinger and Corl in 1929, made possible for the first time the scientific study of the effect of different cultural methods and soil and climatic conditions on the pyrethrin content of the flowers.

Pyrethrum Investigations in Colorado

The possibilities of pyrethrum as a commercial crop for Colorado were first recognized by Dr. E. P. Sandsten, Director of the Colorado Experiment Station. In 1931 the common interest of Dr. Sandsten and the writer in pyrethrum led to the establishing of a joint project by the Colorado Agricultural College and Experiment Station and McLaughlin, Gormley King Co. of Minneapolis, for the purpose of developing a domestic source of supply for pyrethrum.

The climatic conditions in the producing areas of Japan and Dalmatia are comparable with those in the higher altitude sections of Colorado. The agricultural districts of this state range in elevation from 3,400 to 9,000 feet, with a growing season varying from 90 days to 170 days in length.

Pyrethrum cinerariaefolium has been grown in the experimental plots at Fort Collins, Colorado, since 1929; the plants have withstood low temperatures without winter protection and very little loss has resulted. These plants were exposed to a subzero temperature period of 10 days in which a minimum of $-39°$ was recorded. The average length of the growing season at Fort Collins is 120 days. The elevation is 5,000 feet above sea level. It has been generally observed in this district that winter injury is due to a drying out of the plant rather than extremely low temperatures.

The results of preliminary investigations of factors affecting the pyrethrin content of *P. cinerariaefolium* have been reported by Gnadinger, Evans and Corl *(354, 355, 252, 253)*. The horticultural side of this investigation was handled by L. E. Evans, Assistant in Horticulture, at Fort Collins, while the writer was responsible for the chemical side.

EXPERIMENTAL PLOTS FOR PYRETHRUM CULTURE. (COLORADO EXPERIMENT STATION, FORT COLLINS).

The results secured in a study of the effect on pyrethrin content of drying methods, seasonal variation, and storage treatment of dried flowers were reported. The results of a comparative study, showing the variations that occur in botanical characters and the yield of pyrethrins by individual plants, were also presented. The work on pyrethrum flowers in Colorado is being planned to extend over a period of years. In the following pages some of the results obtained in 1932, 1933, 1934 and 1935 are described.

EFFECT OF DIFFERENT DRYING METHODS ON PYRETHRIN CONTENT

Seven drying methods were used in this study:

1. Full sunlight exposure for 27 hours and remainder of drying period under outdoor shade.
2. Sixty-two hours in outdoor full shade and remainder of drying period under full outdoor sunlight.
3. Full outdoor sunlight.
4. Dried completely in outdoor full shade.
5. Dried in thermostatically controlled drying cabinet at 37° to 40°.
6. Dried at 35° to 40° with forced circulating air currents.
7. Dried in vacuum oven at 35° to 40° and 100 mm. pressure.

The first method consisted of drying on a wire screen tray for two days, during which time 27 hours were full sunlight. The screen permitted a natural circulation of air currents around the sample. After the first 48 hours of outdoor exposure, the samples were placed in full shade.

The second method used was drying in full shade for 62 hours. After this period the samples were placed on a wire tray for exposure to full sunlight until completely dry.

Drying in full sunlight was the third method. The samples were placed on unshaded wire trays outdoors until completely dry.

The drying of samples in method 4 was accomplished by the use of a special full shade cabinet. The mean temperature inside of the cabinet was 26°.

For the fifth method of drying a wooden, asbestos insulated cabinet, 46x92x46 cm. was used. The top and bottom were perforated with holes 5mm. in diameter, to permit natural air circulation. The temperature was regulated by a thermostat which was calibrated to hold the temperature between 37° and 40°. The thermostat controlled a 60 watt, 120 volt electric bulb placed at the bottom of the cabinet and partitioned from the drying chamber by perforated asbestos. The samples were suspended in cheese-cloth bags on hooks placed at intervals of 8 cm.

Drying by the sixth method was accomplished in a cabinet constructed as just described, except that a heating unit and electric fan were placed under the bottom of the cabinet. This forced a continual circulation of warm air around the samples. The flowers were placed in cheese-cloth bags and arranged in the cabinet as described in the fifth method.

The seventh method consisted of drying samples in a vacuum oven in which the temperature was regulated at 35° to 40° by heating units. A pressure of 100 millimeters was maintained in the oven by a motor driven vacuum pump.

Two series of samples were run by each of the drying methods. In the first series, 70 grams of flowers were used in each test, and in the second series 50 grams. Samples were made up of composite pickings from 30 plants. All the flowers in a series were picked on the same date and in the same stage of development, accurately weighed before and after drying, and the per cent moisture loss computed. At no time during the tests was there any cloudy or partially cloudy weather.

The data presented in Table LXXXVII give the results of the

A COLORADO PYRETHRUM FIELD, ELEVATION 8000 FEET. (COURTESY L. E. EVANS, COLORADO EXPERIMENT STATION).

two complete series of tests. The first series was started on the 21st of July, and the last series on the 30th of July, 1932. The drying of all samples in a series was begun simultaneously.

Inspection of the data, in Table LXXXVII, does not show any great difference in the pyrethrin content of the samples dried by the different treatments. The lowest average pyrethrin content was secured on flowers dried by methods 3 and 6. In each case

TABLE LXXXVII. EFFECT OF DIFFERENT DRYING METHODS ON PYRETHRIN CONTENT (GNADINGER, EVANS AND CORL)

Treatment	Fresh weight	Dry weight	Moisture loss on drying	Hours required to dry	Moisture in dried flowers	Pyrethrin content moisture-free basis
	g.	g.	%		%	%
Full sunlight 27 hrs. Remainder shade	60	17.842	70.26	310	7.1	1.25
Shade 62 hrs. Remainder in sunlight	60	18.080	69.86	199	5.9	1.24
Full sunlight	60	17.019	71.63	123	5.8	1.18
Full shade	60	17.306	71.16	355	7.9	1.28
Temp. regulated cabinet 37° to 40°	60	17.747	70.43	53	6.0	1.20
Circulated air current cabinet 35° to 40°	60	17.448	70.92	89	9.1	1.18
Vacuum oven 35° to 40°	60	16.584	72.36	19	5.6	1.21

this amounted to 1.18 per cent. The highest average yield of pyrethrins, 1.28 per cent, was secured by method 4. The difference between the highest and lowest average pyrethrin content was 0.10 per cent.

Effect of Shading the Plant on Pyrethrin Content

The purpose of this study was to determine the relationship of the pyrethrin content of flowers produced in full sunlight, partial sunlight, and in absence of sunlight. The partial shade was accomplished by covering the plant with a muslin frame. Absence of sunlight was obtained by the use of a canvas covered frame which did not admit sunlight. Three plants of similar foliage characters, size, and number of flower buds were selected and treated as mentioned above. The yield of flowers and pyrethrin content was determined upon each plant.

The results, presented in Table LXXXVIII, include a summary of data secured on each individual studied. The pyrethrin content of flowers produced in absence of sunlight was 0.42 per cent, which is 0.57 per cent less than for flowers produced in full sunlight. The absence of sunlight affected the normal appearance of the plant, and the flowers were malformed and very small. The difference in the pyrethrin content of flowers produced in full sunlight and in partial sunlight may possibly be due to differences in individuals.

TABLE LXXXVIII. EFFECT OF SHADING ON PYRETHRIN CONTENT (GNADINGER, EVANS AND CORL)

Treatment	No. of flowers picked	Fresh weight g.	Dry weight g.	Moisture in dried flowers %	Pyrethrin content moisture-free basis %
Full sunlight	325	169.44	39.89	9.1	0.99
Partial sunlight	250	119.90	24.85	7.9	1.22
Absence of sunlight	93	.....	7.69	9.0	0.42

Effect of Different Treatments Previous to Drying Flowers

In some sections of Japan, the flowers with stalks are suspended on a frame during the drying period. The object of this study was to note the effect on pyrethrin content of holding flowers, with 7-inch stems, in water 48 hours before drying; also, drying flowers with 7-inch stems as compared with flowers dried without stems. A composite lot of flowers with 7-inch stems was gathered from 30 plants, thoroughly mixed, and divided into three samples. Sample 1 was placed in water, immediately after picking, for a period of 48 hours. At the end of this period, the sample was placed in the thermostatically controlled drying cabinet. Sample 2, with stems, and sample 3, in which the flower heads were removed from the stems, were immediately placed in the thermostatically controlled cabinet. The stems were, of course, removed from samples 1 and 2 before assaying.

The results, shown in Table LXXXIX, are the averages of two complete tests. From the data it is noted that there is only 0.02 per cent difference in the pyrethrin content of flowers dried with stems and without stems. The difference in the pyrethrin content between sample 1 and sample 3 was 0.08 per cent.

TABLE LXXXIX. EFFECT OF TREATMENT GIVEN FLOWERS PREVIOUS TO DRYING ON THE PYRETHRIN CONTENT (GNADINGER, EVANS AND CORL)

Treatment	Weight before drying	Dry weight of flowers without stems	Moisture loss on drying	Moisture in dried flowers	Pyrethrin content moisture-free basis
	g.	g.	%	%	%
Sample No. 1 7-inch stems placed in water 48 hrs. previous to drying	89.52	13.05	77.16	7.7	1.22
Sample No. 2 Dried with 7-inch stems	77.20	13.07	72.21	7.0	1.16
Sample No. 3 Control; dried without stems	45.79	12.90	71.82	7.9	1.14

VARIATION IN *P. cinerariaefolium*

Two hundred and fifty plants, four years old, were available for securing information on the botanical characters and pyrethrin content of *P. cinerariaefolium*. Thirty-nine plants were selected to secure information on the variation in yield of pyrethrins in individuals. The following botanical characters were studied upon each individual plant to furnish information on the possible use of such characters as an index to low or high pyrethrin content.

A. Foliage characters
1. Growth habit
2. Type of foliage
3. Shape
4. Amount of leaf bloom
5. Color
6. Leaf size.

- B. Flower stalk characters
    1. Relative strength of flower stalk
    2. Pubescence
    3. Color
    4. Length
- C. Flower head characters
    1. Bracts
        - a. Relative length and width
        - b. Pubescence
        - c. Imbrication
        - d. Scariousness
    2. Ray flowers
        - a. Entire or toothed
        - b. Color
        - c. Length
    3. Receptacle
        - a. Flat or convex
    4. Disk
        - a. Width of disk
        - b. Shape of achene
        - c. Color of achene
        - d. Length of achene

Each individual plant was carefully studied and the average for each particular characteristic was computed from a large number of observations. Measurements of flower heads from individual plants were made upon a large number of flowers picked at the same stage of maturity, that is, when the first three rows of disk florets were open. Achene measurements were made on the outside row of open disk florets.

The daily yield of flowers from individual plants was carefully recorded. Any deterioration in the pyrethrin content that may have taken place in the first picked flowers was held to a minimum by storing in an airtight, screw top, metal container. Further, the 10-gram sample for analysis was made up from the composite yield of all flowers picked from each plant. This procedure was followed so that any variation in yield of pyrethrins would be due entirely to differences in individual plants.

Non-variable Botanical Characters

From the detailed study of individuals, it was found that certain botanical characters remained constant. The pubescence, imbrication, scariousness and the relative length and width of the

HOTBED AND GREENHOUSE EQUIPMENT FOR PYRETHRUM CULTURE.
(COLORADO EXPERIMENT STATION, FORT COLLINS).

bracts were not variable. The ray flower was found to be creamy white and toothed for all individuals. The slightly convex receptacle and light yellow-green, ribbed achenes were uniformly present in all plants.

Variable Botanical Characters

Growth habit.—A comparison of the growth habits of individual plants revealed striking differences ranging from prostrate and spreading to erect and bushy.

Foliage.—There was considerable difference in foliage, ranging from a light feathery to a coarse and heavy type. The variable appearance of the plant was due to a decided difference in the size and shape of the leaves. In the feathery foliage type the leaves were deeply pinnatifid with small lobes, while in the heavy foliage type the lobes were larger and not as deeply pinnatifid. Some of the plants were glaucous or covered with a whitish bloom, which varied in intensity in individual plants from a light to a heavy covering, giving the plant a silvery appearance. In other plants the whitish bloom was entirely absent, in which case the foliage color varied from a pale yellow-green to a deep, dark green. The variation in the leaves of individual plants is shown on page 291.

Flower stalks.—The relative strength of flower stalks ranged from prostrate to erect. This characteristic of flower stalks is very important from the standpoint of picking the flowers and quality of the dried flower. Those plants with prostrate flower stalks made harvesting difficult. The soil adhering to the floral parts materially reduced the quality of the product. The color and pubescence of stem varied in the same manner as the leaves. The average height of flower stalks for individual plants varied from 37.08 cm. to 67.05 cm. Two typical plants illustrating the variation in foliage and stem are shown on page 291.

Flower head.—The ray flower varied in length from 12.1 mm. to 46.8 mm. The width of disk varied from 6.95 mm. to 14.1 mm., with achenes ranging in length from 1.75 mm. to 2.45 mm. The variation in the width of disk and length of ray flower is shown on page 291.

Variation in Yield of Flowers and Pyrethrins

The variation in the yield of flowers from the 39 individuals under trial in 1932 is shown in Table XC. The yield of dried flowers from individual plants varied from 70 flowers weighing

9.24 grams, to 543 flowers weighing 62.66 grams. It is also shown in Table XC that the percentage of moisture lost on drying is quite constant.

TABLE XC. VARIATION IN YIELD OF FLOWERS FROM INDIVIDUAL PLANTS (GNADINGER, EVANS AND CORL)

Number	Total number bloom picked	Fresh weight g.	Dry weight g.	Moisture lost in drying %
A9-81	70	42.20	9.24	78.10
A9-53	72	44.97	9.71	78.41
A9-153	82	42.68	9.86	76.92
A9-175	97	39.41	9.88	74.87
A9-199	77	40.28	10.15	74.82
A9-90	77	46.48	10.42	77.58
A9-237	85	47.91	10.81	77.43
A9-225	152	88.58	11.08	87.49
A9-187	84	53.21	11.63	78.14
A9-240	98	60.30	12.02	80.06
A9-208	99	41.72	12.28	70.56
A9-233	105	61.59	12.68	79.41
A9-140	100	44.38	13.26	70.12
A9-212	104	52.31	13.33	74.51
A9-86	96	60.60	13.70	77.39
A9-14	121	66.48	15.24	77.07
A9-190	102	66.63	15.53	76.69
A9-162	171	69.40	16.34	76.45
A9-241	125	70.50	16.64	76.39
A9-115	132	66.95	16.96	73.62
A9-75	122	57.21	17.11	71.10
A9-236	131	75.79	17.56	78.83
A9-69	144	71.51	18.99	73.44
A9-77	164	77.28	19.12	75.25
A9-197	152	93.25	19.27	79.21
A9-59	189	81.19	19.55	77.32
A9-18	221	71.04	20.43	71.25
A9-61	187	82.14	21.04	74.50
A9-161	181	101.50	21.69	78.63
A9-210	186	84.50	22.19	73.51
A9-5	127	76.70	22.19	71.06
A9-228	162	100.63	25.35	74.81
A9-239	209	116.76	26.59	77.21
A9-73	215	105.48	27.82	73.62
A9-200	182	111.83	30.30	72.90
A9-202	284	143.61	35.93	74.81
A9-151	283	159.78	38.51	75.89
A9-149	422	185.28	51.95	71.96
A9-150	543	286.76	62.66	78.14

The pyrethrin contents in 1932 of the 39 individual plants are given in Table XCI; they varied from 0.90 per cent to 2.07

per cent. The average pyrethrin content for the 39 individuals was 1.27 per cent.

Similar data on some of these individuals were obtained in 1933 and 1934, together with data from 11 additional plants which were set out in 1932. The analyses, presented in Table

TABLE XCI. VARIATION IN PYRETHRIN CONTENT OF INDIVIDUAL PLANTS (GNADINGER, EVANS AND CORL)

Number	Moisture in dried flowers %	Pyrethrin content moisture-free basis %
A9-115	7.8	0.90
A9-175	6.8	0.92
A9-161	8.0	0.99
A9-162	7.2	1.00
A9-153	7.9	1.02
A9-59	8.6	1.04
A9-86	7.0	1.08
A9-190	5.4	1.09
A9-73	7.2	1.09
A9-212	6.7	1.11
A9-197	8.2	1.11
A9-81	10.1	1.14
A9-69	7.4	1.14
A9-202	7.4	1.16
A9-225	6.6	1.17
A9-149	7.9	1.19
A9-150	7.6	1.19
A9-187	5.6	1.20
A9-18	7.1	1.22
A9-239	7.5	1.24
A9-199	7.6	1.24
A9-210	10.3	1.24
A9-208	5.9	1.25
A9-53	7.1	1.26
A9-151	10.3	1.26
A9-61	7.6	1.27
A9-76	6.4	1.28
A9-200	8.1	1.29
A9-240	7.7	1.34
A9-228	8.7	1.43
A9-90	9.8	1.43
A9-5	8.2	1.53
A9-233	5.8	1.53
A9-14	8.1	1.54
A9-241	6.5	1.56
A9-236	7.0	1.63
A9-140	9.2	1.66
A9-77	9.3	1.95
A9-237	5.4	2.08

VARIATION IN LEAF, FOLIAGE AND FLOWER HEADS.

XCII, show a considerable variation in the pyrethrin content of these individual plants from year to year.

TABLE XCII. VARIATION IN PYRETHRIN CONTENT OF INDIVIDUAL PLANTS FROM YEAR TO YEAR (GNADINGER, EVANS AND CORL)

Plant No.	Pyrethrin content (air-dried basis) 1932 %	1933 %	1934 %
A9-86	1.00	...	0.89
A9-90	1.29	1.11	1.57
A9-150	1.10	1.10	1.05
A9-151	1.13	...	1.07
A9-153	0.94	...	0.89
A9-187	1.13	...	1.05
A9-206	0.90	0.95	1.03
A9-237	1.96	...	1.39
A2-117	...	1.21	0.99
A2-118	1.26	1.07	...
A2-125	...	1.08	0.90
A2-138	...	1.25	0.97
A2-282	...	1.11	1.01
A2-355	1.29	1.22	0.99
A2-361	...	0.94	0.89
A2-369	...	1.04	1.13
A2-706	1.03	1.03	...
A2-927	0.94	0.79	0.80
A2-1052	1.12	0.86	0.81

The A9 plants were from seed planted in 1929; the A2 series was planted in 1932. The 1932 analyses of the A2 series are somewhat abnormal because the plants did not produce sufficient flowers for analysis until autumn; it has since been found that flowers harvested in the autumn are somewhat richer in pyrethrins than spring flowers from the same plants.

These results (Table XCII) are at variance with those of Martin and Tattersfield *(591)*, who state that a given plant yielded flowers containing approximately the same percentage of pyrethrin I for three successive years, while the total pyrethrin content of flowers from eight individual plants was the same for each plant for two successive harvests.

There is also a wide variation in the number and weight of flowers produced by the plants included in Table XCII. This is shown in Table XCIII.

TABLE XCIII. VARIATION IN YIELD OF FLOWERS FROM INDIVIDUAL PLANTS FROM YEAR TO YEAR (GNADINGER AND CORL)

Plant No.	1932 crop		1933 crop		1934 crop	
	No. of flowers	Air-dried weight g.	No. of flowers	Air-dried weight g.	No. of flowers	Air-dried weight g.
A9-86	96	13.70	...	....	106	14.30
A9-90	77	10.42	80	9.50	178	22.90
A9-150	543	66.15	354	43.30	346	46.07
A9-151	283	28.51	...	....	200	26.94
A9-153	82	9.86	...	....	190	20.90
A9-187	84	11.63	...	....	123	16.91
A9-206	325	39.89	184	19.48	131	16.20
A9-237	85	10.81	...	....	221	26.07
A2-117	...	....	217	26.30	113	11.73
A2-118	108	10.61	115	10.80	...	....
A2-125	...	....	169	20.81	468	52.99
A2-138	...	....	259	28.56	399	53.22
A2-282	...	....	...	....	300	31.19
A2-355	57	9.15	195	20.23	439	46.92
A2-361	...	....	...	....	342	51.36
A2-369	...	....	...	....	357	47.79
A2-706	49	8.33	195	22.92	...	....
A2-927	85	10.76	259	24.53	454	47.98
A2-1052	65	8.57	170	22.70	388	57.29

The pyrethrin content of the flowers from a given plant appears to be unrelated to the number and weight of flowers produced by that plant from year to year.

The open fertilized seed from a number of individual plants of the A9 series were sown early in 1933 in the hotbeds and the plants were transplanted to the field in May. Only a few of the

TABLE XCIV. COMPARISON OF PYRETHRIN CONTENT OF MOTHER PLANTS AND PROGENY

Mother Plant A9-237 (F. C.)
1932—1.96; 1933—no samples; 1934—1.39; 1935—no samples

First Generation A-5 (F. C.)

Open pollinated

No.	1	3	5	6	8	9	11	12	13	16	20	30	34	36	41	45	47	48	Comp.
1934	1.54	1.12	1.39	0.96	1.13	1.18	1.13	1.05	1.18	1.18	1.15	0.89	1.13	1.15	1.10	0.94	1.02	1.07	1.18
1935	...	...	...	...	...	...	...	...	...	...	...	...	...	...	...	...	...	...	1.01

Close pollinated

No.	50	53	54	55	56	58	61	62
1934	...	...	...	...	...	...	...	...
1935	1.06	1.28	1.27	1.08	1.21	1.21	1.01	1.19

Second Generation A-5-1 (F.C.)
Open pollinated

No.	4	7
1935	1.11	1.08

Second Generation A-5-45 (Avon)
Open pollinated

No.	1	2	3	6
1935	1.24	1.42	1.40	1.14

Mother Plant A9-77 (F. C.)
1932—1.77; 1933-4-5—no samples

First Generation A-4 (F. C.)

Open pollinated

No.	1	3	4	7	8	11	12	13	15	26	31	32	Comp.
1934	0.97	1.28	1.03	1.05	1.08	1.11	0.94	0.90	1.01	1.58	1.55	1.48	1.20
1935	...	...	...	...	...	...	...	...	...	...	...	...	1.01

Close pollinated

No.	55
1934	...
1935	1.30

Second Generation A-4-11
Open pollinated

	Fort Collins			Avon		
No.	3	14	16	1	3	4
1935	1.02	1.17	1.00	1.28	1.24	1.79

Mother Plant A9-140 (F. C.)
1932—1.51; 1933-4-5 no samples

First Generation A-2
Open pollinated

Fort Collins

No.	15	24	32	40	42	58	60	78	79	Composite
1934	0.83	0.89	0.87	0.89	0.55	0.87	0.73	1.08	1.22	1.18
1935	...	...	...	...	...	...	...	...	...	...

Avon

No.	61
1934	...
1935	1.27

progeny produced sufficient flowers for analysis in 1933, but in 1934 a large number of samples was available. In Table XCIV the pyrethrin contents of some of the mother plants are compared with those of the progeny. The number of high test progeny from the rich mother plants was smaller than anticipated.

In general the physical characteristics noted in the mother plants, such as color of leaf, size and form of leaf, width of floral disk and size of achenes, were found to recur in the progeny.

Martin and Tattersfield *(591, 592)* found that rooted shoots from parent plants rich in pyrethrins continue to yield flowers of high pyrethrin content. They also noted that certain plants give flowers which have short ray florets, or which mature at an earlier or later date than normal and these characteristics are again found in plants produced from these individuals by vegetative propagation.

Drain and Shuey *(237)* used crown division to develop progeny rows of selected individual plants. Crown divisions were more successful when made in the fall than when made in the spring. A single plant, rich in pyrethrins, was vegetatively propagated in 1932 and its progeny yielded flowers assaying 2.05 per cent pyrethrins, at the rate of 1088 pounds per acre, in 1934.

Both Tattersfield and Drain agree that pyrethrin content, although influenced by environmental factors, is genetic in nature, as Evans has suggested *(354)*. No morphological characteristic of the plant has been correlated with pyrethrin content by any investigator.

Effect of Fertilizer on Yield of Flowers and Pyrethrins.

In order to determine the effect of nitrogen, phosphorus and potassium on the yield of flowers and pyrethrins, experiments were conducted in 1932 and 1933. The plants were set out in 1932, spaced 18 inches in the row and 30 inches between rows. The treatments were repeated seven times using a Latin square random distribution.

The fertilizers used were: ammonium sulfate (25 per cent available nitrogen), superphosphate (45 per cent available phosphorus pentoxide), and potassium chloride (50 per cent available potassium oxide). These were combined in the ratios shown on page 294.

The six fertilizers were applied at the rate of 39 g. per plant, or 1000 pounds per acre. The soil was not deficient in any

Treatment No.	N %	P_2O_5 %	K_2O %
1	4	12	4
2	0	12	0
3	4	12	0
4	0	12	4
5	0	0	4
6	4	0	0
7	0	0	0

of the elements. The fertilizer was placed in a furrow around the plant and then covered with soil.

Due to the immaturity of the plants, the yield of flowers was scanty in 1932 and these did not mature until fall. Samples were picked August 27th, September 7th, September 17th and October 1st. Analyses were made of composite samples from the seven treatments. They gave little information excepting that the October 1st samples were appreciably richer in pyrethrins than those collected August 27th.

In 1933, samples were collected from the 9 plants in the center row of each replicate; no additional fertilizer was applied in 1933.

None of the fertilizer treatments caused a significant increase in pyrethrin content. The yield of flowers was not greatly affected.

Ripert *(746)* reports that in favorable years the use of fertilizer causes large increases in the pyrethrum yields, while in poor years it maintains a satisfactory yield. A complete fertilizer is necessary for regular results. Nitrogen is the most active ingredient. Fertilizers with rapid action, such as ammonium phosphate and superphosphate, are the best in humid years, and fertilizers with slow action, cyanamide and slags, are preferable for the dry years. Fertilizers affect the yield of flowers but not the pyrethrin content. The nitrate fertilizers are never to be recommended, according to Ripert.

Martin and Tattersfield *(592)* found no significant difference in the pyrethrin I content of potted plants subjected to the following treatment:

No fertilizer
Farmyard manure
Complete nitrogen, phosphorus, potassium, calcium artificial fertilizer
Complete artificial fertilizer, excluding potash
Complete artificial fertilizer, excluding nitrogen
Complete artificial fertilizer, excluding phosphorus
Complete artificial fertilizer, excluding calcium

The plants were grown in heavy soil and the experiments extended over three years. Plants grown in sand with deficiencies of nitrogen, phosphorus and potash had substantially the same pyrethrin I content as fully manured plants.

WINTER HARDINESS OF PYRETHRUM

No serious loss of plants from winter injury occurred in the 1929 planting during the period of 1929 to 1932. The location of these plants and the conditions during this period were such that the information secured may not be applicable to adverse conditions in the field. In 1933 a further study of winter hardiness was made. Three treatments were included in this study, as follows:

Treatment No. 1. Check plot, foliage acting as a natural covering.
Treatment No. 2. Covering of wheat straw to a depth of 12 inches.
Treatment No. 3. Removal of leaves and exposure of crown.

The leaves of the plants in treatment No. 3 were cut back

TABLE XCV. EFFECT OF COVERING AND EXPOSURE ON YIELD OF FLOWERS AND PYRETHRINS (1933). (GNADINGER, EVANS AND CORL)

Plot No. and treatment	Number flowers harvested*	Dry weight g.	Mean dry weight of individual flower. g.	Pyrethrin content dry basis %	Remarks
1A Natural covering	814	105.77	.1299	1.00	Water applied 7/2/33
1B Natural covering	887	111.22	.1253	1.18	Water applied 6/25/33 and 7/2/33
2A 12″ covering of straw	1165	130.18	.1117	1.27	Water applied 6/25/33 and 7/2/33
2B 12″ covering of straw	927	144.52	.1559	1.15	Water applied 7/2/33
3A Leaves cut back to crown	707	93.47	.1322	1.27	Water applied 6/25/33 and 7/2/33
3B Leaves cut back to crown	563	80.04	.1421	.86	Water applied 7/2/33

* Flowers picked from five plants in center row of each plot.

on December 15, and the straw covering in treatment 2 was applied on January 5, and removed April 25.

The lowest temperature recorded during these trials was –33° C. which continued for a period of five days. During these trials it was noted that the moisture in the soil and the straw material in treatment No. 2 was sufficient to prevent drying out of the plants. The amount of snowfall was not sufficient to affect the data collected from the treatments that received no covering.

In Table XCV are shown the differences in total yield of flowers, dry weight of flowers and yield of pyrethrins for each treatment. The yield of flowers and dry weight of flowers increased with the amount of covering. There were considerable differences in the yield of pyrethrins of each treatment and replication. The highest yield of pyrethrins was recorded from plots 1B, 2A and 3A, which may possibly be due to the additional supply of moisture through irrigation on June 25th. These trials indicated a partial injury to the flower buds or subsequent year's crown growth when covering was insufficient.

TABLE XCVI. PYRETHRIN CONTENT OF FOREIGN AND DOMESTIC STRAINS OF PYRETHRUM GROWN IN COLORADO. (GNADINGER, EVANS AND CORL)

No.	Source of Seed	Pyrethrins %		Flower production
		1934	1935	
E4	U. S. Dept. Agr.	0.96	0.97	Fair
E5	California	...	1.05	Fair
E10	England	0.89	0.94	Poor
E11	Japan	0.93	0.92	Fair
E12	Rose, Yugoslavia	1.09	1.06	Poor
E13	Brac, Yugoslavia	0.92	0.87	Poor
E14	Okayama, Japan	0.90	0.97	Poor
E15	Hiroshima, Japan	0.87	0.92	Poor
E16	Hokkaido, Japan	0.94	0.80	Poor
E17	Yugoslavia	...	0.97	Fair
E18	Switzerland	...	0.96	Fair
E19	Switzerland	...	1.10	Fair
E20	England	...	1.10	Poor
E22	California	...	0.86	Poor
E23	Japan	...	0.82	Fair
E24	Manchuria	...	0.82	Fair
E25	England	...	1.02	Fair
E26	Yugoslavia	...	0.92	Poor
E27	Yugoslavia	...	1.09	Poor
E28	Yugoslavia	...	0.99	Poor
E29	Yugoslavia	...	0.99	Poor
E30	Yugoslavia	...	1.01	Poor

Foreign and Domestic Strains of Pyrethrum

Twenty-nine different strains of pyrethrum have been collected from foreign countries and the United States and planted in Colorado. Analyses from those strains which have produced flowers are given in Table XCVI.

Test Plots Under Varied Conditions

In 1933, 1934 and 1935, pyrethrum test plots were located in different sections of Colorado to determine the effect of altitude and environmental conditions on the growth of the plants, yield of flowers and pyrethrins.

Large composite samples were collected at Rocky Ford, Fort Lupton, Fort Collins, Avon and Vanadium for a comparative study on weight of dried flowers and yield of pyrethrins.

The analyses, presented in Table XCVII, show a great variation in the percentage of pyrethrins. Information was also obtained regarding production costs under different conditions.

Table XCVII. Variation in Pyrethrin Content of Pyrethrum Grown Under Different Conditions in Colorado (Gnadinger, Evans and Corl)

Source	Altitude ft.	Crop year	Pyrethrins %
Rocky Ford	4330	1933	0.93
Rocky Ford	4330	1934	0.87
Fort Lupton	4900	1935	1.05
Fort Collins	5000	1933	0.96
Fort Collins	5000	1934	1.03
Fort Collins	5000	1935	1.05
Avon	7680	1933	1.26
Avon	7680	1934	1.26
Avon	7680	1935	1.37
Vanadium	9000	1934	1.23
Vanadium	9000	1935	1.34

The production of pyrethrum flowers in the high mountain valleys is more successful than in the lower altitudes or plains area. A higher pyrethrin content and greater yields of flowers are obtained in the higher elevations.

Gnadinger, Evans and Corl *(355)* have presented data indicating that the pyrethrin content may be closely related to the temperature during the growing period, low temperatures producing high pyrethrin content.

The cost of handling in the higher elevations is less than in the lower elevations, due to the higher yields of flowers.

A survey of the possible acreage available in the mountain

valleys for commercial production shows that there is plenty of land suitable for the crop.

Effect of Irrigation on Plant Losses.

Pyrethrum, grown under irrigated conditions in Colorado, is subject to a disease (Phytophthora sp.) that causes a gradual decay of the epidermal tissues of the crown stalks. In 1933 and 1934 a preliminary study was made by Evans, Gnadinger and Corl *(253)* to determine the effect of irrigation practices as means of reducing the plant losses due to crown rot.

The results of the study showed that the greatest loss of plants occurred where the irrigation furrow was immediately adjacent (3 to 5 inches) to the rows and where the moisture was excessive in the soil around the plants during the irrigation. Further the soil was not ridged and the excessive moisture did not drain towards the irrigation furrow, and, as a result, was retained in the soil around the plants for several days.

The plant losses were not great where a deep irrigation furrow (6 to 8 inches) was placed at a distance of 12 inches or more from the row and the soil around the crowns of the plants remained dry during the irrigation. In addition the soil was slightly ridged around the plants and the excessive moisture, following periods of precipitation, drained away from the plant and the surface soil dried quickly.

No significant differences were obtained in the yield of flowers or pyrethrins between the different irrigation treatments.

Effect of Spacing Interval on Yield of Flowers

Previous to 1933 the spacing intervals in the pyrethrum experimental and commercial plantings in Colorado varied from 16 to 21 inches between the plants and 30 to 36 inches between the rows. A study was made in 1934 by Evans, Gnadinger and Corl *(252)* to determine the effect of six spacing intervals (9, 12, 15, 18, 21, and 24 inches) between the plants on the mean yield of flowers per plant and yield in pounds of flowers per acre.

The 21 or 24 inch spacing interval between plants yielded a significantly greater mean number of flowers per plant than did the 9 or 12 inch spacing interval.

Significant differences in the mean dry weight of flowers per plant were obtained between the six spacing intervals. The 9-inch spacing interval, on the basis of a perfect stand and calculated yield per acre, gave a significantly greater yield in pounds of dry flowers than either the 18, 21, or 24 inch spacing interval.

The results of the preliminary study show that in experimental work with pyrethrum, where the individual plant is considered as a unit, the worker should consider the spacing interval between plants as a possible source of error in the interpretation of the results.

OTHER EXPERIMENTAL PLANTINGS

Work on pyrethrum was begun at the Tennessee Agricultural Experiment Station in 1928. Plantings were made with seed from Dalmatia, Japan, Switzerland, France and England. Crown division tests were started in 1932. The larger and more easily propagated plants, when divided, yielded over 100 plants each. Propagation by crown division was most successful when made in the fall, managed in a frame during the winter and set in the field during the spring. The cost of this type of propagation compared favorably with propagation from seed in seed beds, transplanting the seedlings to the field.

Drain and Shuey *(237)* reported the analyses of 39 samples grown at the Tennessee Station in 1934. Thirteen samples contained more than 0.90 per cent pyrethrins; eighteen samples contained less than 0.70 per cent. The lowest pyrethrin content was 0.51 per cent, the highest, 2.05 per cent.

In 1935, through the cooperation of McLaughlin Gormley King Co. of Minneapolis, Drain, Gnadinger, Corl and Shuey *(236)* reported a large number of analyses of Tennessee pyrethrum. Twenty-five selected strains were found to contain from

TABLE XCVIII. PYRETHRIN CONTENT WITHIN A VEGETATIVE STRAIN (ANALYSTS, GNADINGER AND CORL)

Strain	Plant	Pyrethrins %	Strain	Plant	Pyrethrins %
M-279	A	1.14	R1S1-1	A	1.18
M-279	B	1.90	R1S1-1	B	1.18
M-279	C	1.57	R1S1-1	C	1.14
M-279	E	1.80	R1S1-1	D	1.17
M-279	F	1.79	R1S1-1	F	1.15
M-279	H	1.83	R1S1-1	G	1.12
M-279	J	1.73	R1S1-1	H	1.16
M-279	K	1.81	R1S1-1	J	1.16
M-279	L	1.73	R1S1-1	K	1.14
M-279	N	1.60	R1S1-1	L	1.10
M-279	O	1.81	R1S1-1	N	1.18
M-279	P	1.73	R1S1-1	P	1.12

0.56 per cent to 1.80 per cent pyrethrins; 10 of these strains contained more than 1.50 per cent pyrethrins.

Samples were also taken to determine the variation in pyrethrin content of the flowers from individual plants within a vegetative strain when these plants were grown in a single progeny row. The pyrethrin content was quite uniform within a given strain, as the analyses show (Table XCVIII).

Sixty-nine samples of flowers grown in Tennessee from Swiss seed were also analyzed. The pyrethrin content varied from 0.80 per cent to 1.76 per cent; six of the samples contained more than 1.40 per cent pyrethrins.

In 1934 the writer assayed a number of samples taken from individual plants grown near Corvallis, Oregon. The variation in pyrethrin content of these plants is shown below:

Plant	Pyrethrins %	Plant	Pyrethrins %
41	1.65	69	1.15
10	1.60	54	1.13
67	1.58	6	1.12
61	1.53	31	1.12
2	1.52	26	1.10
42	1.50	39	1.10
70	1.50	38	1.09
1	1.43	27	1.08
9	1.40	43	1.08
35	1.40	59	1.08
40	1.40	66	1.08
3	1.36	7	1.05
16	1.36	13	1.05
37	1.36	15	1.05
51	1.36	28	1.04
62	1.36	33	1.03
18	1.34	63	1.03
30	1.29	53	1.02
21	1.28	23	1.01
24	1.28	58	1.01
71	1.25	29	0.99
14	1.24	44	0.99
34	1.24	17	0.97
64	1.21	36	0.97
60	1.20	65	0.97
5	1.18	11	0.96
68	1.18	22	0.95
4	1.17	12	0.93
19	1.15	57	0.83
32	1.15	25	0.82
56	1.15	52	0.68
		Avg.	1.18

Cultivation of pyrethrum in Pennsylvania was begun in 1929. Culbertson *(204)* states that more than 150 acres had been planted by 1935. Fields having 7000 to 14,000 plants to the acre yielded from 400 to 1600 pounds of flowers to the acre. The toxicity equalled that of Japanese flowers.

Culbertson *(205)*, after numerous experimental plantings in different parts of the United States, concluded that pyrethrum behaves as an annual in the far south, as a biennial a little farther north, and, in general, as a perennial lasting at least seven years. Culbertson found that pyrethrin content is little influenced by:

Geographical location of area where grown
pH of the soil
Soil type
Fertilizer treatment.

The pyrethrin content reported by Culbertson varied from 0.41 to 0.84 per cent. He considers propagation on a commercial scale by crown division impractical.

The writer has also examined samples from other experimental plantings, as shown in Table XCIX.

TABLE XCIX. PYRETHRIN CONTENT OF FLOWERS FROM EXPERIMENTAL PLANTINGS (ANALYSTS, GNADINGER AND CORL)

Source	Crop year*	Pyrethrins %	Remarks
Argentine (Tucuman)	1934	0.79	....
California (El Monte)	1930	1.10	Semi commercial planting.
Cyprus	1932	0.75	....
Idaho (Twin Falls)	1935	1.13	....
Michigan (Mentha)	1933	0.89	Grown in peat from Dalmatian seed.
Michigan (Mentha)	1933	0.93	Grown in peat from U. S. Dept. Agr. seed.
Minnesota (Minneapolis)	1933	0.76	Grown in sandy soil from Colorado seed.
Montana	1934	0.87	....
Oregon (Corvallis)	1933	1.13	U. S. Dept. Agr. seed.
Oregon (Corvallis)	1933	1.45	U. S. Dept. Agr. seed.
Oregon (Corvallis)	1934	1.17	U. S. Dept. Agr. seed.
Oregon (Albany)	1935	1.13	....
Virginia (Arlington)	1929	1.11	U. S. Dept. Agr. seed.

* All samples assayed shortly after harvesting.

INSTRUCTIONS FOR GROWING PYRETHRUM

The following instructions for first year handling of pyrethrum plantings were issued by Evans to growers cooperating in the experiments conducted by the Colorado Experiment Station:

"Choice of soil. Pyrethrum has been grown on all types of soil. A well-drained, sandy loam soil is preferable to a heavy, clay loam that is poorly drained. The grower should select the lighter and better drained soils, and avoid the creek or river bottom soils that usually are poorly drained. A soil should be selected that retains moisture well during the spring of the year.

"It is extremely important to select ground that has been clean cultivated and kept free of weeds the previous season. Weeds are a serious factor in establishing a pyrethrum planting.

"It is also suggested that, in selecting a location for pyrethrum on the farm or in a particular field, the grower should consider the possibility of putting up a temporary fence to keep out livestock during the winter. Livestock will feed on the leaves of the plant and damage the crowns and roots. Pyrethrum is non-poisonous to livestock.

"Preparation of land and seed bed. The preparation of the ground for pyrethrum is about the same as that for lettuce. Fall plowed ground is preferable to spring plowed ground, especially if it is worked down early in the spring so as to retain the moisture. If the ground is to be spring plowed, it should be done early. Early preparation means early planting and less loss of moisture. The ground should be firmed and harrowed after plowing. An ideal seed bed is one with a finely prepared surface soil that is firmed to the depth of the new plowing.

"Preparation of field for seeding. After the seed bed has been firmed the field is marked off with either a sled marker or a cultivator equipped with ditchers. The marks or furrows are spaced 28 to 30 inches apart and made deep enough to carry water. In some of the lettuce growing areas of Colorado the lettuce is planted in more or less of a bed with the rows spaced 14 to 20 inches apart and the beds or double rows spaced 26 to 28 inches apart. This method can be used very satisfactorily where two or four row planters are available. Growers intending to use this method with single row planters should mark the field first. These furrows enable irrigating for germination. The marking of the field affords a guide for planting the seed and marks the seed row for blind cultivation and also gives protection against blowing.

"Planting rate and equipment. At the present time we do not have definite information regarding the planting rate for pyrethrum. Experiments are being run at the high-altitude station to determine the proper planting rate. Until this information is available the growers will be advised as to the acreage the seed lot will plant.

"The ordinary hand garden drill or lettuce drill is best suited to planting pyrethrum. A two or four row lettuce planter may be used if the ditchers are attached in order to mark the rows for irrigation and cultivation. The Planet Jr. or Iron King garden drills are recommended. At a planting rate of 4 to 6 pounds (planting rate for 1936) per acre, the Planet Jr. is set at the third notch higher than the notch for lettuce; with the Iron King, the No. 7 notch on the regular plate is used.

"Planting. Pyrethrum should be planted early in order that it may have the advantage of the spring moisture. Planting in April is suggested if the ground can be worked. Planting before May 1 is strongly advocated, and it is not advisable to plant after May 15-20.

"The seed should be planted at a distance of 2 or 3 inches or less from the edge of the furrow that has been previously made by the marker or cultivator ditcher. If the field is marked off in 20 and 28 inch spacings, the seed should be planted on both sides of the furrow at a distance great enough to make a total of 14 to 20 inches from row to row on the bed. If the seed row is close to the edge of the furrow, it will be possible to 'sub' to the seed more quickly and thoroughly than where the seed is at a distance of 7 to 10 inches from the furrow.

"The recommended planting depth for pyrethrum is ¼ to ½ inch, preferably ¼ inch. The drill shoe should be carefully regulated for this depth. It is suggested that the planter be tested before planting to be sure that the seed is being properly distributed and that the drill shoe is set for the proper depth.

"Care of planting during germination period. Pyrethrum seed is slow in germinating, requiring 7 to 14 days and even longer periods if the conditions for germination are not favorable. If there is not sufficient spring moisture to germinate the seed, then it will be necessary to irrigate. The irrigation necessary to bring up the seed should be handled very carefully in order that the moisture 'subs' to the seed and also that the surface soil does not completely dry out during the time that the seed is germinating. Regulate and time the irrigation so as to keep the surface soil moist but not wet, and also not to form a crust. As soon as the seed is germinated the irrigations can be discontinued until the soil begins to dry out to a depth of 2 or 3 inches.

"Often there are spring snows and rains that will supply

the soil with sufficient moisture to germinate the seed. The early plantings will have the advantage of this moisture. In some instances it will be necessary to follow light snows or rains with an irrigation to maintain a moist surface soil for proper germination and to prevent drying out.

"Care of planting during the seedling stage. Pyrethrum grows very slowly during the seedling stage and it is approximately 3 to 4 weeks before the plants form a dense crown of foliage. During the seedling stage weeds may cause considerable trouble and damage to the pyrethrum unless they are hoed out. It is suggested that at the first sign of weeds the grower should cultivate the middles. If the pyrethrum is up, and weeds are coming up in the seed row, then it will be necessary to hoe by hand. A short-handled hoe or onion weeder hoe is adapted to hoeing around the seedlings. It is important that the weeds be removed when they are very small, otherwise difficulty will be encountered in removing them and also some damage will be done to the pyrethrum. A pyrethrum seedling can be distinguished by the notched edges of the leaves.

"Pyrethrum is not thinned out but allowed to form a matted row.

"The grower should instruct the weeding crew as to: (1) The appearance of the seedlings; (2) not to thin out the plants; (3) not to disturb the root system any more than is necessary.

"Cultivation. The regular one or two horse cultivators can be used in cultivating pyrethrum. Each row should be cultivated separately, especially if one row is seeded at a time. Deep cultivation is not advisable close to the plants since it has a tendency to disturb or injure the shallow root system. In the early stages of growth the weeds are removed from around the plants by hand. However, as the plants begin to 'stool-out' and crowd the weeds there is less need for hand hoeing.

"Irrigation after the plants 'stool-out'. Pyrethrum, like any other crop, should be irrigated whenever it is needed. There is no definite rule as to irrigating pyrethrum; however, it has been found at the high-altitude station that the plants respond very well to few and short irrigations during the season. Watering can be governed to a certain extent by the amount of moisture that is in the soil at a depth of 4 to 6 inches below the crown of the plants. The soil at this depth should form a moist ball when squeezed in the hand.

"When irrigating, the water should not be allowed to 'sub' beyond the shoulder of the ridges. Keep the surface of the soil around the plants dry during the irrigation. 'Sub' to the roots rather than to the ridges of the row. The long and heavy irrigations at Fort Collins were found to be detrimental as the plants developed a 'crown rot' disease due to the excessive moisture that accumulated in the surface soil around them. The length of the run should be short to prevent the upper end from becoming too wet and the lower end not wet enough.

"As the plants 'stool-out' and get larger, the irrigation furrow should be placed further away from the row and made a little deeper and the soil thrown towards the row to form a slight ridge. The slight ridging of the plants will permit better drainage of the excessive water away from the crowns and will hasten drying of the surface soil around the plants. If the planting is made with double rows, 14 to 20 inches apart, it is suggested that the irrigation should be discontinued between the double rows after the plants become established and begin to interfere with the shovels on the cultivator. The irrigation furrow can then be placed between sets of double rows.

"Irrigations should cease about the first of September in order that the plants may mature properly before going into the resting stage. If the soil is dry late in the fall, after the plants have ceased to grow, an irrigation should be given to prevent drying out during the winter."

The instructions issued to growers in Pennsylvania by Culbertson are very complete. It should be remembered, however, that some adjustments may be necessary to meet local conditions. Culbertson's instructions follow:

Seed-Bed Practice:

"The Site. Choose a rich, well drained plot of land, preferably with a southern exposure and accessible to water.

"Preparing the Bed. Prepare the bed as you would for the growing of tobacco, cabbage, celery, lettuce, etc., getting the soil level and in a fine condition. Beds may be thrown up to a height of 6 to 8 inches. This insures good drainage and root development.

"Sterilizing the Soil. Where steaming is available, use two pans. Steam 15 to 20 minutes and leave the pan on while the next pan is run. The object is to kill weed seeds and promote more rapid growth. The burning of old brush on the bed serves the same purpose. Sterilization is not necessary, but it does cut down labor costs in weeding. For fall sowing, beds may be prepared 4 to 6 weeks previous to seeding and harrowed to kill the weeds.

"Fertilizing. One to two days previous to sowing, apply 2

to 3 lbs. of a 4-8-4 fertilizer to each 100 sq. ft. of bed and rake in thoroughly. (Fertilizers on hand for truck or potatoes will do just as well.)

"Sowing the Bed. Broadcast the seed on sterilized beds; otherwise sow in rows as an aid in weeding. Allow one ounce to 100 to 150 sq. ft. of plant bed; and from 200 to 250 sq. ft. of bed per acre. Cover lightly with soil and roll or firm with a board. Germination may be hastened by soaking the seed 2 to 3 days before sowing.

"As soon as sown, tobacco cloth should be spread over the bed. Beds may be boxed in and the cloth attached to the boards by the use of small nails. The simpler method of fastening the cloth along the edges with wire pins is frequently employed. Some growers prefer placing long ryc straw directly over the bed (between the ground and the muslin) until the plants are up. If the muslin is pinned, after the plants get several true leaves, wire loops similar to a croquet wicket are made and stuck in the bed at frequent intervals to keep the cloth off the plants. In case straw is not used, hog bristles, paper, burlap bags, etc., are helpful in obtaining good germination. It is important that the upper layer of soil should not dry out until the seed has germinated, and these materials help hold the moisture.

"Seed Treatment. Under certain conditions it is advisable to treat the seed with Semesan or a similar preparation.

"Time of Seeding. Depends on the transplanting plan. (1) For June transplanting sow seeds as early in the spring as possible. (2) For August 15—Sept. 15 transplanting sow seed May 1-30. (3) For April-May transplanting sow seed in July and August.

"Care of Plants in Seed-beds. The seed will germinate in 7 to 21 days. The first pair of true leaves have notched margins. These are soon followed by more true leaves; and when there are 7 to 8 of these leaves and the plants are 4 to 7 inches high, they are ready to transplant.

"In case the plants are growing slowly, water frequently and apply sodium nitrate at the rate of one teaspoon to 6 quarts of water once a week. On a large scale, dissolve 6 lbs. in a barrel of water (50 gal.). Apply 1 gallon to every 5 sq. ft. of bed, and follow with clean water to prevent burning. For fall grown plants, there is no need of forcing, since the plants will not be set out until Spring.

"If any disease appears in the bed, spray with Bordeaux.

This may be purchased or prepared by dissolving 1 lb. of copper sulphate in 5 gallons of water and adding to this (after slaking) 1 lb. of stone lime. Then add enough water to make 10 gallons in all. Occasionally damping off occurs. For this trouble, keep the bed as dry as possible and throw it open to the sunshine and air. Semesan sometimes gives good results. For Spring sowing, leave the muslin on the bed until about two weeks previous to transplanting.

"Lifting the Plants. Unlike tobacco, pyrethrum has a fibrous root system so the plants cannot be pulled. They should be raised with a fork, taking the individual plants out from the loose dirt. Soaking makes the process easier. Place in baskets, crates, etc., and cover with burlap. Do not hold the plants in tight containers for more than ½ day or they are apt to heat and spoil.

Planting Instructions:

"Pyrethrum does not like wet feet so do not choose a low field for the planting. Gently rolling land is best. A field with too great a slope is apt to wash and gully. As a means of keeping the planting free from weeds, surround the patch with grain, alfalfa or a clean cultivated crop. Blue grass, dandelion and chick weed may blow in the crowns, get a start and be difficult to eradicate.

"Land Preparation. Prepare the land as for any other crop, except that it is advisable to work it more often to get rid of weeds so as to cut down amount of cultivation. Some growers like to plant pyrethrum after a clean tilled crop as corn, tobacco or potatoes, but as the plants will be in the ground for 6 to 7 years, it is a good plan to plow under sod in order to maintain the organic matter (preferably in the fall). Of course, if the other ground is manured, there is little difference. Manure may likewise be applied to the field from time to time after the crop is established, but should be as free from weed seeds as possible.

"A short time before the plants are ready to be set, harrow, and roll or drag. Many growers do this right ahead of the planter. If set by hand, mark out the rows 30 to 34 inches apart. This may be done with a corn planter, at the same time applying fertilizer in the rows, or by the use of a sled type marker. Plants are not set in furrows, so do not use the usual big shovel marker. In case the plants are to be set by hand, the soil should be ridged slightly. This may be done with:

A two-horse cultivator with hillers attached.

A potato planter.

One horse cultivator with hillers turned in.

Regular hiller.

After the ground is laid out with the ridge, wait for a rain to settle and firm the ground. In the meantime, purchase or make several dibbers or dibbles to use in setting. These may be purchased from any seed house. They may be made by cutting 8 to 9 inches off a broom stick, and rounding one end and pointing the other.

"Transplanting by Hand. After a rain, during a slight drizzle, or any time moisture conditions are right, lift the plants with a fork, place the plants in containers and proceed to the field. Each person transplanting should take a bunch of plants in his left hand and the dibble in the right hand. Punch a hole in the row with the dibble, slip in the roots of the plant with the left hand, and use dibble again to close the dirt around the roots. This is done by shoving the dibble in the ground right beside the roots and should be done so as to close the air space clear to the bottom of the roots. In setting, use care so that the crown is not covered. A boy with a basket of plants moves among the setters and supplies them with plants.

"Transplanting by Machine. As a cup of water is placed with each plant, the moisture conditions are not as important as in the case of hand setting. However, it is, nevertheless, advantageous that the soil is not too dry. Usually, it is best to wait for a rain and let the soil dry off to the point where the rollers on the planter will move readily. As in the case with hand planting, harrow and roll or drag immediately ahead of the planter. This firms the soil while still permitting the shoe to penetrate to the desired depth. It is also easier to mark for the next row. Horses or a tractor may be used to pull the planter. In case fertilizer is to be applied, use about 300 to 500 lbs. of a 4-8-4 per acre. This may be drilled in with a corn planter or seed drill shutting off the hoes which come between the rows. If the men setting will use care, there is little danger of covering the crown. Hold the plant until the rollers press around the roots and they may be set as carefully with a machine as by hand.

"As a rule two men will lift enough plants in 2 to 3 hours to run the planter ½ day. The planter capacity is about two acres per day if the plants and water are supplied to the field. A simple carrier may be rigged on the rear of the machine to carry extra plants. To insure a good stand, have a boy with plants follow the planter. He may fill in where missing and at the same time straighten out any plants improperly set.

"Replanting. There is bound to be some replanting and this may be done any time there is sufficient moisture in the soil. Generally, stick plants just after a rain when the soil will adhere to the roots.

"Planting Distances. This depends somewhat on the method to be employed in cultivating. Usually the rows are run 27 to 33 inches apart with the plants spaced 15 inches apart in the rows. For those wishing to cultivate both ways, plants may be set 21 by 21 inches, 24 by 24 inches, 30 by 21 inches, etc.

Planting Distances	Plants per acre
27 inches x 15 inches	15,560
30 inches x 15 inches	13,942
30 inches x 18 inches	11,614
30 inches x 21 inches	10,000
24 inches x 24 inches	10,890
36 inches x 15 inches	11,614
27 inches x 18 inches	12,922
21 inches x 21 inches	14,235
33 inches x 15 inches	12,663

"Fertilizing. Pyrethrum is not classed as a heavy feeder like tobacco, but the crop responds well to both manure and commercial fertilizers, and the added yield more than offsets the cost of the materials. If the land is in good shape the crop may not need any fertilizing materials applied, at least the first year. The growth of the plants should be taken as the basis for application. Normally, under good growing conditions, Spring set plants will be 15 to 18 inches in diameter by October 1st. The general recommendations include the application of 5 tons of manure per acre. This may be supplemented with 300 lbs. of a 4-8-4 fertilizer applied in the row under the plants. In regard to fertilization from year to year, side applications may be made with fertilizer attachments to the cultivator. Materials should be applied in the early Spring about the time the new growth starts or immediately following harvest. The formula and the ratio will vary according to the soil and climate but one is safe in the application of 400 to 500 lbs. of a 4-8-4 fertilizer each season. On small tufts, apply in the Spring, and for large tufts apply in the early spring or following harvest. Some growers split the application and apply at both periods. In case the soil is quite poor, or the plants are making poor growth, a side dressing of 75 to 100 lbs. nitrate of soda may be made.

"Cultivation. In cultivating do not throw the dirt up over the crowns. The regular one or two horse cultivators may be used but where possible, use scrapes instead of shovels. Deep cultivation injures the root system, and as the main object in any cultivating is merely to get rid of weeds, scrapes do a much better job. Depending on the season, the patch should be gone through 3 to 5 times. Hoeing should follow each cultivation. Use a regular narrow tobacco hoe rather than the usual garden hoe, and scrape along in and out between the plants in the row rather than chop up the ground all around the plant. As the weeds are kept down and the plantation becomes older with a tendency for the plants to fill up the row and crowd out weeds, the amount of cultivation is lessened. Usually stop cultivating about September 20th so as to permit the plants to harden before cold weather.

"Harvesting. Harvest pyrethrum when the stems are still upstanding and most of the flowers are separated from the stems on the plant and the stems mowed off later; and the various methods hereafter enumerated whereby the plant is cut near the ground and the flowers stripped off later by means of a stationary stripper.

"Cutting may be done:

"1. By hand, using sickles or right angled corn knives. Use the left arm and leg in holding the plant, place the sickle (seemingly backwards) between the leg and the lower stems, and with an upward pull of the right hand, sever the entire plant as close to the ground as possible. The bunch is placed across the row and one proceeds to cut at a slow walk. It requires about 30 man hours to cut an acre by this method.

"2. By mowing machine or self reaper. Cut two rows at a time. The plants fall in bunches and may be bound in sheaves or picked up loose and hauled to the stripper.

"3. With a binder. In case the stand is uniform and the stems are upstanding, the crop may be harvested with a binder making small loose sheaves. It is advisable to use a mower blade and to cut only two rows at a time. In case the heads are badly tangled or the plants are down badly, do not use this method of harvesting.

"Stripping. The stripper consists of a cylinder, 12 inches in diameter, and 20 inches wide with 12 rows of teeth, 1½ inches long, $\frac{3}{16}$ in. thick, ¾ in. wide, spaced with the rows 3½ in. and the teeth 1½ in. apart.

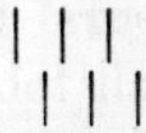

"Thus in line but staggered, and mounted so as to feed readily and with a chute to carry the flowers away. It is really a modified broom corn cleaner and there is probably one in every community which could be purchased or rented for a small sum. Cylinders may be purchased direct from Paris Foundry & Machine Works, Paris, Ill., or Landis Bros., Lancaster, Pa. In case the machine is home-built or a broom corn cleaner is adapted, fix it for under feeding.

"The machine is set up on the barn floor near the doors so as to let out the exhaust from the motor. The legs are braced with boards and an old door or similar feeding table is fastened to the side of the machine. In case the bunches are not tied, they are piled evenly on the wagon or truck with the heads out, hauled in, removed carefully and stripped right off the bed. Bunches are placed on the feeding table and redivided so they may be held well with two hands. The heads are held under and in contact with the cylinder and with a little practice the flower heads separate freely from the stems. The flowers shoot through and are forked over to take out excess stems.

"The capacity is about 1 acre per day and the job requires three men, one to feed, one to separate the bundles on the feeding table and one to supply bundles and take away flowers. A seat may be placed for the feeder to sit on.

"A field stripper is being developed which we hope will do away with hand stripping and strip an acre in several hours. It is at present about 85% efficient.

"Drying. Drying may be accomplished in the sun or shade, but care must be used to prevent heating and discoloration of the flowers. Generally pile flowers about 1 inch deep and turn frequently with a rake for several days. As they become dryer, they may be piled deeper. The following methods of drying are suggested:

"1. On a barn floor or other flooring suitable. 2. On canvases, or tarpaulins, out in the sun and pulled under shelter at night. 3. On wire screens (on a frame 6 by 12 ft. and 4 to 6 inches high). These may be set out in the sun during the day and racked at night on the barn floor or in the open and covered with canvas. They may also be placed in brooder houses, etc., where artificial drying may be used. The latter is especially recommended. 4. In a sweet corn dryer, flowers may be dried beyond the stage where they will heat in 2 to 3 hours' time. 5. On tobacco lath latticed across the tiers over which tobacco muslin is placed. When harvested with a binder, rank the sheaves like oats and dry in the field. If the weather is threatening, the sheaves may be hung on tobacco lath or 5 penny nails in a shed on the barn floor.

"Baling. The flowers are thoroughly dry when they will snap when broken between the fingers. When dry place in large beet pulp sack, or, preferably, bale. Hydraulic presses may be located in the community and in some instances perhaps the local cider press may be used. Any means of reducing the bulk will be satisfactory. When the whole crop is used, the sheaves may be baled with a regular hay baler.

"A few common mistakes noted that you may avoid: 1. having plants too thick in the bed, resulting in spindly and not hardened off plants. 2. Setting out in dry hot weather. If plants are not ready by July. 1, wait until August 15-September 15. 3. Setting out plants too small. 4. Setting too late in fall, resulting in heaving. 5. Failure to cultivate when weeds are small. 6. Piling flowers too deep, with failure to turn frequently. Flowers must not heat and be discolored."

The comments of Sievers *(812)* on the cultivation of pyrethrum are:

"It has been observed that the pyrethrum plant, at least the species worked with by the U. S. Department of Agriculture *(Chrysanthemum cinerariaefolium)*, is quite resistant to drought. However, during the period when the flowers are being formed, sufficient moisture must be available or the yield is affected to an important extent. After the flowers are removed the plants will maintain themselves throughout a long, hot summer with very limited rainfall. When the young seedlings are first set out in the field prolonged dry weather is detrimental. The actual loss of plants may not be excessive but their development is retarded.

"In view of the fact that favorable moisture conditions are essential during the time when the flowers are forming, it is evident that the culture of the crop under irrigation would have distinct and important advantages.

"Under such conditions a maximum growth of flowers can be brought about by applying water at the time when most needed. Furthermore, on the large level tracts usually available in irrigated districts, machinery can be employed to the best advantage although it does not follow that smaller areas on undulating or even hilly ground could not be used. It is also true that in irrigated regions there is greater likelihood that good drying weather will prevail at harvest time. Uninterrupted clear, sunny weather would assure rapid and thorough curing or drying of the flowers in the field, thus eliminating the additional cost involved in removing the harvested crop to barns or sheds for drying. It is possible that flowers dried under such conditions might be adversely affected as to color but while this would be undesirable in the production of the powder their usefulness for spray extracts would not likely be impaired thereby.

"Pyrethrum seed does not germinate quickly nor evenly and the seedlings do not grow rapidly. Therefore, in order to insure a satisfactory stand the plants must be grown in prepared seed beds and later transferred to the field. The seed should be sown in a carefully prepared seed bed, with a southern exposure, as early in April as possible. If the bed is covered with cheese cloth and kept moist, germination should begin in two or three weeks. By early June the plants should be from 4 to 6 inches high and ready for transplanting. The regular tobacco-setting equipment has been used very successfully. The plants should be set in rows 30 inches apart and at intervals of about 15 inches in the rows. Care should be taken not to set the plants too deep, as covering the crown buds may cause the plants either to die or require considerable time to adjust themselves. During the summer and fall the plants make good root growth, and such plants will produce a moderate crop of flowers the following season and will for several years produce a maximum crop. Thereafter, when the yields decline noticeably, it is best to abandon the planting. A practical plan would be to start a new planting each season to replace the acreage discontinued. Such procedure maintains the distribution of labor and the same productive acreage from year to year.

"Experiments have been made during the past few seasons at the Arlington Experiment Farm at Rosslyn, Va., on the cutting and threshing of pyrethrum by machinery. The ordinary grain binder has been used to good advantage in cutting the crop where there is not too much lodging."

Attempts at Threshing Pyrethrum

In an earlier chapter it was shown that most of the pyrethrins in the flower head are present in the achenes. This has suggested the possibility of harvesting the flowers with stems and separating the achenes by threshing, using machinery at present available. Unfortunately, it is impossible to get a complete separation of the achenes without considerable loss, because all of the flowers are not equally mature and the immature achenes are quite small. Even more important is the fact that the pyrethrins in the achenes seem to decompose more rapidly when the achenes are detached from the receptacles. The analyses of a number of separations, obtained by threshing flowers harvested with stems, are given in Table C; these results are not very promising.

TABLE C. PYRETHRIN CONTENT OF SEPARATIONS MADE BY THRESHING

No.	Description	Pyrethrins %
1	Whole flowers, without stems	0.72
2	Immature achenes, disk florets, involucral scales and leaf particles	0.89
3	Ray florets, parts of receptacles and leaves	0.35
4	Involucral scales, immature achenes	0.26
5	Clean, mature achenes	0.45

So called pyrethrum seed offered for import into the United States have, in some cases, proved to be ground stems, leaves and sweepings.

REFERENCES†

Patents are indicated by an asterisk.—*

1. Abbott, W. S.
A study of the effect of storage, heat and moisture on pyrethrum. U. S. Dept. Agr. Bull. **771**, 1-8, 1919.

2. Abbott, W. S.
Experiments with miscellaneous substances against chicken lice and the dog flea. U. S. Dept. Agr. Bull. **888**, 1-5, 1920.

3. Abel, J.
Insect powder. Am. J. Pharm., **32**, 520, 1860.

4. Abelous.
Meeting at Bedarieux on April 15th. Ann. soc. hort. hist. nat. Herault, **54**, 163, 1923.

5. *Adams, E. W.
Pyrethrum and alkyl phthalates used in fly sprays. U. S. Patent 1,957,429. May 8, 1934.

6. *Adams, E. W.
Insecticide comprising an alkyl benzoate and pyrethrum in mineral oil. U. S. Patent 2,000,004. May 7, 1935.

7. Adrien.
Notes on pyrethrum. J. pharm. chim., **25**, 43, 1892.

8. Alessandrini, G.
Malaria in Pontine district during 1933. Conquist. Terra (Rome), No. 2, 1-17, 1934.

9. Allen, T. C.
New spray simplifies potato leaf hopper control. Wisc. Agr. Exp. Sta. Bull. **428**, 107, 1934.

10. Allen, T. C.
Application of atomized oil sprays to truck crop insects. J. Econ. Entomol., **28**, 496, 1935.

11. Alvord, E. B., and Dietz, H. F.
Control of insects on plants by chemical means. Ind. Eng. Chem., **25**, 631-2, 1933.

12. Alwood, W. B.
Pyrethrum powder. Va. Agr. Exp. Sta. Bull. **24**, 22-23, 1893.

13. Andre, E.
Insects and pyrethrum powder. Rev. hort., **76**, 457-8, 1904.

14. Andrus, A. U.
Insect flowers and powders. Pharm. Record, **8**, 316, 1888.

15. Anglade, M. and Gaudin, O.
Three cases of intestinal parasites cured by pyrethrins. Bull. sci. pharmacol., **39**, 480-3, 1932 and Bull. acad. med. **108**, 917-21, 1932.

† The writer is indebted to the excellent bibliography in U. S. Department of Agriculture Bulletin 824 for many of the earlier references.

16. Anglade, M., Gaudin, O. and Arcony, R.
Clinical use of pyrethrins against intestinal parasites and secondary disturbances caused by them. Bull. sci. pharmacol., **39**, 23-6, 1932.
17. Anonymous.
Persian camomile and Caucasian insect powder. Arch. Pharm., **136**, 375-6, 1856.
18. Anonymous.
Persian insect powder. Wochbl. Land-Forstwirt., **8**, no. 47, 256, 1856.
19. Anonymous.
Caucasian insect powder. Arch. Pharm., **144**, 378-9, 1858.
20. Anonymous.
Persian insect powder. Oesterr. Z. Pharm., **11**, 437, 1858.
21. Anonymous.
Persian insect powder. Bonplandia, **10**, 342, 1862.
22. Anonymous.
Insect powder. J. pharm. soc. Anvers, 1862.
23. Anonymous.
Insecticides. Pharm. J. Trans., **31**, 530, 1871.
24. Anonymous.
Insect powder. Pharm. Ztg., **25**, 300, 1880.
25. Anonymous.
Insect powder. Pharm. Ztg., **29**, 443, 1884.
26. Anonymous.
Leucanthemum vulgare as insect powder. Pharm. Ztg., **34**, 118, 1889.
27. Anonymous.
Caucasian insect flowers. Abstract in Chemist Druggist, **34**, 403, 1889.
28. Anonymous.
Poisoning with insect powder. Chemist Druggist, **35**, 557, 1889.
29. Anonymous.
Cultivation of medicinal plants in Japan. Ann. Rept. Cent. Sanitary Bureau, Japan, 1888-9.
30. Anonymous.
Insect powder. Chemist Druggist, **36**, 402-3, 1890.
31. Anonymous.
Purity of insect powders. Pharm. Ztg., **37**, 310, 1892.
32. Anonymous.
Insect flower stems. Drug. Circ. Chem. Gaz., **38**, no. 7, 146, 1894.
33. Anonymous.
Anthelmintic properties of pyrethrum. Chemist Druggist, **52**, 165, 1898.
34. Anonymous.
Cultivation of insect flowers in Algeria. Pharm. J., **63**, 549, 1899.
35. Anonymous.
Pyrethri flores. Pharm. J., **66**, 831-2, 1901.

36. Anonymous.
Wild forms of chrysanthemum. Pharm. J., **67**, 695, 1902.

37. Anonymous.
Insect powder. Insecticide Decision 1. U. S. Dept. Agr., 1-3, 1911.

38. Anonymous.
Cultivation of insect powder plants. Pharm. Era., **51**, no. 3, 61-3, 1918.

39. Anonymous.
Notice to manufacturers of insect powder. U. S. Dept. Agr. S. R. A. Insecticide, **22**, item 54, 1918.

40. Anonymous.
U. S. Navy, Bur. Med. and Surg. Dept. Bull. **86**, 6, 1919.

41. Anonymous.
Notice relative to insect flower stems. U. S. Dept. Agr. S. R. A. Insecticide, **31**, item 56, 1920.

42. Anonymous.
A comparison of fly spray formulas. Soap, **2**, no. 2, 49-53, 1926.

43. Anonymous.
Fly sprays not jeopardized by Terry patent. Soap, **2**, no. 3, 51, 1926.

44. Anonymous.
Pyrethrum. Texas Agr. Exp. Sta. Ann. Rept., p. 22, 1929.

45. Anonymous.
Federal specifications for pyrethrum powder. Federal Standard Stock Catalogue, Sec. IV, part 5, 1930.

46. Anonymous.
Labeling fly sprays for animals. U. S. Dept. Agr., F. and D. Administration Circ., Jan. 31, 1930. Revised Dec. 16, 1935.

47. Anonymous.
The pyrethrum industry of Japan. Bull. (British) Imperial Institute, **28**, 300-42, 1930.

48. Anonymous.
Pyrethrum flowers from Kenya and Cyprus. Bull. (British) Imperial Institute, **28**, 425-9, 1930.

49. Anonymous.
Pyrethrum investigations. Ky. Agr. Exp. Sta. Ann. Rept., p. 26, 1930.

50. Anonymous.
Skinner to help rehabilitate agriculture in Virgin Islands. U. S. Dept. Agr. Off. Rec., **10**, 92, 1931.

51. Anonymous.
Mosquito extermination. Ind. Eng. Chem. News Ed., **10**, 245, 1932.

52. Anonymous.
Toxicity of pyrethrum extracts by the roach test method. Soap, **8**, no. 1, 107-9, 1932.

53. Anonymous.
Insecticidal plants. Ann. Rept. (British) Imperial Institute, p. 30, 1932.

54. Anonymous.
Pyrethrum. Tenn. Agr. Exp. Sta. Ann. Repts., p. 40, 1929; p. 41, 1930; p. 49-50, 56-57, 1931; p. 35, 41, 1932.

55. Anonymous.
History of pyrethrum in Kenya and its possibilities. Kenya Colony Dept. Agr. Mimeographed pamphlets, 1932.

56. Anonymous.
Urge pyrethrum. Lancaster, Penna. Intelligence J., Mar. 10, 1933.

57. Anonymous.
Aeroplanes fighting forest insects. Hamburger Fremdenblatt, May 29, 1933. Also Hamburger Nachrichten, June 14, 1933.

58. Anonymous.
The steam sprayer for insecticides. Soap, **9**, no. 3, 83-4, 1933.

59. Anonymous.
Allergy to pyrethrum. J. Am. Med. Assoc., **101**, 729, 1933.

60. Anonymous.
Controlling the brown dog tick. North Am. Veterinarian, **14**, no. 8, 35, 1933.

61. Anonymous.
Developments in economic entomology. Ind. Eng. Chem. News Ed., **2**, 362, 1933.

62. Anonymous.
Advances in entomology. Ind. Eng. Chem. News Ed., **12**, 447-8, 1934.

63. Anonymous.
Empire pyrethrum. Chemist Druggist, **121**, 338, 1934.

64. Anonymous.
Kerosene-pyrethrum sprays against *Antestia* and *Lygus*. Kenya Colony Dept. Agr. Leaflet 1, 1934.

65. Anonymous.
Report on the bedbug. Ministry of Health (London) Rept. Pub. Health Med. Subj., **72**, 1-46, 1934.

66. Anonymous.
Bedbug liquids. Soap, **11**, no. 3, 87, 1935.

67. Anonymous.
Bulgarian pyrethrum. "Eildienst" No. 235, Oct. 9, 1935.

68. Anonymous.
Control of insects. U. S. Dept. Agr. Bur. Entomol., E-343, April, 1935.

69. Anonymous.
Developments in economic entomology. Ind. Eng. Chem. News Ed., **13**, 475, 1935.

70. Anonymous.
Directory of insect pest control. East. Br. Am. Assoc. Econ. Entomol., New Brunswick, N. J., 1935.

71. Anonymous.
Harvesting pyrethrum. Soap, **11**, no. 8, 107, 1935.

72. Anonymous.
Limit insecticide safety claims. Soap, **11**, no. 8, 107, 1935.

73. Anonymous.
Report of council on pharmacy and chemistry. J. Am. Med. Assoc., **105**, 119, 1935.

74. Anonymous.
Answers Ruzicka on pyrethrum. Soap, **11**, no. 11, 115, 1935.

75. Anonymous.
Pyrethrum powder and pyrethrum products. Calif. Dept. Agr. Special Bull. **138**, 6, 1935.

76. Anonymous.
Pyrethrum tests unsound, says Ruzicka. Soap, **11**, no. 10, 111, 1935.

77. Anonymous.
Advocates pyrethrum subsidy for U. S. Soap, **11**, no. 12, 103, 1935.

78. Anonymous.
Moth sprays. Soap, **12**, no. 3, 95, 1936.

79. Anonymous.
Stock sprays evaluated. Soap, **12**, no. 3, 101, 1936.

80. Anonymous.
Yugoslav pyrethrum. Soap, **12**, no. 3, 117, 1936.

81. Anonymous.
Pyrethrum culture in Bulgaria. U. S. Dept. Commerce, World Trade Notes, **10**, no. 1, 5, 1936.

82. Anonymous.
Production of pyrethrum. U. S. Dept. Commerce, World Trade Notes, **10**, no. 2, 3, 1936 and **10**, no. 18, 5, 1936.

83. Argall, C. B. K.
Japanese pyrethrum. Chemist Druggist, **121**, 750-4, 1934.

84. Arie, J.
Pyrethrum. Octavo, 118 pp. Inst. Anna Rosa, Sao Paulo, 1935.

85. Audiffren, M.
Color reaction of chrysanthemum monocarboxylic acid. J. pharm. chim., **19**, 535-6, 1934.

86. Austin, M. D.
Pyrethrum experiments in 1928. J. So. E. Agr. Coll., Wye, Kent, **26**, 124-35, 1929.

87. Austin, M. D.
Sciara in mushrooms. Ann. Appl. Biol., **21**, 167-71, 1934.

88. Austin, M. D. and Martin, H.
Incorporation of contact insecticides with protective fungicides. J. So. East. Agr. Coll., Wye, Kent, **32**, 49-58, 1933.

89. Aymard, J.
The culture of pyrethrum. Ann. soc. hort. hist. nat. Herault, **51**, 162, 1920.

90. Baber, E.
Insect spray for destruction of mosquitoes and flies. J. Roy. Army Med. Corp. (London), **62**, 411-18, 1934.

91. Back, E. A.
Clothes moths and their control. U. S. Dept. Agr. Farm. Bull. **1353**, 27, 1931.

92. Back, E. A., and Cotton, R. T.
Control of moths in upholstered furniture. U. S. Dept. Agr. Farm Bull. **1655**, 32, 1931.

93. Badertscher, A. E.
Effect of soap on the toxicity of a pyrethrum product. J. Econ. Entomol., **24**, 268-77, 1931.

94. Badertscher, A. E.
Rotenone in fly sprays. Soap, **9**, no. 8, 85-6, 1933.

95. *Badertscher, A. E.
Horticultural spray containing pine oil, dipentene, pyrethrum and soap. U. S. Patent 1,938,652. Dec. 12, 1933.

96. *Badertscher, A. E.
Pyrethrum dust. U. S. Patent 1,940,899. Dec. 26, 1933.

97. Badertscher, A. E. and Wotherspoon, R.
Derris and pyrethrum powders. Soap, **11**, no. 4, 87, 1935.

98. Bailey, J. F.
Pyrethrum on cabbage and cotton worms. Am. Entomol., **3**, 296, 1880.

99. Bake, L. S.
Effect of lead on pyrethrum extracts. Soap, **11**, no. 11, 111, 1935.

100. Baldacci, A.
Relation of Dalmatian pyrethrum to *P. cinerariaefolium*. Ann. soc. agr. prov. Bologna, **34**, 209, 1894.

101. Baldacci, A.
Vegetable products used in Albania and Epiro. Atti accad. Georgofili, **19**, 58, 1896.

102. Baldi.
Pyrethrum. Boll. soc. ort. Palermo, p. 17, 1915.

103. Ballard, E., Nustikawi, A. M. and El Zoheiry, M. S.
The desert locust in Egypt. Ministry Agr., Egypt Tech. Sci. Serv. Bull. **110**, 1-149, 1932.

104. Bandt, H. I.
Fish poisoning caused by pyrethrum. Mitt. Fischereiver. Ost-Ausgabe, **37**, 165-6, 1933.

105. Basinger, A. J. and Boyce, A. M.
Orange worm control. Calif. Citrograph, **20**, 158, 178-9, 1935.

106. Beckwith, C. S.
Control of false blossom of cranberries. Am. Cranberry Growers' Assoc. Proc., 64th Ann. Conv., 8-9, 12-13, 1933.

107. Beckwith, C. S.
Dusting cranberry bogs from the air. Am. Cranberry Growers' Assoc. Proc. 69th Ann. Conv., 14-16, Aug. 1935.

108. Beckwith, M. H.
Rept. Entomologist. Del. Agr. Exp. Sta., 109-36, 1889.

109. Beckwith, M. H.
Rept. Entomologist. Del. Agr. Exp. Sta., 97-8, 1891.

110. Bedell, C. S.
Exterminating mosquitoes without harm to wild life. Game Breeder, **36**, no. 8, 1932.

111. Beringer, G. M.
The Hungarian daisy as an adulterant of insect powder. Am. J. Pharm., **61**, 1-4, 1889.

112. Birdsall, R. W.
Commercial aspect and future possibilities of rotenone. Ind. Eng. Chem., **25**, 643, 1933.

113. Bishop, W. D.
Rept. Commissioner of Patents for 1859, Agriculture, 13-19, Washington, 1860.

114. Bishopp, F. C.
The pigeon fly. J. Econ. Entomol., **22**, 974-80, 1929.

115. Bishopp, F. C.
The stable fly. U. S. Dept. Agr. Farm. Bull. **1097**, 12, 1931.

116. Bishopp, F. C., Laake, E. W., Brundrett, H. M. and Wells, R. W.
The cattle grubs or ox warbles. U. S. Dept. Agr. Bull. **1369**, 1926.

117. Bishopp, F. C., Roark, R. C., Parman, D. C., and Laake, E. W.
Repellents and larvicides for the screw worm and other flies. J. Econ. Entomol., **18**, 776-8, 1925.

118. Bissell, T. L.
Fly control. Ga. Agr. Exp. Sta. Circ. **102**, 1-5, 1934.

119. Blin, H.
Culture of pyrethrum in Southern France and Algeria. Sec. de Agr., Viacao, Industria e Obras Publicas do Estado da Bahia (Brazil) Boletim, **2**, no. 4, 279-83, 1903.

120. Bohmer, K.
Chrysanthemum corymbosum. Pharm. Ztg., **40**, 523, 1895.

121. Bojtschinow, A.
Methods for the examination of pyrethrum and its preparations. Pharm. Presse, Wiss.-prakt. Heft, 61-63, **1933**.

122. Borzi, A.
The culture of pyrethrum. Boll. Studi Giard. Colon. Palermo, **5**, 45, 1919.

123. Bosredon, Dr.
Poisoning by pyrethrum powder. Bull. gen. therapeutique, medicale, chirurgical, obstetrical et pharmaceutique, **132**, 275-6, 1897.

124. Boucard, J.
Pyrethrum caucasicum. L'Union pharmaceutique, **57**, **1858**.

125. *Bouma, J. A. L.
Pyrethrum is added to an oil containing dissolved sulfur. Dutch Patent 30,941. Sept. 15, 1933.

126. Bourcart, E.
Diseases of plants (pyrethrum). Pp. 182, 440, 543. Paris **1910**.

127. *Bousquet, E. C.
Insecticide comprising soya bean lecithin, with or without pyrethrum or other contact insecticides. U. S. Patent 2,006,227. June 25, 1935.

128. Boyce, A. M.
Codling moth in Persian walnuts. J. Econ. Entomol., **28**, 869, 1935.

129. Braquehaye.
Pyrethrum powder as an insecticide. Rev. hort., **76**, 508, 1904.

130. Breakey, E. P.
Halowax as a contact insecticide. J. Econ. Entomol., **27**, 393-8, 1934.

131. Breakey, E. P. and Miller, A. C.
Halowax as an ovicide. J. Econ. Entomol., **28**, 358-65, 1935.

132. *Breuer, A.
Steam sprayer. U. S. Patent 1,989,912. Feb. 5, 1935.

133. Britton, W. E.
Springtail troublesome in houses. Conn. Agr. Exp. Sta. Bull. **368**, 253, 1935.

134. Britton, W. E. and Friend, R. B.
Insect pests of elms. Conn. Agr. Exp. Sta. Bull. **369**, 1935.

135. Browne, D. J.
Persian insect powder. Rept. Commissioner of Patents for 1857, Agriculture, 129-30. Washington, 1858.

136. Browne, D. J.
Persian insect powder for *Coccus hesperidum*. Rept. Commissioner of Patents for 1858, Agriculture, 270, Washington, 1859.

137. Buchmann, W.
Pyrethrum for combating insects injurious to health. Z. Desinfekt. Gesundh., **21**, no. 9, 1929.

138. Buchmann, W.
Physiological action of pyrethrum. Z. Desinfekt. Gesundh., **21**, 14-6, 37-9, 72-6, 1929.

139. Buchmann, W.
Evaluation of pyrethrum insect powder through physiological tests on animals. Z. Desinfekt. Gesundh., **22**, 414-8, 1930.

140. Buchmann, W.
Powders valuable as insecticides. Kleine mitt. Mitglieder Ver. Wasser-Boden-Lufthyg., **7**, 1931.

141. Buchmann, W.
Chironomous damage, its prevention and control. Z. Gesundh. Städtehyg., **24**, 1932.

142. Buchmann, W.
Particle size of active principles in aqueous suspensions of pyrethrum powder. Z. angew. Entomol., **20**, 136-49, 1933.

143. Buhach Producing and Mfg. Co.
Buhach, the great California insecticide. 1-32, Stockton, Calif., 1879.

144. Burdette, R. C.
Sprays for squash vine borer. J. Econ. Entomol., **28**, 229, 1935.

145. *Burgess, W. S. and Golley, E. K.
Method of destroying insects with pyrethrum. U. S. Patent 1,926,579. Sept. 12, 1933.

146. *Burgess, W. S. and Golley, E. K.
Apparatus for diffusing pyrethrum solutions. U. S. Patent 2,006,437. July 2, 1935, and Canadian Patent 359,106. July 14, 1936.

147. Caesar and Loretz.
Chrysanthemum leucanthemum. Pharm. Ztg., **34**, 396-7, 1889.

148. Caesar and Loretz.
Powdered flowers of *Chrysanthemum cinerariaefolium*. Pharm. Zentralhalle, **39**, 727, 1898.

149. Caesar and Loretz.
Insect powder. Pharm. Ztg., **43**, 325, 1898.

150. Caesar and Loretz.
Insect flowers. Jahresbericht, p. 110, Aug., 1911.

151. Caldairou, A.
Note on attempts to cultivate pyrethrum. Bull. soc. centre. agr. Aude, **102**, 478, 1921.

152. Campbell, F. L., Sullivan, W. N. and Jones, H. A.
Derris in fly sprays I. Tests of derris and cube root. Soap, **10**, no. 3, 81-87, 1934.

153. Campbell, F. L., Sullivan, W. N. and Jones, H. A.
Derris in fly sprays II. Comparative tests of extracts of derris and pyrethrum. Soap, **10**, no. 4, 83, 1934.

154. Cannon, C. W.
Introduction of pyrethrum culture in Bulgaria. Rept. U. S. Consul, Sofia, June 26, 1935.

155. Carlos, A. S.
Manufacture and use of insecticides and fungicides. Fertilizer, Feeding Stuff and Farm Supplies J., **16**, 222-3, 1931.

156. Carpenter, W. L.
Experiments with *Pyrethrum roseum*. Am. Naturalist, **13**, 176-7, 1879.

157. Cates, J. S.
Something new in insecticides. Country Gentleman, 18, Oct., 1935.

158. *Cattanach, J. R.
Non-poisonous insecticide. Canadian Patent 327,739. Nov. 22, 1932.

159. *Cattanach, J. R.
Pyrethrum insecticide. Canadian Patent 329,936. Feb. 7, 1933. Also British Patent 401,519, Nov. 16, 1933.

160. Cazeneuve, P.
Proposed French law for suppression of fraud in insecticides. Ann. fals., **21**, 261-6, 1928.

161. Cecuk, S.
Dalmatian pyrethrum. Agronomski Glasnik (Belgrade), **4**, no. 11, 272-9, 1933.

162. Chevalier, J.
Pyrethrum insecticide, its preparations, its activity. Rev. vit. **58**, 320-22, 1923.

163. Chevalier, J.
Insecticidal pyrethrum. Recherches et inventions, **9**, 143-52 1928.

164. Chevalier, J.
Present status of the evaluation of pyrethrum. Ann. fals., **21** 318-23, 1928.

165. Chevalier, J.
Pyrethrins from *Chrysanthemum cinerariaefolium* as anthelminthics. Bull. acad. med., **99**, 446, 1928.

166. Chevalier, J.
A new anthelminthic. Rev. vet. et J. med. vet. Zootech., 80, 1928.

167. Chevalier, J.
Pyrethrum. The pharmacodynamic and therapeutic activity. Bull. sci. pharmacol., **37**, 154-65, 1930.

168. Chevalier, J.
Pyrethrum. Culture and economic future. Bull. sci. pharmacol., **37**, 235-9, 1930.

169. Chevalier, J.
Pyrethrum. Industrial and pharmaceutical preparations. Bull. sci. pharmacol., **37**, 422-31, 1930.

170. *Chevalier, J.
Insecticide; pyrethrum in oil is washed with permanganate solution and dehydrated. French Patent 724,216. Dec. 18, 1930.

171. Chevalier, J.
Ashes of medicinal plants. IV Cong. intern. plantes med. essences (Paris), 173-6, 1931.

172. Chevalier, J.
Pharmacologic and therapeutic researches on the pyrethrins. Bull. soc. therapeut., **37**, 6-26, 1932.

173. Chevalier, J.
The pyrethrum problem. Ann. fals., **25**, 525-8, 1932.

174. Chevalier, J. and Dantony, E.
Toxic action of the insecticidal principle of pyrethrum flowers. Compt. rend., **177**, 1077-79, 1923.

175. Chevalier, J. and Mercier, F.
Pharmacodynamic action of the insecticidal principle of pyrethrum flowers. Compt. rend., **176**, 1847-8, 1923.

176. Chevalier, J. and Mercier, F.
The pyrethrins. Compt. rend. soc. therapeut., **34**, 222, 1928.

177. Chevalier, J. and Ripert, J.
Pharmacodynamic action and physiological titration of pyrethrum preparations. Compt. rend., **184**, 776-8, 1927.

178. Childs, L.
Substitutes for lead arsenate in codling moth control. Oregon State Hort. Soc. 26th Ann. Rept., 85-9, 1934.

179. Chittenden, F. H.
Insects injurious to beans and peas. U. S. Dept. Agr. Yearbook for 1898, p. 254, 1899.

180. Chou, T. Q. and Chu, J. H.
Chrysantin and chrysanthene, crystalline neutral principles from *C. cinerariaefolium.* Chinese J. Physiol., **8**, 167-70, 1934.

181. *Cleveland, C. R.
Insect repellent. U. S. Patent 1,963,955. June 26, 1934.

182. Collin, E.
Anatomy of insect flowers. Pharm. J., **67**, 474-6, 503-6, 601-5, 1901.
183. Collin, E.
Pyrethrum and insect powders. J. pharm. chimie, **15**, 409-19, 1902.
184. *Compagnie Universelle Négoce.
Pyrethrins are extracted with halogenated hydrocarbons. French Patent 761,228. Mar. 14, 1934.
185. Comstock, J. H.
Pyrethrum. Rept. Commissioner Agr. for 1880, 289-90, Washington, 1881.
186. Conroy, M.
Insect powder and insect flowers. Pharm. J. Trans., **42**, 869, 1883.
187. Cook, A. J.
17th Ann. Rept. Mich. Board Agr., 431-8, 1878.
188. Cook, A. J.
Pyrethrum. 22nd Ann. Rept. Mich. Board Agr., 244-5, 1884.
189. Cook, A. J.
Buhach. 25th Ann. Rept. Mich. Board Agr., 136-7, 1886.
190. Cook, A. J.
Buhach or California pyrethrum. Mich. Agr. Exp. Sta. Bull. **58**, 6-7, 1890.
191. Cook, A. J. and Davis, G. C.
Kerosene emulsion. Mich. Agr. Exp. Sta. Bull. **76**, 3-16, 1891.
192. Coquillet, D. W.
Production and manufacture of buhach. U. S. Dept. Agr. Div. Entomol. Bull. **12**, 7-16, 1886.
193. Corbett, L. C.
Pyrethrum. S. Dak. Agr. Exp. Sta. Bull. **35**, 89, 1893.
194. Cornu, C.
Insecticidal pyrethrum; its cultivation and future possibilities. Chimie et industrie, special no., 849-58, Apr., 1928.
195. Cory, E. N.
Notes on a new and promising insecticide. J. Econ. Entomol., **14**, 345-7, 1921.
196. Cory, E. N., and Eaton, N. A.
Insecticidal value of pyrethrum soaps. Md. Agr. Exp. Sta. Bull. **308**, 417-28, 1929.
197. Cory, E. N. and Langord, G. S.
Sulfated alcohols in insecticides. J. Econ. Entomol., **28**, 257-60, 1935.
198. Costa, D.
Research on pyrethrum powder. Giorn. chim. ind. applicata, **4**, 91-3, 251-3, 1922.
199. Cotes, E. C.
Pyrethrum. Indian Museum Notes, **3**, no. 4, 64, 1896.
200. Couperot, E.
Some plants with hydrocyanic acid. J. pharm. chim., **28**, 542, 1908.

201. Covell, G.
Anti-mosquito measures. Govt. India Health Bull. **11**, Delhi, 1934.

202. Crane, D. B.
Cornell Vet. **23**, 15, 1933.

203. Cruickshank, R. D.
Persian insect powder plant. Calif. Agr. Exp. Sta. Rept. for 1891-2, p. 202, 1893.

204. Culbertson, R. E.
Pyrethrum, a new crop for Pennsylvania. Pennsylvania Farmer, **112**, no. 5, 5, 1935.

205. Culbertson, R. E.
Points for pyrethrum buyers. Chemical Industries, **37**, 341-4, 1935.

206. Dal Sie, G.
The active principle of insect powder. Bull. soc. chim., **31**, 542-3, 1879.

207. Darley, M. M.
Comparative tests with rotenone, nicotine and pyrethrum. J. Econ. Entomol., **24**, 111-5, 1931.

208. Daskaloff, C.
Importance and culture of pyrethrum. State Agr. Exp. Sta. Plovdiv, Bulgaria, Bull. **4**, 1935.

209. Daveau, J.
Culture of pyrethrum. Ann. soc. hort. hist. nat. Herault, **51**, 47, 1919.

210. Davin, V.
The pyrethrum of Dalmatia. Agr. Provence Colon., **8**, 3, 1922.

211. Davis, G. C.
31st Ann. Rept. Mich. Board Agr., 215-7, 1892.

212. Davis, G. C. and Woodworth, C. W.
Some new insecticides and their effect on cotton worms. Ark. Agr. Exp. Sta. Bull. **15**, 1890.

213. Davy, J. B.
Natural vegetation of alkali lands. Univ. Calif. Agr. Exp. Sta. Rept. for 1895-7, 63-75, 1898.

214. De Bellesme, J.
An alkaloid from *Pyrethrum carneum*. J. pharm. chim., **24**, 139, 1876.

215. De Boisse, P.
Extraction of the yellow resin of pyrethrum of Dalmatia. Rev. sci. nat. appl., **42**, 589-90, 1895.

216. *De Fenq, E.
Process for extracting pyrethrum. Australian Patent 20, 203/29 May 21, 1929.

217. Degrully, L.
Destruction of flea beetles with pyrethrum powder. Prog. agr vit., **27**, 375-7, 1897.

218. Degrully, L.
Destruction of flea beetles with pyrethrum powder. Prog. agr vit., **27**, 407-8, 1897.

219. Delong, D. M.
Biology and control of the potato leaf hopper. J. Econ. Entomol., **21**, 183-8, 1928.

220. DeLong, D. M.
Experiments with sulfur and pyrethrum. Crop Protect. Inst. Bull. **44**, 1934.

221. DeLong, D. M.
Potato leafhopper control. J. Econ. Entomol., **27**, 525-33, 1934.

222. Demetriou, A. G. and Chatzenikolaou, I. N.
Greek pyrethrum. Mpenakeion Phytopathol. Institute, June, 1933.

223. Deshusses, L. and Deshusses, J.
Determination of pyrethrins in pyrethrum base insecticides. Mitt. Lebensm. Hyg., **21**, 304-12, 1930.

224. Deshusses, L. and Deshusses, J.
Determination of the power of pyrethrum as an insecticide. Compt. rend. acad. agr. France, **17**, 517-20, 1931.

225. De Visiani.
The insecticidal plants *Pyrethrum roseum* and *Pyrethrum cinerariaefolium*. Ann. chim. applicata medicina, **19**, 84-8, Milan, 1854.

226. Dibble, C. B.
Pyrethrum a safe winter control for insect pests of livestock. Mich. Agr. Exp. Sta. Quart. Bull. **15**, 219-21, 1933.

227. Dibble, C. B.
Home made cattle sprays. Mich. Agr. Exp. Sta. Quart. Bull. **16**, 243-4, 1934.

228. Dibble, C. B.
Home made cattle spray. Mich. Agr. Exp. Sta. Quart. Bull. **18**, 5-7, 1935.

229. Dickover, E. R.
Japan increases output of insect powder material. U. S. Dept. Commerce, Bur. For. Dom. Comm., Rept. **30**, 476, 1918.

230. Dieterich, E.
Powdered pyrethrum flowers. Helfenberger Annalen, pp. 263, 360, 1896; 273-4, 1897; 321, 1900; 195, 1901; 186, 1902; 280, 1903; 209, 1904. Erst. Dezenn. Helfenb. Ann., p. 420, 1886-1895.

231. Dietze, F.
Evaluation of insect powder. Pharm. Ztg., **44**, 196-7, 1899.

232. Doehlert, C.
Compatibility of pyrethrum insecticides with Bordeaux mixture. Am. Cranberry Growers' Assoc. Proc. 66th Ann. Convention, 6-10, Aug., 1935.

233. Donovan, W.
Report of dominion laboratory. New Zealand Dept. Sci. Research, Ann. Rept., 36-7, 1934.

234. Doty, A. E.
Cattle spray tests. Soap, **12**, no. 4, 97, 1936.

235. Dowzard, E.
Valuation of insect powder. Chemist Druggist, **54**, 936, 1899.

236. Drain, B. D., Gnadinger, C. B., Corl, C. S., and Shuey, G. A. Further experiments with high test pyrethrum strains. Proceed. Am. Soc. Hort. Sci., **33**, 201, 1936.

237. Drain, B. D. and Shuey, G. A. Isolation and propagation of high pyrethrin strains of pyrethrum. Proc. Am. Soc. Hort. Sci., **32**, 190-1, 1934.

238. Driggers, B. Experiments with white oil-pyrethrum for control of oriental peach moth. J. Econ. Entomol., **24**, 319-25, 1931.

239. Driggers, B. F. and Pepper, B. B. The spittle insect. N. J. Agr. Exp. Sta. Bull. **593**, 1-4, 1935.

240. Duchartre, P. Note on *P. willemoti* or Caucasian pyrethrum. J. soc. imper. centr. hort., **5**, 206, 1859.

241. Duchartre, P. Supplementary note on Caucasian pyrethrum. J. soc. imper. centr. hort., **6**, 673, 1860.

242. Dufour, J. Destruction of cabbage worms. Chroniq. agr. Canton de Vaud, **6**, 196-7, 1893.

243. Dufoux, A. Pyrethrum and the vine worms. Rev. vit., **59**, 99-101, 1923.

244. Dufoux, A. Culture of pyrethrum and the fight against the vine worms. Rev. vit., **61**, 129-38, 1924.

245. Durrant, G. R. Insect powders of commerce. Pharm. J., **58**, 505-7, 1897.

246. Dustan, A. G. Gladiolus thrips and its control. Quebec Soc. Protection Plants, Rept. 1930-32, 32-7, 1932.

247. Edmonton, W. E. Pyrethrum insecticides. Am. J. Pharm., **104**, 793-8, 1932.

248. Elliott, A. T. Pyrethrum for grain weevils. Am. Entomol., **3**, 178-9, 1880.

249. Elliott, D. C. The squash bug in Connecticut. Conn. Agr. Exp. Sta. Bull. **368**, 224, 1935.

250. Evans, J. W. Control of thrips. J. Council Sci. Ind. Research **6**, 99-102, 1933.

251. Evans, L. E. Variation in *P. cinerariaefolium*. Thesis Colorado Agr. Coll., 1934.

252. Evans, L. E., Gnadinger, C. B., and Corl, C. S. Pyrethrum plant investigations in Colorado. III. Effect of spacing interval in the row on yield of flowers. Proceed. Am. Soc. Hort. Sci., **33**, 205, 1936.

253. Evans, L. E., Gnadinger, C. B., and Corl, C. S. Pyrethrum plant investigations in Colorado. IV. Effect of irrigation practices on plant losses due to crown rot. Proceed. Am. Soc. Hort. Sci., **33**, 207, 1936.

254. Evans Sons, Lescher and Webb.
Analytical notes: insect powder, p. 20, 1906; 25, **1907**; 35, 1909; 36, 1910.

255. Eyer, J. R.
Tests of insecticides in control of grape leaf hopper and oriental fruit moth. J. Econ. Entomol., **20**, 253-61, 1927.

256. Eyer, J. and Crawford, R. F.
Codling moth investigations. New Mex. Agr. Exp. Sta. Ann. Rept., 38-42, 1932.

257. Eymard, L.
Researches on powdered Caucasian pyrethrum. L'Union pharmaceutique, **31**, 357-61, 1890.

258. Faes, H.
Pyrethrum and the fight against the vine worm. La Terre Vaudoise, **6**, 57-9, 71-3, 91-3, 97-100, 1914.

259. Faes, H.
Pyrethrum and its culture. J. suisse pharm., **54**, 670-74, 1916.

260. Faes, H.
Pyrethrum and its culture. Schweiz. Apoth. Ztg., **56**, 429-31, 447-50, 1918.

261. Faes, H.
The cultivation of *Pyrethrum cinerariaefolium*. Prog. agr. vit., **78**, 394-400, 1922.

262. Faes, H.
The cultivation of pyrethrum in Dalmatia. Schweiz. Apoth. Ztg., **61**, 323-5, 335-8, 1923.

263. Faes, H. and Staehelin, M.
Treatments against cochylis (vine worm). Prog. agr. vit., **78**, 326-30, 350-53, 1922.

264. Feinberg, S. M.
Allergy in general practice. Lea and Febiger, Philadelphia, 1934.

265. Feinberg, S. M.
Pyrethrum sensitization; its importance and relation to pollen allergy. J. Am. Med. Assoc., **102**, 1557, 1934.

266. Fernald, M. L.
Chrysanthemum leucanthemum and American white weed. Rhodora, **5**, 177-91, 1903.

267. Ferrand.
A case of poisoning by insect powder. Abstract in Jahr. Pharm., **32**, 92, 1897.

268. Ficht, G. A.
Insecticides used against the European corn borer. J. Econ. Entomol., **26**, 747-54, 1933.

269. Finzelberg.
Tincture of insect powder. Pharm. Zentralhalle, p. 118, 1880.

270. Fisher, H. V. and Bailey, E. M.
Composition of commercial insecticides. Conn. Agr. Exp. Sta. Bull. **300**, 1929; Bull. **346**, 1933, and Bull. **398**, 1937.

271. Fleming, W. E.
Contact sprays for Japanese beetle. U. S. Dept. Agr. Circ. **280**, 1-4, 1933.

272. Fleming, W. E., Metzger, F. W. and Osburn, M. R.
Protecting plants from Japanese beetle. U. S. Dept. Agr. Circ., **326**, 1934.

273. Flückiger, F. A.
Pharmakognosie des pflanzenreiches. 3rd ed. 825-9, 1891.

274. *Fly Tox, Le.
Insecticide comprising extract of pyrethrum in glycol. French Patent 722,325. Nov. 26, 1930.

275. Francis, J. M.
The folly of buying powder ground from hand-picked closed flowers. Bull. Pharm., **12**, 5-6, 1898.

276. Freeborn, S. B. and Regan, W. M.
Fly sprays for dairy cows. J. Econ. Entomol., **25**, 167-74, 1932.

277. Freeborn, S. B., Regan, W. M. and Berry, L. J.
Effect of petroleum oil fly sprays on dairy cattle. J. Econ. Entomol., **27**, 382-8, 1934.

278. Freeborn, S. B., Regan, W. M. and Folger, A. H.
Relation of flies and fly sprays to milk production. J. Econ. Entomol., **18**, 779-80, 1925.

279. Freeborn, S. B., Regan, W. M. and Folger, A. H.
Relation of flies and fly sprays to milk production. J. Econ. Entomol., **21**, 494-501, 1928.

280. Fremd.
Cannot use gasoline in fly sprays. Soap, **2**, no. 3, 49, 1926.

281. Fromme, G.
Modification of Durrant's method of assay of insect powder. Proc. Am. Pharm. Assoc., **49**, 697, 1901.

282. Fromme, G.
Insect powder. Apoth Ztg., **28**, 721-2, 1913.

283. Fromme, G.
Evaluation of insect powder. Apoth. Ztg., **30**, 405-6, 1915.

284. Frontali, F.
Chrysanthemum cinerariaefolium as a remedy against *Oxyuris*. Bull. Sci. Mediche (Bologna), **9**, 333-5, 1858.

285. Frydlender, J. H.
Pyrethrum and its derivatives. Rev. prod. chim., **36**, 1-6, 65-70, 161-3, 1933.

286. Fryer, J. C. F. and Gimingham, C. T.
Pyrethrum as an insecticide and its cultivation in England. Nature, April 11, 1931.

287. Fryer, J. C. F., Tattersfield, F. and Gimingham, C. T.
English grown pyrethrum as an insecticide. Ann., Appl. Biol., **15**, 423-45, 1928.

288. *Fulton, S. C.
Insecticide comprising pyrethrins and rotenone. U. S. Patent 1,967,024. July 17, 1934.

289. Fujitani, J.
Chemistry and pharmacology of insect powder. Arch. exp. Path. Pharm., **61**, 47-75, 1909.

290. Gadd, H. W. and Gadd, S. C.
Testing of drugs, chemicals and galenicals by dispensing chemists: insect powder. Pharm. J., **75**, 902, 1905.

291. Garman, P.
Tests with insecticides designed to do the work of nicotine sulfate. Conn. Agr. Exp. Sta. Bull. **294**, 277-8, 1928.

292. Garman, P.
Study of aphicides. Conn. Agr. Exp. Sta. Bull. **360**, 458, 1934.

293. Garman, P.
White apple leafhopper control. Conn. Agr. Exp. Sta. Bull. **368**, 239, 1935.

294. Garman, P. and Townsend, J. F.
Control of white apple leafhopper. Conn. Agr. Exp. Sta. Bull. **360**, 449-51, 1934, also J. Econ. Entomol., **27**, 361, 1934.

295. Garratt, J. R. and Bigger, J. W.
Asthma due to insect powder. Brit. Med. J., **2**, 764, 1923.

296. Garrigues, S. S.
On insect powder. Proc. Am. Pharm. Assoc., **19**, 505-6, 1871.

297. Gattefosse, J.
Cultivation of Dalmatian pyrethrum. Rev. bot. appl. agr. colon., **2**, 397, 1922.

298. Gattefosse, R. M.
Oleoresin of Pyrethrum. Rev. vit., **58**, 11-13, 1923.

299. Gattefosse, R. M. and Gattefosse, J.
New vehicle for pyrethrum. J. agr. prat., **37**, 349, 1922.

300. Gattefosse, R. M. and Gattefosse, J.
Dalmatian pyrethrum. Parfumerie moderne, **16**, 107, 1923.

301. Gaudin, O.
Notes on pyrethrum. Ann. fals., **26**, 26-8, 1933.

302. Gaudin, O.
Action of pyrethrins on marine animals. Bull. sci. pharmacol., **42**, 145, 222, 1935.

303. Gaudin, O.
Comparative toxicities of pyrethrins. Compt. rend., **201**, 356-7, 1935.

304. Gaudin, O. and Carron, B.
Action of pyrethrins on the musculature of helminths. Bull. sci. pharmacol., **38**, 627-31, 1931 and Bull. acad. méd. **106**, 226-9, 1931.

305. Gaudin, O. and Carron, B.
Application of method of Lopez-Lomba to the physiological determination of pyrethrins. Bull. soc. therapeut., **38**, 75, 1933.

306. Gehe and Co.
Pharm. Zentralhalle, **12**, 137, 1891.

307. Gerard, R.
Pyrethrum powders and their botanical origin. J. pharm. chimie, **8**, 383-4, 1898.

308. Gerard, R.
The culture of pyrethrum in Dalmatia. Hort. nouv., no. 3, 1919.

309. Gersdorff, W. A.
Method for the study of toxicity using goldfish. J. Am. Chem. Soc., **52**, 3440-5, 1930.

310. Gersdorff, W. A. and Davidson, W. M.
New solvents for the active principles of pyrethrum. Ind. Eng. Chem., **21**, 1251-3, 1929.

311. Gervais, P.
The fight against the flea beetle. Prog. agr. vit., **27**, 442-6, 1897.

312. Gibson, A.
Report of committee on policy. J. Econ. Entomol., **21**, 19-22, 1928.

313. Gibson, A.
Mosquito supression in Canada. Proc. 16th Ann. Meet. N. J. Mosquito Exterm. Assoc., 102-8, 1929.

314. Giemsa, G.
Arch. Schiffs.-Tropen.-hyg. **15**, 533, 1911; **16**, 565, 1912; **17**, 181, 1913; **18**, 26, 1914.

315. Gieseler.
Flowers used as insecticides. Proc. Am. Pharm. Assoc., **10**, 112, 1862.

316. Gillette, C. P.
Experiments with pyrethrum. Iowa Agr. Exp. Sta. Bull. **5**, 161-96, 1889.

317. Gillette, C. P.
Experiments with injurious insects and insecticides. Iowa Agr. Exp. Sta. Bull. **12**, 535-49, 1891.

318. Gillette, C. P.
Pyrethrum experiments. Insect Life, **3**, 259-60, 1891.

319. Gillies, K. D.
Insecticidal properties of nicotine, pyrethrum, rotenone and penetrol in codling moth control. Thesis, State Coll. Wash. 1932.

320. Gilmer, P. M.
Derris as a parasiticide. Minn. State Entomol. Rept., 41-9, 1922.

321. Ginsburg, J. M.
Insecticide investigations. 49th Ann. Rept. N. J. Agr. Exp. Sta. 1928.

322. Ginsburg, J. M.
Test to determine toxicity of pyrethrum vapors to honey bees. J. Agr. Research, **40**, 1053-7, 1930.

323. Ginsburg, J. M.
Pyrethrum as a mosquito larvicide. Proc. 17th Ann. Meet. N. J. Mosquito Exterm. Assoc., p. 57, 1930.

324. Ginsburg, J. M.
Larvicide to exterminate mosquitoes without injuring water-fowl. Game Breeder, March, 1930.

325. Ginsburg, J. M.
Pyrethrum larvicide to exterminate mosquitoes without injury to waterfowl, fish and vegetation. N. J. Agr. Exp. Sta. Ann. Rept. 1930.

326. Ginsburg, J. M.
Experiments with waxes as possible carriers of insecticides. J. Econ. Entomol. **25**, 599-607, 1932.

327. Ginsburg, J. M.
Pyrethrum; its value in exterminating insects. N. J. Agr. Exp. Sta. Circ. **272**, 1-4, 1933.

328. Ginsburg, J. M.
Mosquito larvicides. Proc. 21st Ann. Meet. N. J. Mosquito Extermination Assoc., 121-7, 1934.

329. Ginsburg, J. M.
New wetting agents for old insecticides. J. Econ. Entomol., **28**, 224, 1935.

330. Ginsburg, J. M. and Forman, L.
Causes and remedies for mosquito breeding in sewage disposal plants. Sewage Works J., **2**, 412, 1930.

331. Ginsburg, J. M. and Schmitt, J. B.
Comparison between rotenone and pyrethrum as contact insecticides. J. Econ. Entomol., **25**, 918-22, 1932.

332. Ginsburg, J. M., Schmitt, J. B. and Granett, P.
Petroleum distillate in controlling insects infesting plants. J. Econ. Entomol., **28**, 236, 1935.

333. Glaser, R. W.
Notes on continuous breeding of *Musca domestica*. J. Econ. Entomol., **20**, 432-3, 1927.

334. Glassford, J.
Economics of pyrethrum. J. Econ. Entomol. **23**, 874-7, 1930.

335. Glassford, J.
Report on the chemical evaluation of pyrethrum. Soap, **5**, no. 5, 123, 1930.

336. Glassford, J.
Evaluation of pyrethrum. Soap, **7**, no. 1, 107-9, 1931.

337. Glover, T.
Report of U. S. Commissioner of Agriculture for 1864, p. 544, 1865.

338. Glover, T.
Report of U. S. Commissioner of Agriculture for 1874, p. 133, 1875.

339. *Gnadinger, C. B.
Pyrethrum insecticide. U. S. Patent 1,773,102. Aug. 19, 1930.

340. Gnadinger, C. B.
Review of methods for evaluating pyrethrum. Soap, **7**, no. 1, 97-101, 117, 1931.

341. *Gnadinger, C. B.
Ethylene dichloride as solvent for pyrethrum. U. S. Patent 1,915,662. June 27, 1933.

342. *Gnadinger, C. B.
Pyrethrum insecticide. Canadian Patent 355,978. Feb. 18, 1936.

343. Gnadinger, C. B. and Corl, C. S.
Studies on pyrethrum flowers. I. Quantitative determination of the active principles. J. Am. Chem. Soc., **51**, 3054-64, 1929.

344. Gnadinger, C. B. and Corl, C. S.
Studies on pyrethrum flowers. II. Relation between maturity and pyrethrin content. J. Am. Chem. Soc., **52**, 680-4, 1930.

345. Gnadinger, C. B. and Corl, C. S.
Studies on pyrethrum flowers. III. Pyrethrin content of commercial varieties. J. Am. Chem. Soc., **52**, 684-8, 1930.

346. Gnadinger, C. B. and Corl, C. S.
Studies on pyrethrum flowers. IV. Relative toxicity of pyrethrins I and II. J. Am. Chem. Soc., **52**, 3300-7, 1930.

347. Gnadinger, C. B. and Corl, C. S.
Effect of storage on pyrethrum flowers. Ind. Eng. Chem., **24**, 901-3, 1932.

348. Gnadinger, C. B. and Corl, C. S.
Manufacture of concentrated pyrethrum extract. Ind. Eng. Chem., **24**, 988-91, 1932.

349. Gnadinger, C. B. and Corl, C. S.
Relative toxicity of pyrethrins and rotenone as fly spray ingredients. J. Econ. Entomol., **25**, 1237-40, 1932.

350. Gnadinger, C. B. and Corl, C. S.
Studies on pyrethrum flowers. V. Presence of pyrethrolone and methyl pyrethrolone in the flowers. J. Am. Chem. Soc., **55**, 1218-23, 1933.

351. Gnadinger, C. B. and Corl, C. S.
Studies on pyrethrum flowers. VI. Dalmatian *Pyrethrum cinerariaefolium*. Soap, **9**, no. 7, 82-5, 1933.

352. Gnadinger, C. B. and Corl, C. S.
Comparison of methods for assaying pyrethrum. Soap, **10**, no. 9, 89, 1934.

353. Gnadinger, C. B., Corl, C. S., and Clark, C. A.
Effect of antioxidants on highly concentrated pyrethrum extracts. Soap, **11**, no. 10, 95-9, 1935.

354. Gnadinger, C. B., Evans, L. E. and Corl, C. S.
Pyrethrum investigations in Colorado. I. Factors affecting pyrethrin content. Colo. Agr. Exp. Sta. Bull. **401**, 1-19, 1933.

355. Gnadinger, C. B., Evans, L. E. and Corl, C. S.
Pyrethrum plant investigations in Colorado. II. A review of progress since 1932. Colo. Agr. Exp. Sta. Bull. **428**, 1936.

356. Goessmann, C. A.
7th Ann. Rept. Mass. Agr. Exp. Sta., p. 172, 1890.

357. Goessmann, C. A.
Dalmatian insect powder. 9th Ann. Rept. Mass. Agr. Exp Sta., pp. 295, 317, 1892.

358. Goessmann, C. A.
10th Ann. Rept. Mass. Agr. Exp. Sta., p. 190, 1893.

359. Goester, L. E.
Pharmacognostic themes. V. Mutation, heredity and selection. Pharm. Weekblad, **68**, 41-59, 1931.
360. Goetze, G.
Two new methods for determining the toxicity of insecticides. Anz. Schädlingskunde, **8**, 54-7, 1932.
361. Goff, E. S.
Experiment with home-grown pyrethrum. 7th Ann. Rept. N. Y. Agr. Exp. Sta., pp. 151-2, 1889.
362. Gösswald, K.
Pyrethrum. Anz. Schädlingskunde, **4**, 56, 1933.
363. Gösswald, K.
Control of chloropiden with pyrethrum. Anz. Schädlingskunde, **9**, 84, 1933.
364. Gösswald, K.
Effect of pyrethrum on forest insects. Z. angew. Entomol., **20**, 489-530, 1934.
365. Gösswald, K. and Zwölfer, W.
Pyrethrum as a weapon against forest insects. Anz. Schädlingskunde, **9**, 73-7, 1933.
366. Gothard, N. J.
What is a proper insecticide base? Soap, **8**, no. 6, 103-107, 1932.
367. Gothard, N. J.
Soap, **8**, no. 6, 113-15, 1932.
368. *Graafland, A. J. N.
Portland cement mixed with pyrethrum is used as an antifouling composition to prevent growth of organisms in water. British Patent 362,526. Dec. 10, 1931.
369. Grady, A. G.
Studies in breeding insects throughout the year for insecticide tests: house flies. J. Econ. Entomol., **21**, 598-604, 1928.
370. Graine, M. P.
Clinical study of pyrethrins. Thesis, Paris, 1933.
371. *Grant, D. H.
Insecticide containing pyrethrum and sulfonated mineral oil. U. S. Patent 1,940,646. Dec. 19, 1933.
372. Graybill, N. W.
Repellents for protecting animals from the attacks of flies. U. S. Dept. Agr. Circ. **131**, 22, 1914.
373. Green, S. B.
Ornamental and timber trees, shrubs and herbaceous plants in Minnesota. Minn. Agr. Exp. Sta. Bull. **24**, 226, 1892.
374. Greenish, H. G. and Collin, E.
Anatomical atlas of vegetable powders, 102-3, London, 1904.
375. Grieb, C. M. W.
Valuation of insect powder. Chemist Druggist, **72**, 848, 1908.
376. Griffiths, T. H. D. and Michel, C.
Destruction of mosquitoes in aircraft. Bull. Office Internat. Hyg. Publ., **27**, 550, 1935.
377. Grote, C.
Colored insect powder. Pharm. Ztg., **25**, 208, 1880.

378. Guillaume, A.
Use of pyrethrum against bed bugs. Bull. sci. pharmacol., **38**, 80-4, 1931.

379. *Guillissen, J.
Pyridine and ethyl lactate are used as solvents for pyrethrum. British Patent 354,680. Jan. 26, 1931.

380. Haas, A.
Insect powder. Pharm. Zentralhalle, **25**, 19-20, 1884.

381. Haegele, R. W.
Pyrethrum in the control of *Mineola scitulella*. J. Econ. Entomol., **25**, 1073-7, 1932 and Idaho Hort. Assoc. 38 Ann. Convention, 83-89, 1933.

382. Haensel, H.
Pyrethrum oil. Pharm. Ztg., **43**, 760, 1898.

383. Hager, H.
Insecticidal part of the flowers of *P. carneum* and *roseum*. Pharm. Zentralhalle, **19**, 74-5, 1878.

384. Hall, M. C. and Wigdor, M.
J. Amer. Vet. Med. Assn. **69**, 19, 1926.

385. Haller, H. L. and Acree, F.
Constituents of pyrethrum flowers. Determination of pyrethrin II. Ind. Eng. Chem. Anal. Ed., **7**, 343, 1935.

386. Haller, H. L. and LaForge, F. B.
Non-crystalline constituents of derris root. J. Am. Chem. Soc., **56**, 2415, 1934.

387. Halsted, B. D.
28th Ann. Rept. N. J. Agr. Exp. Sta. p. 376, 1908.

388. Halsted, B. D.
30th Ann. Rept. N. J. Agr. Exp. Sta. pp. 326-7, 1910.

389. Hamilton, C. C.
Mexican bean beetle and how to control it. N. J. Agr. Exp. Sta. Circ. **216**, 1-16, 1929.

390. Hamilton, C. C.
Control of leaf hoppers infesting dahlias and asters. J. Econ. Entomol., **25**, 304-11, 1932.

391. Hamilton, C. C.
Control of gladiolus thrips. J. Econ. Entomol., **26**, 555-64, 1933.

392. Hamilton, C. C. and Gemmell, L. G.
Comparative efficiency of derris, pyrethrum and hellebore powder on different insects. J. Econ. Entomol., **27**, 446-53, 1934.

393. Hamilton, C. C. and Henderson, K.
Control of orchid weevil. Am. Orchid Soc. Bull. **2**, 43-50, 1933.

394. Hanamann, J.
Persian insect powder. Vierteljahr. prakt. pharm., **12**, 522-5, 1863.

395. Hanausek, T. F.
Microscopic characteristics of pyrethrum flowers. Pharm. Post, **25**, 18-24, 177-83, 717-23, 829-31, 1892.

396. Hanausek, T. F.
Insect powder. Lehrbuch der technischen Mikroskopie, pp. 293-306, Stuttgart, 1901.

397. Hanausek, T. F.
Real-Enzyklopadie der gessamten Pharmazie, 2nd ed., **7**, 49-54, 1906.

398. Hanausek, T. F. and Winton, A. L.
Microscopy of technical products, 1st ed., 310-23, 1907.

399. Hansberry, T. R.
Tracheal penetration in aphis with reference to efficiency of pyrethrum insecticides. Thesis, State Coll. Wash. 1932.

400. Hanson, A. J.
The blackberry mite and its control. Wash. Agr. Exp. Sta. Bull. **279**, 1-15, 1933.

401. Hanson, A. J.
The potato flea beetles. Wash. Agr. Exp. Sta. Bull. **280**, 1-22, 1933.

402. Harder, H.
Determination of the active principles of insect powder. Dissertation, Tech. Hochsch. Zurich, 46-79, 1927.

403. Hart, J.
Insect powder. Abstract in Pharm. Record, **8**, 315-6, 1888.

404. Hartzell, A.
Pyrethrum fields of Dalmatia. J. Econ. Entomol., **26**, 583-6, 1933.

405. Hartzell, A.
Histopathology of insect nerve lesions caused by insecticides. Contrib. Boyce Thompson Institute, **6**, no. 2, 211-23, 1934.

406. Hartzell, A. and Wilcoxon, F.
Chemical and toxicological studies of pyrethrum. Contrib. from Boyce Thompson Inst., **4**, no. 1, 107-17, 1932.

407. Harukawa, C., Takato, R. and Kumashiro, S.
Corn seed maggot control. Ber. Ohara Inst. landw. Forsch., Japan, **6**, 83-111, 1933.

408. Harvey, F. L.
Rept. Me. Agr. Exp. Sta. for 1892, pp. 132, 143-4, 1893.

409. Haseman, L.
Report of subcommittee on insecticides and appliances. J. Econ. Entomol., **22**, 13-5, 1929.

410. Haseman, L. and Jones, E. T.
Greenhouse pests and their control. Mo. Agr. Exp. Sta. Bull. **342**, 1-32, 1934.

411. Haywood, J. K.
Chemical composition of insecticides and fungicides. U. S. Dept. Agr. Bur. Chem. Bull. **68**, 38-40, 1902.

412. Haywood, J. K.
Pyrethrum powders containing poisonous metals. U. S. Dept. Agr. Bur. Chem. Bull. **76**, 7-21, 1903.

413. Headlee, T. J.
Effect of pyrethrum extract on wireworms and upon plants infested by them. J. Econ. Entomol., **23**, 251-9, 1930.

414. Headlee, T. J.
Sprays for codling moth control. New Jersey Agr., **15**, no. 2, 4, 1933.

415. Headlee, T. J.
Organic insecticides as substitutes for arsenicals. J. Econ. Entomol., **27**, 148-154, 1934.

416. Headlee, T. J., Ginsburg, J. M. and Filmer, R. S.
Substitutes for arsenic in control of codling moth. J. Econ. Entomol., **23**, 45, 1930.

417. Heckel, E.
Dalmatian pyrethrum, its culture in Provence. Abstract in J. soc. nat. hort. France, 14, 270, 1913.

418. Heinrich, R.
Analysis of the ash of *Chrysanthemum segetum.* 2nd Bericht der landwirthschaftlichen Versuchs-Station zu Rostock, pp. 227-8, 1894.

419. Hellier, J. B.
Pyrethrum. Agr. J. (Cape Colony), **3**, 4-5, 1890.

420. Hellier, J. B.
Pyrethrum as an insecticide. Agr. J. (Cape Colony), **3**, 93, 1890.

421. Hely, P. C.
Control of green peach aphid. Agr. Gaz. N. S. Wales, **44**, 751-8, 1933.

422. Henri, A. and Ripert, J.
Pyrethrum; its employment in veterinary medicine. Rev. med. vet., **103**, no. 9, 1927.

423. Henry, W. A.
Buhach or pyrethrum. 2nd Ann. Rept. Wisc. Agr. Exp. Sta., pp. 45-6, 1885.

424. Herrera, A. L.
Destruction of mosquitoes with pyrethrum powder. Bol. Com. Parasitologia Agricola (Mexico), **2**, 139-42, 1905.

425. Herrera, A. L.
Pyrethrum powder and the plant which produces it. Circ. Com. Parasitologia Agricola (Mexico), **61**, 1-24, 1907.

426. Herrick, G. W.
Fly control on dairy farm and in milk plant. N. Y. State Assoc. Dairy Milk Inspectors, 7th Ann. Rept., 112-24, 1933.

427. Hervey, G. E. R. and Palm, C. E.
Non-arsenical dusts for cauliflower worm control. N. Y. Agr. Exp. Sta. Bull. **640**, 1934.

428. Hess.
Treatment of helminthiasis. Thesis, Strasbourg, 1933.

429. Hilgard, E. W.
Examination of Persian insect powder and buhach. Suppl. Bienn. Rept. Univ. Calif. Coll. Agr., pp. 68-71, 1879.

430. Hilgendorf, G.
Pyrethrum and pyrethrum preparations. Nachr. Deut. Pflanzenschutzdienst, **14**, 11-13, 1934.

431. Hill, J. R.
Insect powder. Abstract in Proc. Am. Pharm. Assoc., **42**, 874. 1894.

432. Hirschsohn, E.
The active part of insect powder. Pharm. Z. Russ., **29**, 209-13, 1890.

433. Hobson, R. P.
Calcium and hydrogen ion concentration and the interfacial tension of pyrethrum extracts. J. Agr. Sci., **21**, 101-14, 1931.

434. Hockauf.
Z. österr. Apoth.-Ver., p. 81, 1903.

435. Holmes, E. M.
Cultivation of the insect powder plant. Pharm. J., **98**, 6-7, 1917.

436. Holt, R. L. and Kintner, J. H.
Anti-mosquito sprays. Philippine J. Sci., **47**, 433-8, 1932.

437. Hooker, J. D.
Chrysanthemum cinerariaefolium. Curtis' Botanical Magazine, **40**, tab. 6781, 1884.

438. *Hopkins, M. B.
Extraction of pyrethrum with petroleum oils. U. S. Patent 1,877,875. Sept. 20, 1932.

439. Horsfall, J. L.
Practical moth control. Soap, **11**, no. 2, 93, 1935.

440. Howard, L. O.
Effect of pyrethrum on heartbeat of *Plusia brassicae*. Am. Naturalist, **16**, 1015, 1882.

441. Howard, L. O. and Bishopp, F. C.
Mosquito remedies and preventives. U. S. Dept. Agr. Farm. Bull. **1570**, 1932.

442. Howard, N. F., Brannon, L. W. and Mason, H. C.
Insecticides for the control of the Mexican bean beetle. J. Econ. Entomol., **26**, 123-9, 1933.

443. Howard, N. F., Brannon, L. W. and Mason, H. C.
Control of Mexican bean beetle. J. Econ. Entomol., **28**, 444-8, 1935.

444. Howie, W. L.
Colored insect powder. Pharm. J. Trans., **13**, 939-40, 1883.

445. Hoyer, D. G.
Effect of sunlight on fly sprays. Soap, **11**, no. 12, 105, 1935.

446. Hoyer, D. G.
Effect of metals on fly sprays. Soap, **12**, no. 1, 105-7, 1936.

447. Hoyer, D. G. and Weed, A.
Pyrocatechin for protecting pyrethrins. J. Econ. Entomol., **28**, 1074, 1935.

448. Hoyer, D. G., Weed, A., Schmidt, S. and Leonard, M. D.
Seasonal quality of Japanese pyrethrum. J. Econ. Entomol., in press (1936) abstract **28**, no. 6, 25, Dec., 1935.

449. Hubbard, H. G.
Ants vs. cotton worms. Am. Entomol., **3**, 249, 1880.

450. Hubbard, H. G.
Effects of pyrethrum on different insects. Am. Entomol., **3**, 250-1, 1880.

451. Huber, E. J.
Adulteration of insect powder. Bull. Pharm., **13**, 253, 1899.

452. Hubert, A.
Tests for the control of cochylis. Prog. agr. vit., **76**, 12, 1921.

453. Hubert, A.
Pyrethrum and cochylis. Prog. agr. vit., p. 248, 1921.

454. Hübscher, J.
Evaluation of pyrethrum flowers and extracts. Süddeut. Apoth. Ztg., **75**, 655-6, 1935.

455. Huckett, H. C.
Use of arsenicals and arsenical substitutes. J. Econ. Entomol., **27**, 156-161, 1934.

456. Huckett, H. C.
Derris for cabbage worms. J. Econ. Entomol., **27**, 441, 1934.

457. Huckett, H. C. and Hervey, G. E. R.
Arsenical substitutes for vegetable pest control. J. Econ. Entomol., **28**, 602, 1935.

458. *Hughes, W. A.
Preserving pyrethrum flowers with kerosene or cymene. U. S. Patent 1,945,148. Jan. 30, 1934.

459. Hutchinson, H. P. and Kearns, H. G. H.
Control of *Galerucella lineola* and *Phyllodecta vitellinae*. Univ. Bristol Agr. Hort. Research Sta. Ann. Rept., 108-26, 1930.

460. Hutson, R.
Leaf-tier damages celery. Mich. Agr. Exp. Sta. Quart. Bull. **14**, no. 2, 1931.

461. Hutson, R.
Pyrethrum for determining insect infestations in golf greens. J. Econ. Entomol., **26**, 1171, 1933.

462. Hutson, R.
Control of strawberry leaf roller. Mich. Agr. Exp. Sta. Quart. Bull. **16**, 7-8, 1933.

463. Hutson, R.
Pyrethrum controls cabbage worms. Mich. Agr. Exp. Sta. Quart. Bull. **16**, 100-1, 1933.

464. Hutson, R.
Controlling raspberry beetle and saw fly. Mich. Agr. Exp. Sta. Quart. Bull. **16**, 183-5, 1934.

465. Hutson, R.
Pyrethrum and derris as arsenical substitutes. Mich. Agr. Exp. Sta. Quart. Bull. **16**, 241-2, 1934.

466. Hutson, R.
Mexican bean beetle control. Mich. Agr. Exp. Sta. Quart. Bull., **18**, 7-9, 1935.

467. Hutson, R.
Strawberry leaf roller control. J. Econ. Entomol., **28**, 388-90, 1935.

468. *Iiba, T. and Nakatuka, S.
Extraction of pyrethrum with a mixture of trichloroethane and methanol. Japanese Patent 91,478. May 30, 1931.
469. *Interessengemeinschaft Farbenindustrie A.-G.
Pyrethrum extract emulsified with sulfonated mixture of aromatic compounds. German Patent 506,889. Oct. 27, 1928
470. *Interessengemeinschaft Farbenindustrie A.-G.
Pyrethrum extract emulsion. German Patent 525,821. Jan. 29, 1929.
471. *Interessengemeinschaft Farbenindustrie A.-G.
Hydrogenated tar oil used with pyrethrum as an insecticide. German Patent 556,507. Dec. 14, 1930.
472. *Interessengemeinschaft Farbenindustrie A.-G.
Benzophenone or other aromatic ketone is used with pyrethrum as an insecticide. German Patent 494,463. Mar. 24, 1930.
473. *Interessengemeinschaft Farbenindustrie A.-G.
Hydrogenated or synthetic oils used as solvents for insecticides. French Patent 726,992. Nov. 27, 1931.
474. *Interessengemeinschaft Farbenindustrie A.-G.
Hydrogenated petroleum oil is used as a base for pyrethrum sprays. German Patent 555,117. Nov. 26, 1930 and British Patent 377,069. July 21, 1932.
475. Inverni, C.
Pyrethrum insecticide. Riv. ital. essenze profumi piante offic. Milan, **16**, 1934.
476. Isachenko, V. B. and Goritzkaya, O. V.
Toxicological analysis of pyrethrum. Bull. Plant Protection (U. S. S. R.), **3**, no. 1, 165-74, 1931.
477. Issoglio, G.
Activity of pyrethrum powders. Ann. Schiapparelli, **3**, no. 5, 10-12, 1929.
478. Jaeger, P.
Roseum against insect bites. Pharm. Z. Russ., **27**, 191, 1888.
479. Jager, H.
Insecticidal plants worth cultivating. Natur (Ger.), **29**, 458, 1880.
480. Jagor, F.
Insect powder and tincture of insect powder. Abstract in Neu. Jahr. Pharm., **30**, 226, 1868.
481. Jary, S. G.
Pyrethrum. J. So. E. Agr. Coll. Wye, Kent, **30**, 183-5, 1932.
482. Jean, F.
Analysis of pyrethrum powder. Ann. chim. anal., **8**, 285, 1903.
483. Jelliffe, S. E.
Adulteration of insect powder by addition of powdered stems. Drug. Circ. Chem. Gaz., **39**, 4-5, 1895.
484. Jones, H. A., Campbell, F. L. and Sullivan, W. N.
Cracca, a source of insecticides. Soap, **11**, no. 9, 101, 1935.
485. Jones, R. W.
Effect of pyrethrum on boll worm. Am. Entomol., **3**, 252-3, 1880.

486. Juillet, A.
Medicinal plants cultivated in the Pyrenees. Com. plantes med. reg. Montpellier, Roumegous et Dehan, Montpellier, 1920.

487. Juillet, A.
Cultivation of pyrethrum and the insecticidal properties of soap-pyrethrum. Rev. vit., **57**, 310-3, 1922.

488. Juillet, A.
Culture of pyrethrum and properties of soap-pyrethrum. Petite rev. agr. hort., **28**, 186, 1922.

489. Juillet, A.
Industrial culture of Dalmatian pyrethrum. Rev. bot. appl. agr. colon., **2**, 402, 1922. Bull. sci. pharm., **29**, 431, 1922. Ann. soc. hort. hist. nat. Herault, **53**, 151, 1922.

490. Juillet, A.
Regarding oleoresin of pyrethrum and soap-pyrethrum. Rev. vit., **58**, 169-76, 1923.

491. Juillet, A.
Reply to Chevalier and Mercier on the pharmacodynamic action of pyrethrum. Bull. sci. pharmacol., **30**, 533-5, 1923.

492. Juillet, A.
Extraction of the active principles of Dalmatian pyrethrum. Bull. sci. pharmacol., **30**, 592-604, 1923.

493. Juillet, A.
The pyrethrum insecticide of Dalmatia. Ministere du commerse et de l'industrie, Travaux de l'office national des matieres premieres vegetales, Notice **16**, 1-263, 1924.

494. Juillet, A. and Dalmier, E.
Anatomy of *P. cinerariaefolium*. Bull. sci. pharm., **31**, 3, 1924.

495. Juillet, A. and Daveau, J.
Pyrethrum in grape culture. Rev. vit., **52**, 92, 1920.

496. Juillet, A., d'Everlange, M. and Ancelin, M.
The pyrethrum insecticide of Dalmatia. 1-236, Roumegous and Dehan, Montpellier, 1924.

497. Juillet, A. and Diacono, H.
Destruction of lice on the body or clothing by soapy emulsions of olecresin of pyrethrum. Bull. sci. pharmacol., **32**, 413-28, 464-78, 1925.

498. Juillet, A. and Galavielle, L.
New treatment of lice with soap pyrethrum. Bull. sci. pharm., 29, 233, 1922.

499. Juillet, A., Galavielle, L. and Ancelin, M.
Preparation of soap-pyrethrum. Rev. vit., **55**, 201-6, 1921.

500. Juillet, A., Galavielle, L. and Ancelin, M.
Secreting system of the vegetative organs of Dalmatian pyrethrum. Bull. sci. pharmacol., **28**, 449-59, 1921.

501. Juillet, A., Galavielle, L. and Ancelin, M.
Insecticidal properties of soap-pyrethrum. Rev. vit., **55**, 301-6, 1921.

502. Juillet, A. and Pasquet, C.
Pyrethrum; culture, harvesting and preparation. Notice publiee office agr. dept. Herault. Roumegous et Dehan, Montpellier, 1921.
503. Juillet, A. and Roucher, P.
The culture of Dalmatian pyrethrum. Notice publiee office agr. dept. Pyrenees-Orientales, Roumegous et Dehan, Montpellier, 1923. Bull. sci. pharm., **30**, 289, 1923.
504. Jüttner, E. and Siedler, P.
Production, trade, adulteration and testing of Dalmatian and Montenegrin pyrethrum. Ber. pharm. Ges., **22**, 397-417, 1912.
505. Kalbruner, H.
Insecticidal action of several kinds of pyrethrum. Z. österr. Apoth.-Ver., **12**, 542-3, 1874.
506. Karsten, G. and Oltmanns, F.
Lehrbuch der Pharmakognosie, 2nd ed., 223-5, Jena, 1909.
507. Kazanetski, N.
Compt. rend. acad. agr. France, **30**, 1080, 1927.
508. Kazanetski, N.
The enemies of pyrethrum. Rev. Appl. Entomol. (A), **16**, 216, 1928.
509. Kearns, H. G. H.
Insect pests of basket willows. Univ. Bristol Agr. Hort. Research Sta. Ann. Rept., 126-46, 1934.
510. Kearns, H. G. H. and Swarbrick, T.
Control of the apple sawfly. Univ. Bristol Agr. Hort. Research Sta. Ann. Rept., 90-4, 1932.
511. Kearns, H. G. H. and Walton, C. L.
Control of loganberry and raspberry beetle. J. Pomology Hort. Sci., **11**, 39-52, 1933.
512. Kearns, H. G. H. and Walton, C. L.
Control of chrysanthemum eelworm. Agr. Hort. Research Sta. Bristol, Ann. Rept. 66-73, 1933.
513. Keimatsu, S.
Chrysanthemum oil. J. Pharm. Soc. Japan, no. 326, 1, 1909.
514. Kemper, H.
Spray materials for combating insects injurious to health. Z. Desinfekt. Gesundh. **21**, no. 6, 1929.
515. Kemper, H.
Pyrethrum for combatting woodlice. Z. Gesundh. Städtehyg., **24**, 1932.
516. Kemper, H.
Action of pyrethrum on different classes of animals. Anz. Schädlingskunde, **9**, 104-6, 1933.
517. Kemper, H.
Use of pyrethrum powder for combatting animal pests in water supplies, drainage and sewage disposal plants. Z. Gesundh. Städtehyg., **25**, 681-90, 1933.
518. Kemper, H.
Physiological evaluation of pyrethrum powder. Z. angew. Entomol., **21**, 208-23, 1934.

519. Kew Royal Gardens.
Insect powders. Bull. Misc. Information, **143**, 297-301, 1898.

520. King, W. M.
Rept. U. S. Commissioner Agr. for 1886, p. 53, 1887.

521. Kirkby, W.
Note on insect powder. Pharm. J. Trans., **19**, 239-41, 1888.

522. Kisskalt, K.
The fight against lice. Deut. Med. Wochschr., **41**, 154, 1915.

523. Klee, W. G.
The insect powder plant. Rept. Prof. in Charge to Pres. Univ. Calif. Coll. Agr. for 1882, 112-4, 1883.

524. Kline, M. N.
Report of committee on drug market. Proc. Am. Pharm. Assoc., **33**, 371, 1885.

525. Kline, M. N.
Report of committee on drug market. Proc. Am Pharm. Assoc., **34**, 3, 1886.

526. Knight, H. G.
Report of chief of bureau of chemistry and soils. U. S. Dept. Agr. Aug. 31, 1933.

527. Koch, W.
Report on an oriental flower. Linnaea, **24**, 329-30, 1851.

528. Kraemer, H. A.
Textbook of botany and pharmacognosy. 4th ed., 395-6, 1910.

529. Kraemer, H. A.
Applied and economic botany. 718-9, 1914.

530. Kral, H.
Colored insect powder. Pharm. Zentralhalle, **21**, 281, 1880.

531. Kritchevsky, W.
Chemistry of modern wetting out agents. Chem. Bull. **21**, no. 10, 251-3, 1934.

532. Kruger, F.
Toxicity of Dalmatian insect powder to the larva of *Corethra plumicornis*. Z. angew. Entomol., **18**, 344-53, 1931.

533. Kunz, E. C.
Insecticide odors to fit uses. Soap, **10**, no. 9, 99-100, 1934.

534. Kuraz, R.
Physiological action of insect powder from Korneuburg. Arch. Chem. Mikros., **8**, 1-17, 1915.

535. Laake, E. W., Parman, D. C., Bishopp, F. C. and Roark, R. C. Chemotropic responses of the house fly, green bottle flies and the black blow fly. U. S. Dept. Agr. Tech. Bull. **270**, 1931.

536. Labler, K.
Cultivation of medicinal plants. Z. österr. Apoth.-Ver., **27**, 232, 1889.

537. Lacroix, D. S.
Flea-beetle control. Conn. Agr. Exp. Sta. Bull. **367**, 137-9, 1935.

538. LaForge, F. B. and Haller, H. L.
Constituents of pyrethrum flowers. II. Isolation of pyrethrin II. J. Am. Chem. Soc., **57**, 1893-6, 1935.

539. Lamiman, J. F.
Control of the grape leaf hopper in California. Univ. Calif. Agr. Exten. Serv. Circ. **72**, 1933.

540. Landerer, X.
Notes on some oriental plants and vegetable products. Am. J. Pharm., **47**, 498-9, 1875.

541. Landerer, X.
Notes on some medicinal and other useful plants. Am. J. Pharm., **49**, 155-6, 1877.

542. Lehmann, E.
Evaluation of insect powder. Münch. Med. Wochschr., **62**, 344, 1915.

543. Lehmann, E.
Evaluation of insect powder. Apoth. Ztg., **30**, 458-9, 1915.

544. Lemaire, A. and Gaudin, O.
Pyrethrins in the treatment of scabies. Bull. sci. pharmacol., **38**, 692-4, 1931.

545. Lemee, H. A.
The question of pyrethrum. Rev. vit., **66**, 373-7, 1927.

546. *Leonardi, P., Zen, P. and Sardi, G.
Improved insect destroyer. British Patent 17,256. Nov. 29, 1890.

547. LePelley, R. H.
Kerosene extracts for the control of *Lygus* and *Antestia.* Kenya Colony Dept. Agr. Bull. **22**, 1932.

548. LePelley, R. H.
Field spraying with undiluted paraffin extracts of pyrethrum against *Antestia* and *Lygus* on coffee in Kenya. Bull. Entomol. Research, **24**, 1-32, 1933.

549. LePelley, R. H.
Pyrethrum extract spraying for control of *Antestia* on coffee. Kenya Colony Dept. Agr. Bull. **8**, 1-15, 1934.

550. Lesne, A.
The vine cochylis. J. agr. prat., **1**, 638-9, 1892.

551. Leubner, B. O.
Valuation of insect powder. Merck's Rept., p. 229, Aug., 1910.

552. Levy, J.
Biological evaluation of extracts of pyrethrum. Bull. soc. therapeut., **38**, 16, Jan., 1933.

553. *Lingler, J.
Pyrethrum used with powdered calcium cyanide as fumigant. U. S. Patent 1,820,394. Aug. 25, 1931.

554. Linke, H.
Insect powder and its testing. Apoth. Ztg., **27**, 775-6, 1912.

555. Linsbauer, K.
Insect flowers. Wiesner's Rohstoffe des Pflanzenreiches, 2nd ed., **2**, 671-8, Leipzig, 1903.

556. Lintner, J. A.
Pyrethrum as an insecticide. 1st Ann. Rept. on Insects of N. Y., pp. 36-40, Albany, 1882.

557. List, G. M.
Psyllid yellows of tomatoes. J. Econ. Entomol., **28**, 431-6, 1935.

558. List, G. M. and Sweetman, L.
Collection and analysis of data on control of cabbage worms. J. Econ. Entomol., **28**, 298-304, 1935.

559. Lodeman, E. G.
The spraying of plants. pp. 52, 116, 156, 163-5, New York, 1896.

560. Loeser.
An old insecticide in a new form. Umschau, **28**, 444-5, 1924.

561. Loring.
Pyrethrum. Its use as an insecticide. Rept. U. S. Dept. Agr., p. 76, 1881-82.

562. Lowe, F. B.
Some facts about insect powder. Bull. Pharm., **24**, 509-10, 1910.

563. Lucas, V.
Analysis and commercial valuation of pyrethrum. Gaz. Pharm. (Brazil), p. 11, June, 1934.

564. Lugger, O.
Classification of insects and their relation to agriculture. Minn. Agr. Exp. Sta. Bull. **28**, 139-40, 1893.

565. Lyons, A. B.
Plant names, scientific and popular. 2nd ed., 118-9, Detroit, 1907.

566. MacDougall, R. S.
Ox warble flies. Trans. Highland Agr. Soc. Scotland, **46**, 1-90, 1934.

567. Mackie, D. B.
Treating balled nursery stock to destroy earwig. Calif. Dept. Agr. Monthly Bull. **23**, 300, 1934.

568. MacLeod, G. F. and Maughan, F. B.
Experiments with onion thrips. J. Econ. Entomol. **28**, 150-3, 1935.

569. MacOwan, L.
Insect powder plants. Abstract in Jahr. Pharm., **31**, 74-5 1896.

570. MacOwan, P.
Memoranda on pyrethrum. Agr. J. (Cape Good Hope), **2**, 270-1, 1889.

571. MacOwan, P.
Cultivation of pyrethrum. Pharm. J. Trans., **20**, 605, 1890.

572. Maercks, H.
Action of nicotine and pyrethrum on eggs of codling moth. Anz. Schädlingskunde, **11**, 13-19, 1935.

573. Magens.
Vet. Med. **27**, 23, 1932.

574. Mail, D. A.
Mosquitoes of Montana. Mont. Agr. Exp. Sta. Bull. **282**, 1-72, 1934.

575. Maisch, J.
Minutes of meeting of pharmaceutical society. Am. J. Pharm., **61**, 50, 1889.

576. Malfatti, J.
Caucasian insect powder. Pharm. Post, **26**, 165-6, 181-2, 193-4, 205-6, 1893.

577. Manget.
A common adulteration of pyrethrum powder. Arch. med. pharm. militaires, **27**, 311-14, Paris, 1896.

578. Mann, D.
Technical preparation of pyrethrum extracts. Chem. Zeit., **39**, 401, 1934.

579. Mann, D.
Pyrethrins, rotenone and nicotine. Seifensieder Zeit., **62**, 185-9, 228-9, 249, 271-2, 290-2, 311, 333-4, 351-3, 373-5, 393-5, 892-3, 913-5, 934-5, 954, 973-5, 994-5, 1016-17, 1035-7, 1935.

580. Mann, D.
Analysis of pyrethrum preparation. Chem. Zeit., **60**, 147-9, 1936.

581. Marchal, P.
Presence of *Eudemis botrana* near Paris. Compt. rend. soc. nat. agr. France, p. 297, 1910.

582. Marchal, P.
Work accomplished against cochylis and eudemis in 1911. Compt. rend. soc. nat. agr. France, 1911.

583. Marchal, P.
Cultivation of Dalmatian pyrethrum in France. Compt. rend. acad. agr. France, **9**, 85-95, 1923.

584. Marchal, P.
Utilization of parasites of cochylis and eudemis. Bull. soc. centr. agr. Aude, **104**, no. 41, 261, 1923.

585. Markoe, G. F. H.
On *Pyrethrum roseum*. Proc. Am. Pharm. Assoc., **12**, 248-9, 1864.

586. Markoe, G. F. H.
Plants useful as insecticides. Proc. Am. Pharm. Assoc., **19**, 115-6, 1871.

587. Marlatt, C. L.
Ann. Rept. Bur. Entomol. U. S. Dept. Agr., 1932.

588. Marpmann.
Insect powder and its use as a fly killer. Pharm. Zentralhalle, **29**, 341-2, 1888.

589. Marsais, P.
The question of pyrethrum. Rev. vit., **64**, 322-4, 1926.

590. Martin, J. T. and Tattersfield, F.
The evaluation of pyrethrum flowers. J. Agr. Sci., **21**, 115-35, 1931.

591. Martin, J. T. and Tattersfield, F.
Effect of environmental conditions on pyrethrum. I. Ann. Appl. Biol. **21**, 670-81, 1934.

592. Martin, J. T. and Tattersfield, F.
Effect of environmental conditions on pyrethrum. II. Ann. Appl. Biol. **21**, 682-690, 1934.

593. Mason, A. H.
Report of committee on the drug market. Proc. Am. Pharm. Assoc., **35**, 382, 393, 1887.

594. Massey, W. F.
Work of the horticultural division. N. C. Agr. Exp. Sta. Bull. **72**, 4-5, 1890.

595. Mattei.
Dalmatian pyrethrum. Boll. soc. ort. Palermo, p. 11, 1906.

596. Maughan, F. B.
Control of onion thrips. J. Econ. Entomol., **27**, 109-12, 1934.

597. Maynard, S. T.
Insecticides, fungicides, spraying calendar. Hatch Exp. Sta. Bull. **60**, 6, 1899.

598. Mazieres.
Industrial culture of pyrethrum. Rev. hort. Tunisienne, p. 294, 1909.

599. McClintock, C. T., Hamilton, H. C. and Lowe, F. B.
Contribution to our knowledge of insecticides. J. Am. Pub. Health Assoc., **1**, 233, 234, 236, 1911.

600. McCord, C. P., Kilker, C. H. and Minster, D. K.
Pyrethrum dermatitis. J. Am. Med. Assoc., **77**, 448, 1921.

601. McDaniel, E. I.
Leaf hoppers and aphids easily controlled. Mich. Agr. Exp. Sta. Quart. Bull. **10**, 171-2, 1928.

602. McDaniel, E. I.
Mushroom maggots. Mich. Agr. Exp. Sta. Quart. Bull. **15**, 72, 1932.

603. McDaniel, E. I.
Spraying to control the gladiolus thrips. J. Econ. Entomol., **26**, 835-6, 1933.

604. McDaniel, E. I.
Michigan termites. Mich. Agr. Exp. Sta. Bull. **150**, 3-14, 1934.

605. McDonnell, C. C., Abbott, W. S., Davidson, W. M., Keenan, G. L. and Nelson, O. A.
Relative insecticidal value of commercial grades of pyrethrum. U. S. Dept. Agr. Tech. Bull. **198**, 1-9, 1930.

606. McDonnell, C. C. and Roark, R. C.
Occurrence of manganese in insect flowers and insect flower stems. J. Agr. Research, **11**, 77-82, 1917.

607. McDonnell, C. C., Roark, R. C. and Keenan, G. L.
Insect powder. U. S. Dept. Agr. Bull. **824**, 1-82, 1920.

608 McDonnell, C. C., Roark, R. C., LaForge, F. B. and Keenan, G. L.
Insect powder. U. S. Dept. Agr. Bull. **824** Revised, 1-78, 1926.

609 McIndoo, N. E. and Sievers, A. F.
Plants tested for or reported to possess insecticidal properties. U. S. Dept. Agr. Bull. **1201**, 1-61, 1924.

610. McIntosh, A.
Brown dog tick. North Am. Veterinarian, **14**, no. 8, 34, 1933.

611. Melvin, R.
Physiological studies on the effect of flies and fly sprays on cattle. J. Econ. Entomol., **25**, 1151-69, 1932.

612. Mendelsohn, M.
Report on the toxicity of insect powder. Charite-Annalen, **12**, 207, 1887.

613. Menke, A. E. and Davis, G. C.
See Davis, G. C. and Woodworth, C. W.

614. Meyer, A.
Insect powder. Dammer's Illust. Lexikon der Verfälschungen, p. 379, Leipzig, 1885-7.

615. Meyer, E. and Berman, P.
Insecticides and their odors. Soap, **9**, no. 12, 105, 1933.

616. Miege, E.
Attempt to cultivate pyrethrum in Morocco. Compt. rend. acad. agr. France, **9**, 780, 1923.

617. Miller, A. W.
Report of committee on adulterations. Proc. Am. Pharm. Assoc., **23**, 522, 1875.

618. *Mills, L. E. and Fayerweather, B. L.
Antioxidant and stabilizer for pyrethrum. U. S. Patent 1,942,827, Jan. 9, 1934.

619. Millspaugh, C. F.
Your weeds and your neighbors. W. Va. Agr. Exp. Sta. Bull. **19**, 123, 1891.

620. Mitlacher, W.
Culture experiments with medicinal plants: *P. cinerariaefolium*. Pharm. Post, **44**, 216, 1911.

621. Moeller, J.
Pyrethrum flowers. Pharmakognostischer Atlas, pp. 158-9, Berlin, 1892.

622. Moore, W.
Esters as insect repellents. J. N. Y. Entomol. Soc., **42**, 185-92, 1934.

623. *Moos, J.
Insecticide comprising tomato juice, water, sulfur, alcohol, pyrethrum and gasoline. Swiss Patent 171,993. Dec. 1, 1934.

624. Morgan, W. L.
The pumpkin beetle. Agr. Gaz. N. S. Wales, **44**, 915-22, 1933.

625. *Mori, K.
Insecticide comprising machine oil, sulfur, pine resin, soap, pyrethrum and eucalyptus oil. Japanese Patent 101,075. May 11, 1933.

626. Mote, D. C. and Thompson, B. G.
Substitutes for lead arsenate. Proc. 5th Pacific Sci. Congr. **5**, 3411, 1934.

627. Moznette, G. F.
Control of pecan nut-case bearer. J. Econ. Entomol., **28**, 791, 1935.

628. Murphy, D. F. and Peet, C. H.
Factors in insecticide performance. Soap, **10**, no. 1, 95, 1934.

629. Murtfeldt, M. E.
Pyrethrum for house plants. Am. Entomol., **3**, 105, 1880.

630. Murtfeldt, M. E.
Entomological notes. U. S. Dept. Agr. Div. Entomol. Bull. **26**, 38, 1892.

631. Nagasawa, T.
Study of pyrethrum flowers; determination of active principle Repts. Imp. Ind. Research Inst., Osaka (Japan), **15**, no. 10 1-18, 1934.

632. Neiswander, C. R.
Control of greenhouse mealybugs. J. Econ. Entomol., **28**, 405, 1935.

633. Nelson, F. C.
Penetration of contact oil spray into breathing system of an insect. J. Econ. Entomol., **20**, 632-5, 1927.

634. Nelson, F. C.
Commercial cattle sprays. Soap, **10**, no. 2, 79-81, 97-103, 1934.

635. Nelson, F. C.
Changes and additions to the Peet-Grady method. Soap, **10**, no. 8, 81, 1934.

636. Nelson, F. C., Buc, H. E., Sankowsky, N. A. and Jernakoff, M. A.
Evaluating liquid insecticides. Soap, **10**, no. 10, 85-91, 105-7, 1934.

637. Neu, R.
Evaluation of pyrethrum extracts. Seifensieder Ztg., **29**, 790-1, 1932.

638. Neu, R.
Combatting insects with chemicals. Chem.-Ztg., **59**, 333-5, 1935.

639. Neumark, E.
Campaign against mosquitoes in Berlin. Z. Desinfekt. Gesund., **19**, 1927.

640. Nijkamp, J. A. and Swellengrebel, N. H.
Pyrethrum preparations in mosquito control. Nederland. Tijdschr. Geneeskunde, **78**, 2327-38, 1934.

641. Niswander, F. J.
Pyrethrum or buhach. Wyo. Agr. Exp. Sta. Bull. **7**, 6, 1892.

642. Nitzschmann, N. J.
Examination of insect powder. Pharm. Era, **6**, 137, 1891.

643. Noble, W. B.
Sod web-worms and their control in lawns and golf greens. U. S. Dept. Agr. Circ. **248**, 1932.

644. Noodt, Dr.
Caucasian insect powder. Buchner's Neues Repertorium für Pharmacie, **7**, 562-4, 1858.

645. North, H. F. A. and Thompson, G. A.
Blue grass webworms in turf. J. Econ. Entomol., **26**, 1117-25, 1933.

646. Nöschel, A.
Caucasian insect powder. Med. Ztg. Russ., 12th yr., no. 9, 70-2, 1855.

647. Notley, F. B.
Control of *Antestia* in wetter districts. Kenya Dept. Agr. Bull. **4**, 1-12, 1933.

648. *O'Kane, W. C.
Oil extract of pyrethrum carrying colloidal cuprous oxide is used as an insecticide and fungicide. U. S. Patent 1,786,125. Dec. 23, 1930.

649. O'Kane, W. C.
Pyrethrum culture in the United States. J. Econ. Entomol., **28**, 3, 1935.

650. O'Kane, W. C., Walker, G. L., Guy, H. G. and Smith, O. J.
Reactions of insects to various chemicals and initial appraisal of proposed contact insecticides. N. H. Exp. Sta. Tech. Bull. **54**, 3-20, 1933.

651. O'Kane, W. C., Westgate, W. A., Glover, L. C. and Lowry, P. R.
Studies of contact insecticides. N. H. Agr. Exp. Sta. Tech. Bull. **39**, 1930.

652. Osborn, H.
Methods of attacking parasites of domestic animals. Insect Life, **6**, 165, 1893.

653. Osborn, H. and Pammel, L. H.
Machinery and methods for spraying: pyrethrum. Iowa Agr. Exp. Sta. Bull. **20**, 709, 1893.

654. *Osita, D.
Pyrethrum insecticide. Japanese Patent 90,865. Mar. 31, 1931.

655. Paillot, A. and Faure, J. C.
Cultivation of pyrethrum and utilization at place of harvesting. Compt. rend. acad. agr. France, **9**, 806, 1923.

656. Pannewitz, E.
Insecticides with petroleum bases. Seifensieder Ztg., **28**, no. 23, 1931.

657. Pannewitz, E.
Toxicity of combinations of mineral oil with ethereal oils. Z. Desinfekt. Gesundh., **23**, 339-42, 1931.

658. Pannewitz, E.
Selection and treatment of petroleum bases for insecticides. Z. Desinfekt. Gesundh., **23**, 465-76, 1931.

659. Pannewitz, E.
Preparation of liquid insecticides with petroleum bases. Seifensieder Ztg., **29**, 365-6, 1932.

660. Pannewitz, E.
Evaluation of pyrethrum. Seifensieder Ztg., **29**, 789-90, 1932.

661. Pannewitz, E.
Standardization of insect powder and liquid insecticides. Seifensieder Ztg., **30**, 391-3, 1933.

662. Parker, W. B.
Development in scientific pest control. J. Econ. Entomol., **26** 718-20, 1933.

663. Parker, W. B.
Recent developments of vapo dust method of pest control. J. Econ. Entomol., **27**, 1036-40, 1934.

664. *Parker, W. B.
Apparatus for treating plants with fog-like insecticidal mists. U. S. Patent 1,980,427. Nov. 13, 1934.

665. Parman, D. C., Bishopp, F. C., Laake, E. W., Cook, F. C. and Roark, R. C.
Chemotropic tests with the screw worm fly. U. S. Dept. Agr. Bull. **1472**, 1927.

666. Parman, D. C., Laake, E. W., Bishopp, F. C. and Roark, R. C.
Blowfly baits and repellents during 1926. U. S. Dept. Agr. Tech. Bull. **80**, 1-14, 1928.

667. Passerini, N.
Combatting the grape worm. Giorn. agr. comm. Firenze, **12**, 227, 1894.

668. Passerini, N.
Insect powder from *P. cinerariaefolium* grown at Florence. Nuovo giorn. bot. italiano, **9**, 30, 1919.

669. Passerini, N.
Pyrethrum powder from different parts of the plant. Nuovo giorn. bot. ital., **27**, 1, 1920.

670. Pauckert, C. A.
Cultivation of *Pyrethrum roseum* and *carneum*. Pharm. Zentralhalle, **7**, 449-52, 1866.

671. Pearson, A. M.
Improved method for determining cattle fly spray repellence. J. Econ. Entomol., **28**, 160-1, 1935.

672. Pearson, A. M.
Role of pine oil in cattle fly sprays. Del. Agr. Exp. Sta. Bull. **196**, 1-63, 1935.

673. Pearson, A. M., Wilson, J. L. and Richardson, C. H.
Methods used in testing cattle fly sprays. J. Econ. Entomol., **26**, 269-74, 1933.

674. Peet, C. H.
Soap, **6**, no. 6, 111, 1930.

675. Peet, C. H.
Testing commercial insecticides by use of composite samples. Soap, **8**, no. 1, 101-5, 1932.

676. Peet, C. H.
The Peet-Grady method revised. Soap, **8**, no. 4, 98-102, 121, 1932.

677. Peet, C. H.
Insecticides. Soap, **9**, no. 1, 87-8, 1933.

678. Peet, C. H.
Soap, **9**, no. 6, 89, 1933.

679. Peet, C. H.
Soap, **10**, no. 1, 85, 1934.

680. Peet, C. H.
Report of committee on insecticide standardization. Soap, **11**, no. 1, 84, 1935.

681. Feet, C. H. and Grady, A. G.
Studies in insecticidal activity: testing insecticides against flies. J. Econ. Entomol., **21**, 612-7, 1928.

682. *Peiser, J. and Karo, W.
Talc impregnated with pyrethrum preparation in vacuum is used as an insecticide. German Patent 501,451. June 3, 1927.

683. Penny, C. L.
C. leucanthemum and *Erigeron annuum.* 8th Ann. Rept. Del. Agr. Exp. Sta., p. 159, 1895-6.

684. Perkins, G. H.
Pyrethrum. 5th Ann. Rept. Vt. Agr. Exp. Sta., p. 159, 1892.

685. Peronne, P.
Dalmatian pyrethrum. Rev. hort. Algerie, **6**, 105, 1921.

686. Perrier, G.
Extract of pyrethrum. Bull. soc. chim., **23**, 216-7, 1900.

687. Perrot, E.
Substitution and adulteration of some medicinal plants. Bull. sci. pharm., **14**, 347, 1907.

688. Perrot, E.
Report of director of national office of vegetable materials. Ministere du Commerce et de l'Industrie, France, 1923.

689. Perrot, E.
The question of pyrethrum. Compt. rend. acad. agr. France, **14**, 539-53, 1928.

690. Perrot, E.
Cultivation of pyrethrum in France. Prog. agr. vit., **92**, 525-8, 1929 and Compt. rend. acad. agr. France, **15**, 885, 1929.

691. Perrot, E.
Insecticides and vermicides; pyrethrum and its applications. Bull., soc. encour. ind. nat., **130**, 709-22, 1931. Also, Pharm. Presse. Wiss.-prakt. Heft, 79-80, 1932, and Bull. sci. pharmacol., **39**, 42-57, 1932.

692. Perrot, E.
The question of the pyrethrins. Bull. agr. soc. nat. encour. agr., **19**, 1932.

693. Perrot, E. and Gaudin, O.
Action of pyrethrins on isolated rabbit intestine. Bull. sci. pharmacol., **40**, 7-13, 1933 and Bull. soc. therapeut. Nov. 9, 1932.

694. Perrot, E., Gaudin, O. and Rondeau Du Noyer, M.
Pyrethrin against helminthiasis in sheep and syngamosis in fowl. Bull. sci. pharmacol. **40**, 13-19, 1933.

695. Petrov, A. D.
Methods for testing pyrethrum. Bull. Plant Protection (U. S. S. R.), **3**, no. 1, 155-8, 1931.

696. Petrov, A. D. and Konen, E. V.
Pyrethrin content of pyrethrum flowers. Zhur. Prikladnoi Khim., **4**, 383-7, 1931.

697. Pierstorff, A. L. and Parks, T. H.
Sod webworms. Ohio State Univ. Plant Diseases and Insect Notes, **2**, no. 9, 1931.

698. Plant, A.
Adulteration of drugs in North America. Deut.-Am. Apoth. Ztg., p. 107, 1894. Also, Jahr. Pharm., **29**, 2-3, 1894.

699. Poole, O. M.
Resume of insecticides. Soap, **11**, no. 7, 91, 1935.

700. Pouchet.
Use of pyrethrins in human and veterinary medicine and in hygiene. Presse médi., **39**, 1931.

701. Pregl, F.
Quantitative organic micro-analysis (Fylemann trans.), pp. 96-9, Churchill, 1924.

702. Price, W. A.
Green June beetle in lawns. J. Econ. Entomol., **27**, 69-72, 1934.

703. Profft, E. and Körting, A.
Pyrethrin content and insecticidal value of Dalmatian pyrethrum. Lanw. Jahrb., **79**, 415-30, 1934.

704. Ramirez, M. A.
Pyrethrum: an etiologic factor in vasomotor rhinitis and asthma. J. Allergy, **1**, 149, 1930.

705. Raupp, O.
Simple titrimetric determination of pyrethrins. Süddeut. Apoth. Ztg., **75**, 216, 1935.

706. Rebrassier, R. E.
Pyrethrum as an anthelminthic for *Ascaridia lineata*. J. Am. Vet. Med. Assoc., **37**, 645-8, 1934.

707. Reeb, E.
Active principles of insect powder. J. Pharm. Elsass-Lothringen, **35**, 267-71, 1909.

708. Regan, W. M. and Freeborn, S. B.
Effect of flies and fly sprays on physiological processes of the dairy cow. J. Dairy Sci., **19**, 11-28, 1936.

709. Richardson, H. H.
Insecticidal method for the estimation of kerosene extracts of pyrethrum. J. Econ. Entomol., **24**, 97-105, 1931.

710. Richardson, H. H.
Research on kerosene extract of pyrethrum. J. Econ. Entomol., **24**, 763-4, 1931.

711. Richardson, H. H.
Pyrethrin I content of pyrethrum powder as an index of insecticidal power. J. Econ. Entomol., **24**, 1098-1105, 1931.

712. Richardson, H. H.
Pyrethrum sprays in the control of greenhouse pests. J. Econ. Entomol., **25**, 415, 1932.

713. Richardson, H. H.
Insecticidal efficiency of the pyrethrins, nicotine and rotenone against the greenhouse red spider mite. J. Econ. Entomol., **25**, 592-9, 1932.

714. Richardson, H. H.
Insecticidal studies of mid-continent distillates as bases for pyrethrum extract. Ind. Eng. Chem., **24**, 1394-7, 1932.

715. Richardson, H. H.
An efficient medium for rearing house flies throughout the year. Science, **76**, no. 1972, 350-1, 1932.

716. Richardson, H. H.
Extractive efficiency of kerosene on pyrethrum powders of varying fineness. J. Econ. Entomol., **26**, 252-9, 1933.

717. Richardson, H. H.
Insecticidal efficiency of kerosene extracts of derris alone and in combination with kerosene extract of pyrethrum. J. Econ. Entomol., **26**, 914-5, 1933.

718. Richardson, H. H.
Poisons with molasses for control of gladiolus thrips. J. Agr. Research, **49**, 370, 1934.

719. Richardson, H. H.
Control of Mexican mealy bug on chrysanthemums. J. Econ. Entomol., **28**, 399-405, 1935.

720. Richardson, H. H. and Nelson, R. H.
Field control of gladiolus thrips. J. Econ. Entomol., **26**, 546-54, 1933.

721. Richmond, E. A.
Collecting hydrophids with pyrethrum dust. J. Econ. Entomol., **27**, 973, 1934.

722. Riedel, J. D.
Insect powder. Pharm. Ztg., **25**, 423, 1880.

723. *Riedel-deHäen Co.
Solutions of pyrethrum, finely atomized by gas or air at high pressure, for destroying insects. French Patent 778,844. Mar. 25, 1935.

724. Ries, D. T.
A new mite on boxwood. J. Econ. Entomol. **28**, 55-62, 1935.

725. Rigal, M.
Action of pyrethrins on isolated organs. Compt. rend. soc. biol., **111**, 687-9, 1932.

726. Riley, C. V.
Use of pyrethrum. Am. Entomol., **3**, 193-5, 1880.

727. Riley, C. V.
Directions for raising pyrethrum. Am. Entomol., **3**, 197, 1880.

728. Riley, C. V.
Pyrethrum for the screw worm. Am. Entomol., **3**, 276, 1880.

729. Riley, C. V.
Pyrethrum in the garden. Am. Entomol., **3**, 296, 1880.

730. Riley, C. V.
Experiments with pyrethrum. Am. Naturalist, **15**, 145-7, 1881.

731. Riley, C. V.
Cultivation of pyrethrum and manufacture of the powder. Am. Naturalist, **15**, 569-72, 744-6, 817-9, 1881.

732. Riley, C. V.
Pyrethrum, its use as an insecticide. Rept. Commissioner Agr., 76-87, Washington, 1882.

733. Riley, C. V.
Rept. Commissioner Agr., 131, 133-5, 163, Washington, 1883.

734. Riley, C. V.
Rept. Commissioner Agr., 326-9, 416, Washington, 1884.

735. Riley, C. V.
Rept. Commissioner Agr., 243-4, 291, Washington, 1885.

736. Riley, C. V.
Pyrethrum. 4th Rept. U. S. Entomological Commission, 164-80, 1885.

737. Riley, C. V.
Rept. Commissioner Agr. for 1886, 500, Washington, 1887.

738. Riley, C. V.
Rept. Commissioner Agr., 259, Washington, 1890.

739. Riley, C. V.
Remedies against flies and mosquitoes. Insect Life, **3**, 470, 1891.

740. Riley, C. V.
Kerosene emulsion and pyrethrum. Insect Life, **4**, 32-3, 1891.

741. Riley, C. V.
Rept. Secretary Agr. for 1891, 259-60, 262, Washington, 1892.

742. Riley, C. V.
Syrian bookworms. Insect Life, **6**, 265-6, 1894.

743. Ripert, J.
Chemical methods of analysis of pyrethrum. Ann. fals., **24**, 323-41, 1931.

744. Ripert, J.
Review of the pyrethrum problem. Ann. fals., **25**, 395-409, 1932.

745. Ripert, J.
New method of analysis for products containing extracts of pyrethrum. Ann. fals., **27**, 580-95, 1934; **28**, 27-38, 1935.

746. Ripert, J.
French pyrethrum. Dehon and Co., p. 1-55, Paris, 1935.

747. Ripert, J.
Pyrethrum cultivation in France. Soap, **11**, no. 8, 97, 1935.

748. Ripert, J. and Gaudin, O.
Relative toxicity of pyrethrins I and II. Compt. rend. Acad. Sci., **200**, 2219, 1935 and Ann. fals. **29**, no. 327, 132, 1936.

749. Roark, R. C.
Pyrethrum and soap, a chemically incompatible mixture. J. Econ. Entomol., **23**, 460-2, 1930.

750. Roark, R. C.
Insecticide possibilities of derris root. Soap, **7**, no. 3, 98-101, 1931.

751. Roark, R. C.
Advantages and limitations of organic insecticides. J. Econ. Entomol., **28**, 211, 1935.

752. Roark, R. C.
Insecticides and fungicides. Ind. Eng. Chem., **27**, 530-2, 1935.

753. Roark, R. C.
Household insecticides. Soap, **11**, no. 11, 101, 1935.

754. Roark, R. C. and Keenan, G. L.
Adulteration of insect powder with powdered daisy flowers. U. S. Dept. Agr. Bull. **795**, 1-9, 1919, Revised 1923.

755. Roark, R. C., Parman, D. C., Bishopp, F. C. and Laake, E. W.
Repellents for blowflies. Ind. Eng. Chem., **19**, 942, 1927.

756. Robinson, R. H.
Sprays, their preparation and use. Ore. Agr. Exp. Sta. Bull. **336**, 3-30, 1935.

757. Rogers, H. D. and Calamari, J. A.
Rotenone determination by colorimetric methods. Ind. Eng. Chem., Anal. Ed., **8**, 135, 1936.

758. Roney, J. N. and Thomas, F. L.
Arsenical substitutes for controlling vegetable insects. J. Econ. Entomol., **28**, 615-620, 1935.

759. Rother, R.
Persian insect powder. Drug. Circ. Chem. Gaz., **20**, no. 7, 118, 1876.

760. Rother, R.
Active principle of Persian insect powder. Drug. Circ. Chem. Gaz., **20**, no. 12, 198, 1876.

761. Rudolfs, W.
Effect of chemicals upon the behavior of mosquitoes. N. J. Agr. Exp. Sta. Bull. **496**, 1930.

762. Runner, G. A.
Controlling grape insects. Ohio State Hort. Soc. Proc., 65th Ann. Meet., 206-11, 1932.

763. *Saito, Y and Koide, T.
Insecticide comprising cresol, casein, soap and pyrethrum. Japanese Patent 101,048. May 11, 1934.

764. Saling, T.
Active principles of pyrethrum and a new biological method for their evaluation. Z. Desinfekt. Gesundh., **20**, 33-42, 1928.

765. Saling, T. and Buchmann, W.
Mode of action of pyrethrum. Kleine Mitt. Mitglieder Ver. Wasser-Boden-Lufthyg., **7**, 1931.

766. Salomon, R.
Treatments against eudemis with soap-pyrethrum. Rev. vit., **57**, 188-90, 1922.

767. *Sanders, G. E.
Making insecticides flotable. U. S. Patent 1,847,540. Mar. 1, 1932.

768. *Sankowsky, N. A.
Method of extracting pyrethrum. U. S. Patent 1,933,077. Oct. 31, 1933.

769. *Sankowsky, N. A. and Fulton, S. C.
Unsaturated hydrocarbons are used as solvents for pyrethrum. British Patent 394,977. July 4, 1933.

770. *Sankowsky, N. A. and Grant, D. H.
Colorless pyrethrum extract is prepared by means of methanol. U. S. Patent 1,945,235. Jan. 20, 1934.

771. Sato, S.
Chrysanthemum from which insect powder is prepared. J. Pharm. Soc. Japan, no. 283, 766-88, 1905.

772. Sato, S.
Effective constituents of insect powder. J. Pharm. Soc. Japan, no. 304, 607, 1907.

773. Sattler.
Adulterated insect powder. Apoth. Ztg., **27**, 627-8, 1912.

774. Saunders, W.
Insect powder. Can. Entomol., **11**, 41-3, 1879.

775. Sayre, L. E.
Am. J. Pharm., **52**, 636, 1880.

776. Sayre, L. E.
Pyrethrum. Trans. Kans. Acad. Sci., **25**, 141, 1913.

777. Schafer, G. D.
How contact insecticides kill. Mich. Agr. Exp. Sta. Bull. **11**, 1911.

778. Schedl, K. E.
Fly sprays. Anz. Schädlingskunde, **11**, 68, 1935.

779. Schenck.
Persian insect powder. Würzburg Wochschr., May, 1859.

780. Schipulinsky, G.
Investigations of the use of *Pyrethrum roseum* and *carneum* against scabies and worms. Med. Ztg. Russ., no. 35. Also Canstatt's Jahresbericht (1854) **4**, 338, 343-4, Würzburg, 1855.

781. Schlagdenhauffen and Reeb.
Pyrethrum flowers, their toxic principle. J. Pharm. Elsass-Lothringen, **17**, 123-6, 1890.

782. Schlagdenhauffen and Reeb.
Powdered pyrethrum flowers. J. Pharm. Elsass-Lothringen, **17**, 273-81, 1890.

783. Schlotshauber.
Persian insect powder. Proc. Am. Pharm. Assoc., **10**, 112, 1862.

784. Schoene, W. J.
Leafhoppers injurious to apple trees. Va. Agr. Exp. Sta. Bull. **283**, 30, 1932.

785. Schoene, W. J. and Jefferson, R. N.
Tests of contact insecticides on eggs of peach moth and codling moth. J. Econ. Entomol., **28**, 182, 1935.

786. *Schotte, H. and Goernitz, K.
Insecticide comprising pyrethrum mixed with a molten carrier, solidified and pulverized. Canadian Patent 348,380.

787. Schrenk, J.
Notes on insect powder. Am. Drug., **16**, 3-4, 1887.

788. Schrenk, J.
Insect flowers. Am. Drug., **18**, 42-3, 1889.

789. Schwartz, B. and Shook, W. B.
Rabbit parasites and diseases. U. S. Dept. Agr. Farm. Bull. **1568**, 1928.

790. Schwartz, L.
Dermatitis due to pyrethrum. U. S. Pub. Health Service Bull. **215**, 51-4, 1934.

791. Schwarz, C.
Colored insect powder. Pharm. Ztg., **33**, 691-2, 1888.

792. Scidmore, G. H.
Insect powder. Rept. U. S. Consul General, Yokohama, Japan, Mar. 31, 1915.

793. Searls, E. M. and Snyder, F. M.
Control of ectoparasites of laboratory rats by pyrethrum in oil. J. Econ. Entomol., **28**, 304-10, 1935.

794. Seil, H. A.
Estimation of pyrethrins. Soap, **10**, no. 5, 89, 1934 and Chem. Trade J. **95**, 168, 1934.

795. Selenski.
Physiological action of Turkish insect powder. St. Petersburg. Med. Wochschr., **10**, 114-5, 1885.

796. Semenoff.
Action of *Pyrethrum roseum* and *carneum* on insects. Just's Botan. Jahresbericht, **6**, 1129-30, 1878.

797. Semichon, L.
Treatment of cochylis and eudemis. Bull. soc. centr. agr. Aude, **104**, no. 41, 254, 1923.

798. Senft, E.
State culture of medicinal plants: *C. cinerariaefolium*. Pharm. Post, **44**, 894, 1911.

799. Shepard, H. H.
Recent usage of the term "pyrethrin." J. Am. Pharm. Assoc., **22**, 479-80, 1933.

800. Shepard, H. H. and Richardson, C. H.
Method of determining the relative toxicity of contact insecticides. J. Econ. Entomol., **24**, 905-14, 1931.

801. Shimkin, M. B. and Anderson, H. H.
Acute toxicities of rotenone and mixed pyrethrins in mammals. Proceed. Soc. Exp. Biol. Med., **34**, 135-8, 1936.

802. Shinn, C. H.
Persian insect powder. Rept. Univ. Calif. Agr. Exp. Sta. for 1895-6, 1896-7, p. 339, Berkeley, 1898.

803. Shinn, C. H.
P. cinerariaefolium and *P. roseum*. Rept. Univ. Calif. Agr. Exp. Sta. for 1895-6, 1896-7, p. 383, Berkeley, 1898.

804. Shropshire, L. H. and Compton, C. C.
Saving garden crops from insect injury. Ill. Agr. Exp. Sta. Circ. **437**, 1935.

805. Shull, W. E.
Control of the cattle louse, *Bovicola bovis*. J. Econ. Entomol., **25**, 1208-11, 1932.

806. Sicard, H.
Destruction of eudemis with soap-pyrethrum. Bull. soc. centr. agr. Herault, **107**, 68, 1920.

807. Sicard, H.
Action of pyrethrum-Bordeaux and nicotine-Bordeaux against first generation of eudemis. Bull. soc. centr. agr. Herault, **108**, June, 1921.

808. Siedler, P.
Insect powder. Riedels Berichte-Riedels Mentor, 13-16, 1913.

809. Siedler, P.
Culture of pyrethrum at Berlin and properties of Dalmatian insect powder. Arb. pharm. Inst. Univ. Berlin, **11**, 69, 1914.

810. Siedler, P.
C. cinerariaefolium. Ber. pharm. Ges., **25**, 287-302, 1915.

811. Sievers, A. F.
Pyrethrum culture in the United States. U. S. Dept. Agr. Bureau Plant Industry, 1931.

812. Sievers, A. F.
Pyrethrum as a crop in the United States. U. S. Dept. Agr. Bur. Plant Ind., Mimeographed. Feb., 1934.

813. Simmonds, P. L.
Medicinal and other useful plants of Algeria. Am. J. Pharm., **63**, 77, 1891.

814. Simmonds, P. L.
Medicinal uses of flowers. Am. J. Pharm., **63**, 200-2, 1891.

815. Sinton, J. A. and Wats, R. C.
Insecticidal sprays in the destruction of adult mosquitoes. Record of Malaria Survey of India, **5**, 275-306, Sept., 1935.

816. Slaus-Kantschieder, J.
C. cinerariaefolium. Z. landw. Versuchsw., **16**, 1-8, 1913.

817. Sleesman, J. P.
Controlling insects on truck crops. Ohio Veg. Growers' Assoc. Proc. 20th Ann. Meeting, 130-36, 1935.

818. Smith, F. F. and Richardson, H. H.
Control of gladiolus thrips on corms in storage. J. Econ. Entomol., **26**, 536-45, 1933.

819. Smith, J. B.
Insecticides and how to apply them. N. J. Agr. Exp. Sta. Bull. **75**, 26-7, 1890.

820. Smith, J. B.
The rose-chafer; pyrethrum kerosene. N. J. Agr. Exp. Sta. Bull. **82**, 20-3, 1891.

821. Smith, J. B.
An experience with the rosebug. Insect Life, **3**, 220-4, 1891.

822. Smith, L. W.
One year's experience with pyrethrum larvicide. Proc. 19th Ann. Meet. N. J. Mosquito Exterm. Assoc., 1932.

823. Smith, R. H
Experiments with toxic substances in spray oils in controlling red scale. Calif. Citrograph, **14**, 315, 1929.

824. Smith, R. H.
Experiments with toxic substances in highly refined oil sprays. J. Econ. Entomol., **25**, 988-90, 193?

825. Smith, R. H.
Experiments with toxic substances in oil sprays. Calif. Citrograph, **17**, 473, 1932.

826. Smith, R. H.
Controlling codling moth. Proc. Wash. State Hort. Assoc., 72-5, 1934.

827. Southall Bros. and Barclay.
Yearbook Pharmacy, p. 172, 1911.

828. Southall Bros. and Barclay.
Chemist Druggist, **82**, 123, 1913.

829. Spratling, W. P.
Inflammation of upper air passages and ears caused by inhalation of Persian insect powder. Med. Record, **44**, 584, 1893.

830. Sprengel, L.
Combating the cherry fruit fly. Anz. Schädlingskunde, **8**, no. 5, 49-51, 1932.

831. Sprengel, L.
Pyrethrum as an insecticide. Anz. Schädlingskunde, **10**, 1-7, 14-21, 111-17, 1934.

832. Spuler, A., Dorman R. and Gillies, K.
Factors influencing the cost of codling moth control. Wash. Agr. Exp. Sta. Sci. Paper **214**, 1-16, 1931.

833. Spuler, A., Strand, T., Dorman, R., Gillies, K. and Brannon, D.
Value of combinations with lead arsenate and summer oils in codling moth control. Northwest Fruit Grower, **2**, 4-12, 1930.

834. Staehelin, M.
Pyrethrum, its culture and application as a vermicide and an insecticide. Heil-Gewürz-Pflanzen, **9**, 39-45, 1926.

835. *Standard Oil Development Co.
Insecticide comprising white petroleum oil, pyrethrum and an emulsifier. French Patent 690,775. Feb. 27, 1930.

836. *Standard Oil Development Co.
Pyrethrum extract in oil boiling between 205° and 280°. French Patent 692,568. Mar. 22, 1930.

837. *Standard Oil Development Co.
Insect repellent comprising terpinyl acetate, oil, soap and pyrethrum. U. S. Patent 1,755,178. Apr. 22, 1930.

838. *Standard Oil Development Co.
Unsaturated hydrocarbons extracted from petroleum used with pyrethrum as an insecticide. French Patent 729,213. Jan. 4, 1932.

839. Stanley, W. W. and Marcovitch, S.
Control of tobacco flea beetle and hornworm. J. Econ. Entomol., **28**, 797, 1935.

840. Staudinger, H. and Harder, H.
Estimation of the pyrethrin content of insect powder. Ann. Acad. Sci. Fennicae, A**29**, no. 18, 1-14, 1927.

841. Staudinger, H., Muntwyler, O., Ruzicka, L. and Seibt, S.
Synthesis of chrysanthemum acid and other trimethylene carboxylic acids. Helv. Chim. Acta, **7**, 390-406, 1924.

842. Staudinger, H. and Ruzicka, L.
Isolation and constitution of the active portion of Dalmatian insect powder. Helv. Chim. Acta, **7**, 177-201, 1924.

843. Staudinger, H. and Ruzicka, L.
Constitution of chrysanthemum mono- and dicarboxylic acids. Helv. Chim. Acta, **7**, 201-11, 1924.

844. Staudinger, H. and Ruzicka, L.
Constitution of pyrethrolone. Helv. Chim. Acta, **7**, 212-35, 1924.

845. Staudinger, H. and Ruzicka, L.
Constitution of tetrahydropyrethrone. Helv. Chim. Acta, **7**, 236-44, 1924.

846. Staudinger, H. and Ruzicka, L.
Synthesis of tetrahydropyrethrone. Helv. Chim. Acta, **7**, 245-59, 1924.

847. Staudinger, H. and Ruzicka, L.
Cyclopentanolone derivatives and comparison with pyrethrolone. Helv. Chim. Acta, **7**, 377-90, 1924.

848. Staudinger, H. and Ruzicka, L.
Preparation of pyrethrolone-like alcohols. Helv. Chim. Acta, **7**, 406-41, 1924.

849. Staudinger, H. and Ruzicka, L.
Further experiments on the preparation of cyclopentanolone derivatives. Helv. Chim. Acta, **7**, 442-8, 1924.

850. Staudinger, H. and Ruzicka, L.
Synthesis of the pyrethrins. Helv. Chim. Acta, **7**, 448-58, 1924.

851. Staudinger, H., Ruzicka, L. and Reuss, E.
Constitution and synthesis of the pyrethrins. Ann. Acad. Sci. Fennicae, A**29**, no. 17, 1927.

852. Stedman, J. M.
Injurious and beneficial insects. Ala. Agr. Exp. Sta. Bull. **45**, 7-11, 1893.

853. *Steindorff, A., Pfaff, K., Krämer, R. and Kern, H.
Urethanes are added to pyrethrum fly sprays. German Patent 564,018. Nov. 12, 1932.

854. Steiner, H. M.
Effect of insecticides on white apple leaf hopper. J. Econ. Entomol., **28**, 385-8, 1935.

855. Stellwaag, F.
Toxicity and value of insecticides. Z. angew. Entomol., **18**, 1931.

856. Stellwaag, F.
Control of the cherry fly. Kranke Pflanze, **10**, 71-2, 1933.

857. Stellwaag, F.
Comparison of pyrethrum and derris. Weinbau Kellerwirtschaft, **13**, 4-5, 14-16, 1934.

858. Stone, E. P.
Composition of ox-eye daisy. 5th Ann. Rept. N. H. Agr. Exp. Sta., p. 139, 1893.

859. Stone, W. E., Boyden, B. L., Wisecup, C. B. and Tatman, E. C. Control of celery leaf-tier in Florida. Fla. Agr. Exp. Sta. Bull. **251**, 1932.

860. Sulzberger, M. B. and Weinberg, C. B. Dermatitis due to insect powder. J. Am. Med. Assoc., **95**, 111, 1930.

861. Sweitzer, S. E. and Tedder, J. W. Pyrethrum in the treatment of scabies. Minn. Medicine, **18**, 793, Dec., 1935.

862. Swellengrebel, N. H. Seven year campaign against mosquitoes at Medemblik. Nederland. Tijdschr. Geneeskunde, **78**, 345-53, 1934.

863. Swellengrebel, N. H. and Nijkamp, J. A. Health Organization, League of Nations, Document No. C. H. (Malaria) **220**, 1934.

864. Swingle, M. C. Difference in action of derris and pyrethrum. J. Econ. Entomol., **27**, 1101, 1934.

865. *Sy, A. P. Extraction of pyrethrum with hot oil. U. S. Patent 1,595,261. Aug. 10, 1926.

866. Symes, C. B. Insect pests of mushrooms. Fruit Grower, **51**, 235, 1921.

867. *Tabar-Nouval, J. M. L. Insecticide comprising pyrethrum, mineral oil and essential oils. French Patent 733,563. Mar. 16, 1932.

868. Taft, L. R. and Davis, G. C. Pests of orchard and garden. Mich. Agr. Exp. Sta. Bull. **121**, 11, 1895.

869. Takai, S. Life history of *Pryeria sinica* Moore. Abstract in Rev. Appl. Entomol., **17**, 565, 1929.

870. Tamaro, D. Pyrethrum and its cultivation. Coltivatore, **60**, 39-76, 1914.

871. Tattersfield, F. Pyrethrum flowers. A quantitative study of their development. Ann. Appl. Biol., **18**, 602-35, 1931.

872. Tattersfield, F. Laboratory methods for evaluating insecticides. Ann. Appl. Biol., **19**, 281-90, 1932.

873. Tattersfield, F. Loss of toxicity of pyrethrum dusts on exposure to air and light. J. Agr. Sci., **22**, 396-417, 1932.

874. Tattersfield, F. Apparatus for testing contact insecticides. Ann. Appl. Biol., **21**, 691-703, 1934.

875. Tattersfield, F. Present status of pyrethrum and derris investigation. Soap, **11**, no. 7, 87, 1935.

876. Tattersfield, F.
Recent developments in pyrethrum and derris products. Chem. Trade J. **96,** 231-2, 251-2, 273-4, 1935.

877. Tattersfield, F., Gimingham, C. T. and Morris, H. M.
Studies on contact insecticides. Ann. Appl. Biol., **12**, 61-76, 1925.

878. Tattersfield, F. and Hobson, R. P.
Pyrethrins I and II. Their estimation in pyrethrum. J. Agr. Sci., **19**, 433-7, 1929.

879. Tattersfield, F. and Hobson, R. P.
Extracts of pyrethrum. Permanence of toxicity and stability of emulsions. Ann. Appl. Biol., **18**, 203-43, 1931.

880. Tattersfield, F., Hobson, R. P. and Gimingham, C. T.
Pyrethrins I and II. Their insecticidal value and estimation in pyrethrum. J. Agr. Sci., **19**, 266-96, 1929.

881. Tattersfield, F. and Martin, J. T.
Loss of activity of pyrethrum. J. Agr. Sci., **24**, 598-626, 1934.

882. Tattersfield, F. and Morris, H. M.
Apparatus for testing the toxic values of contact insecticides. Bull. Entomol. Research, **14**, 223-33, 1924.

883. Tatu, H.
Determination of the active principle of pyrethrum extract. Parfumerie moderne, **24**, 607-15, 1930.

884. Taubenhaus, J. J. and Parks, H. B.
Pyrethrum, a new host for *Phymatotrichum*. Plant Disease Rept., **17**, no. 1, 5, 1934.

885. *Terry, F. D.
Insecticide comprising the active principle of pyrethrum dissolved in kerosene. U. S. Patent 1,599,851. Sept. 14, 1926.

886. Textor, O.
Examination of Persian insect powder for its active principle. Am. J. Pharm., **53**, 491-3, 1881.

887. Thomas, C. A.
Observations on mushroom insects. J. Econ. Entomol., **25**, 322-31, 1932.

888. Thomas, C. A.
Review of research on the control of wireworms. Penn. Agr. Exp. Sta. Bull. **259**, 1930.

889. Thomas, C. A.
Mushroom insects: biology and control. Penn. Agr. Exp. Sta. Bull. **270**, 1931.

890. Thomas, C. A.
Observations on mushroom insects. J. Econ. Entomol., **27**, 200-208, 1934.

891. Thomas, F. L.
Non-poisonous insecticides for use on vegetable crops. Texas Agr. Exp. Sta. Press Release, Nov. 8, 1935.

892. Thomas, W.
Liquid insecticides for preventing moth damage. Soap, **9**, no. 8, 89-91, 1933.

893. Thomason, O.
Botanical insecticides—Kenya, East Africa. American Consular Rept., Nairobi, Kenya Colony, June 13, 1933.

894. Thompson, B. G.
Western willow tingid in Oregon. J. Econ. Entomol., **26**, 1090-5, 1933.

895. Thompson, F. A.
Notes on insect powder. Am. J. Pharm., **63**, 583-4, 1891.

896. Thoms, H.
Insect powder. Pharm. Ztg., **35**, 242, 1890.

897. Thoms, H.
Research on insect powder. Chem. Ztg., **14**, 1284-5, 1372, 1890.

898. Thoms, H.
Croton flavens and *C. cinerariaefolium*. Ber. pharm. Ges., **1**, 241-7, 1891.

899. Tischler, N.
Reproductivity of flies exposed to pyrethrum sprays. J. Econ. Entomol., **24**, 558, 1931.

900. Trabut, L.
Pyrethrum insecticide. Bull. agr. Algerie-Tunisie, **5**, 323, 1899.

901. Trabut, L.
Report on botanical-agricultural enterprises in 1895. Gouv. gen. Algerie, Serv. botan. 1895 and 1901.

902. Trabut, L.
Soap-pyrethrum insecticide. Bull. agr. Algerie-Tunisie-Maroc, **27**, 155, 1921.

903. Trapp, J.
Pyrethrum roseum and *Pyrethrum carneum*. Pharm. Z. Russ., **15**, 326, 1876.

904. Trappmann, W. and Nitsche, G.
Action of rotenone and pyrethrins on different insects. Nachrbl. deut. Pflanzenschutzdienst, **15**, 6-7, 1935.

905. Treherne, R. C.
Bio-chemical aspects of insect control. Sci. Agr., **3**, 109-13, 1922.

906. *Trevillian, W. J.
Process for extracting pyrethrum flowers. U. S. Patent 1,786,967. Dec. 30, 1930.

907. Treviranus, L. G.
Index Seminum Horti Vratislaviae, 1820.

908. Tropea, C.
Pyrethrum insecticide; history, culture and uses. Bol. Giard. Colon. Palermo, **2**, 117, 1915.

909. Tropea, C.
Practical instructions for cultivating, harvesting, and handling pyrethrum. Arch. farmacog. **5**, 149, 1916.

910. Trottner, K.
Evaluation of insect powder. Arch. Pharm., **253**, 92-102, 1915.

911. Tschirch, A.
Angewandte Pflanzenanatomie, **1**, 153, 1889.

912. Tschirch, A.
Cultivation of medicinal plants in Germany. Arch. Pharm., **228**, 685, 1890.

913. Tschirch, A.
Handbuch der Pharmakognosie. Volume 3, p. 197, 1921.

914. Tschirch, A. and Oesterle, O.
Anatomischer Atlas der Pharmakognosie und Nahrungsmittelkunde, pp. 171-4, Leipzig, 1900.

915. Turner, N.
Tests of mosquito light traps and larvicides. Conn. Agr. Exp. Sta. Bull. **349**, 441-5, 1933.

916. Turner, N.
Mosquito larvicides. Conn. Agr. Exp. Sta. Bull. **360**, 445, 1934.

917. Turner, N.
Control of vegetable insects. Conn. Agr. Exp. Sta. Bull. **368**, 245, 1935.

918. Turner, N. and Friend, R. B.
Control of onion thrips. Conn. Agr. Exp. Sta. Bull. **360**, 480, 1934.

919. Turner, N. and Friend, R. B.
Mexican bean beetle control. Conn. Agr. Exp. Sta. Bull. **371**, 423-52, 1935.

920. Tutin, F.
Preparation of a pyrethrum spray fluid. Ann. Rept. Agr. Hort. Sta. Univ. Bristol, 96-102, 1928.

921. Twinn, C. R. and Herman, F. A.
A cheap and effective fly spray. Sci. Agr., **18**, 441-5, 1928.

922. Ueno, S.
Methods for the determination of toxic principles contained in pyrethrum and its products. Manuscript.

923. Unger, H.
Pyrethrum flowers. Pharm. Ztg., **32**, 685-6, 1887; **33**, 81-2, 1888; **33**, 131-2, 1888; **33**, 166-7, 1888.

924. Unger, H.
Colored insect powder. Pharm. Ztg., **33**, 715-6, 1888.

925. Unger, H.
Adulterated insect powder. Chemist Druggist, Mar. 15, 1890.

926. *Union Climique Belge S. A.
Pyrethrum is extracted with a solvent miscible with water, as pyridine or ethyl lactate. French Patent 682,290. Sept. 25, 1929.

927. Urbain, A. and Guillot, G.
The pyrethrins; their employment in veterinary medicine. Rev. pathol. comp. hyg. gen., **31**, no. 416, 493, 1931.

928. Van Laren, A. J.
Cultivation of pyrethrum in France and the manufacture of soap-pyrethrum. Pharm. Weekblad, **60**, 877-92, 1923.

929. Van Leeuwen, E. R.
Sodium oleate-oleoresin of pyrethrum spray. N. J. Dept. Agr. Circ. **92**, 1926.

930. Van Leeuwen, E. R.
Control of Japanese beetle on fruit and shade trees. U. S. Dept. Agr. Circ. **237**, 1932.

931. Vannote, R. L. and Ginsburg, J. M.
Practical application of pyrethrum mosquito larvicide. Proc. 18th Ann. Meet. N. J. Mosquito Exterm. Assoc., 111-20, 1931.

932. Verneau, V.
Study on pyrethrum. These Ecole Superieure Pharmacie Montpellier, Apr. 9, 1892.

933. Vogt, G.
Test for insect powder. Pharm. J., **76**, 23, 1906.

934. Vogtherr, M.
Diagnosis of the medicinal composite flowers. Ber. pharm. Ges., **7**, 47-50, 1897.

935. Vollmar, R. C.
Quantitative determination of pyrethrin I. Ind. Eng. Chem. Anal. Ed., **3**, 110-2, 1931.

936. *Voorhees, V.
Light-stable pyrethrum extract. U. S. Patent 2,011,428. Aug. 13, 1935.

937. Vulpius, G.
Moisture content of vegetable powders. Arch. Pharm., **225**, 1046, 1887.

938. Walker, F. W.
Control of black hickory aphid. Ga. Fla. Pecan Grow. Assoc. Proc., 26th Ann. Conv., 39-43, 1932.

939. Walker, H. G.
Insecticidal value of certain pyrethrum extracts. Va. Truck. Exp. Sta. Bull. **75**, 943-71, 1931.

940. Walker, H. G. and Anderson, L. D.
Report on the control of the harlequin bug. J. Econ. Entomol., **26**, 129-35, 1933.

941. Walker, H. G. and Anderson, L. D.
Derris and pyrethrum dusts for control of insects attacking cruciferous crops. J. Econ. Entomol., **27**, 388-93, 1934.

942. Walker, H. G. and Anderson, L. D.
Tolerance of cabbage seedlings to insecticide dips. Va. Truck Exp. Sta. Bull. **86**, 1205-10, 1935.

943. Walker, H. G. and Anderson, L. D.
Arsenical substitutes for control of vegetable crop insects. J. Econ. Entomol., **28**, 603, 1935.

944. Walton, C. L.
The raspberry and loganberry beetle and its control. J. Pomology Hort. Sci., **8**, 173-83, 1930, and Univ. Bristol Agr. Res. Sta. Ann. Rept., 100-5, 1930.

945. Walton, C. L. and Kearns, H. G. H.
Carabid beetles as strawberry pests. Univ. Bristol Agr. Hort. Research Sta. Ann. Rept. 77-82, 1931.

946. Wasicky, R.
Evaluating insect powder. Münch. Med. Wochschr., **62**, 735. 1915.

947. Watanabe, T.
Control for *Hellula undalis* Fab. Abstract in Rev. Appl. Entomol., **15**, 641, 1927.

948. Webster, R. L., Marshall, J., Miller, C. E. and Hansberry, T. R.
Fish oils, spreaders and non-arsenicals for codling moth control. Proc. Wash. State Hort. Assoc., 28th Ann. Meet. **Apr.**, 1933.

949. Weed, A.
Pyrethrum, its agricultural future. Soap, **6**, no. 3, 107-11, 1930.

950. Weed, A.
Soap, **6**, no. 5, 117, 1930.

951. Weed, A.
Correlating pyrethrum tests. Soap, **6**, no. 7, 105-9, 131-3, 1930.

952. Weed, A.
Problems in the manufacture of liquid household insecticides. J. Econ. Entomol., **24**, 93-7, 1931.

953. Weed, A.
Action of pyrethrum on insects. Soap, **9**, no. 8, 84-5, 1933.

954. Weed, A.
Report of insecticide committee. Soap, **11**, no. 1, 87, 1935.

955. Weed, A.
Commercial aspects of the Peet-Grady test. Soap, **11**, no. 6, 105, 1935.

956. Weed, C. M.
Pyrethrum. 3rd and 4th Ann. Rept. N. H. Agr. Exp. Sta., 246-7, 1892.

957. Weed, H. E.
Injurious insects. Miss. Agr. Exp. Sta. Bull. **14**, 11, 1891.

958. West, A. P. and Russell, P. F.
Various toxic substances partially adsorbed on charcoal as *Anopheles* larvicides. Philippine J. Sci., **49**, 211-5, 1932.

959. West, G. N.
Insect powder industry. Rept. U. S. Consul, Kobe, Japan, Mar. 22, 1915.

960. Wheeler, H. W.
Loss of toxicity of dusts. J. Council Sci. Ind. Research (Aust.), **7**, 70, 1934.

961. White, R. C.
Pyrethrum concentrates. Soap, **9**, no. 7, 91-2, 1933.

962. *White, R. C.
Insecticide comprising pyrethrum, methyl protocatechuic aldehyde and a thiocyanate of the methoxymethyl cyano propionate group. U. S. Patent 1,990,422. Feb. 5, 1935.

963. White, R. C.
Specifications for liquid insecticide. Soap, **12**, no. 1, 94-d, 1936.

964. White, R. P. and Hamilton, C. C.
Diseases and insect pests of rhododendron and azalea. N. J. Agr. Exp. Sta. Circ. **350**, 3-23, 1935.

965. White, W. H.
Arsenical substitutes for cabbage worm control. J. Econ. Entomol., **28**, 607, 1935.

966. Wiebelitz.
The ash residue of insect powder. Apoth. Ztg., **27**, 688, 1912.

967. *Wikström, E. G. A.
Insecticide comprising cymol or its derivatives and pyrethrum. British Patent 434,660. Sept. 6, 1935.

968. Wilcoxon, F. and Hartzell, A.
Further chemical and toxicological studies of pyrethrum. Contrib. Boyce Thompson Inst., **5**, no. 1, 115-27, 1933.

969. Willemot, C.
Destruction of noxious insects by means of pyrethrum. Rept. Commissioner Patents for 1861, Agriculture, 223-31, Washington, 1862.

970. Williams, C. L. and Dreesen, W. C.
Nonflammable pyrethrum spray for fumigating airplanes. U. S. Public Health Reports, **50**, 1401-4, 1935.

971. Wilson, E. D.
Relative toxicities of pyrethrins I and II. Soap, **9**, no. 9, 84, 1933.

972. Wilson, J. L., Pearson, A. M. and Cannon, C. Y.
Effect of fly sprays on hides and body temperatures of dairy cows. J. Dairy Sci., **16**, 427-33, 1933.

973. Windisch, R.
Ash and sand content of insect powder. Z. Nahr. Genussm., **41**, 78-81, 1921.

974. Worsley, R. R.
Household sprays. E. African Agr. J., **1**, 61-2, 1935.

975. Worthley, H. N.
Rept. 3rd Ann. Conference European Corn Borer Research, Washington, 1929.

976. Worthley, H. N.
Codling moth spraying experiments in Pennsylvania. J. Econ. Entomol., 27, 240-4, 1934.

977. *Yamamoto, H. and Inouye, K.
Extract of pyrethrum is treated with alkali to produce a readily soluble material. U. S. Patent 1,595,538. Aug. 10, 1926.

978. Yamamoto, R.
Insecticidal principle in *C. cinerariaefolium*, I. J. Tokyo Chem. Soc., **40**, 126-47, 1919.

979. Yamamoto, R.
Insecticidal principle in *C. cinerariaefolium*, II and III. J. Chem. Soc. Japan, **44**, 311-30, 1923.

980. Yamamoto, R.
Insecticidal principle of insect powder. Inst. Phys. Chem. Research (Tokyo), **3**, 193-222, 1925.

981. Yamamoto, R. and Sumi, M.
Insecticidal principle in *C. cinerariaefolium*, IV. J. Chem. Soc. Japan, **44**, 1070-85, 1923.

982. *Yanagida, T.
Pyrethrum powder is covered with Japanese acid clay. Japanese Patent 90,435. Feb. 24, 1931.

983. *Yates, W. J.
Stabilizers for pyrethrum-oil sprays. Canadian Patent 343,209. July 17, 1934.

984. Yoshimura, K. and Trier, G.
Further reports on the presence of betaine in plants. Z. physiol. Chem., **77**, 292-6, 1912.

985. Zacharewic, E.
Pyrethrum. Prog. agr. vit., **94**, 569-70, 1930.

986. Zeck, E. H.
The green vegetable bug, *Nezara viridula*. Agr. Gaz. N. S. Wales, **44**, 594, 675, 1933.

987. Zeigler, W. H.
Pharmacology of pyrethrum flowers. J. Am. Pharm. Assoc., **12**, 19-26, 1923.

988. Zuco, F. M.
Higher homolog of cholesterine. Gazz. chim. ital., **19**, 209-12, 1889.

989. Zuco, F. M.
A new alkaloid from pyrethrum. Atti. accad. Lincei, **6**, 571-5, 1890.

990. Zuco, F. M.
A new alkaloid from pyrethrum. Gazz. chim. ital., **21**, 516-54, 1891.

991. Zuco, F. M.
Chrysanthemin. Boll. chim. farm., **31**, 203, 1892.

INDEX